GLOBAL AGRICULTURE DEVELOPMENTS

HANDBOOK ON HERBICIDES

BIOLOGICAL ACTIVITY, CLASSIFICATION AND HEALTH AND ENVIRONMENTAL IMPLICATIONS

Global Agriculture Developments

Additional books in this series can be found on Nova's website under the Series tab.

Additional e-books in this series can be found on Nova's website under the e-book tab.

GLOBAL AGRICULTURE DEVELOPMENTS

HANDBOOK ON HERBICIDES

BIOLOGICAL ACTIVITY, CLASSIFICATION AND HEALTH AND ENVIRONMENTAL IMPLICATIONS

DAIKI KOBAYASHI
AND
EITO WATANABE
EDITORS

nova publishers
New York

For permission to use material from this book please contact us:
Telephone 631-231-7269; Fax 631-231-8175
Web Site: http://www.novapublishers.com

Library of Congress Cataloging-in-Publication Data

Handbook on herbicides : biological activity, classification and health & environmental implications / editors: Daiki Kobayashi and Eito Watanabe.
p. cm.
Includes index.
ISBN 978-1-62948-053-4 (hardcover)
1. Herbicides. I. Kobayashi, Daiki. II. Watanabe, Eito.
SB951.4.H333 2013
632'.954--dc23
2013033805

Published by Nova Science Publishers, Inc. † New York

CONTENTS

Preface		**vii**
Chapter 1	Productive Degradation of Dichlorprop by Transconjugant Strains Carrying Plasmid pMC1 of Delftia acidovorans MC1 with Special Focus on the Activity of RdpA *Sabine Leibeling, Karin Glaser, Monika Neytschev, Hauke Harms and Roland H. Müller*	**1**
Chapter 2	Weeds and their Mechanisms of Resistance to Herbicides *Félix Morales-Flores, Beatriz King-Díaz, María Isabel Aguilar and Blas Lotina-Hennsen*	**29**
Chapter 3	Application of ^{31}P-NMR Spectroscopy to Glyphosate Studies in Plants: Insights into Cellular Uptake and Vacuole Sequestration Correlated to Herbicide Resistance *Xia Ge, D. Andre' d'Avignon, Joseph J. H. Ackerman, Elizabeth Ostrander and R. Douglas Sammons*	**55**
Chapter 4	Lethal and Sublethal Glyphosate Effects on Non-Target Fish Species: Biomarkers Responses in Cnesterodon decemmaculatus *Renata J. Menéndez-Helman, Alfredo Salibián and Maria dos Santos Afonso*	**85**
Chapter 5	Phenylurea Herbicides: Chemical Properties and Genotoxic Effects *Concetta Federico, Matteo Pappalardo, Claudia Giovanna Leotta, Zelica Minniti, Vito Librando, and Salvatore Saccone*	**111**
Chapter 6	Action Mode of Triazines and Toxic Effects on Vertebrates *G. D. C. Severi-Aguiar, E. C. M. Silva-Zacarin, F. D. Campos-Pereira and A. A. Alves*	**133**
Chapter 7	Photolysis Experiments on Alloxydim Herbicide and Biological Response of its Transformation Product *P. Sandín-España, B. Sevilla-Morán, M. Villarroya-Ferruz, J. L. Alonso-Prados and I. Santín-Montanyá*	**151**

Chapter 8	Glyphosate-Resistant Weeds in Southern Europe: Current Status, Control Strategies and Future Challenges *Demosthenis Chachalis and Ilias S. Travlos*	**175**
Chapter 9	Herbicide Paraquat Genotoxicity-Enhancement by the Phenolic Antioxidants Dl-α -Tocopherol and 2,6-Di-Tert-Butyl-P-Cresol *Hideki Hanada*	**191**
Chapter 10	Glyphosate Adsorption in a Soil Depth Profile *H. Tévez and M. dos Santos Afonso*	**213**
Chapter 11	Adsorption-Desorption Processes and Mobility of (4-Chloro-2-Methylophenoxy) Acetic Acid (MCPA) in Irish Grassland Soils *A. A. Piwowarczyk and N. M. Holden*	**231**
Chapter 12	Pesticides and Cancer: Studies on the Interaction of Phenoxy acid Herbicides with DNA *Sofia Benfeito, Jorge Garrido, M. J. Sottomayor, Fernanda Borges and E. Manuela Garrido*	**245**
Chapter 13	Synthesis, Structure and Herbicidal Activity of Pyrazole Derivatives *Yuhan Zhou, Xiaoliang Dong and Jingping Qu*	**263**
Chapter 14	Behavior of Herbicides in Paddy Water and Soil after Application *Kuniaki Kawata and Tomohiro Kose*	**277**
Chapter 15	A Laboratory Lysimeter for Pesticide Transport Studies *A. A. Piwowarczyk and N. M. Holden*	**291**
Index		**303**

PREFACE

In this handbook, the authors present current research in the study of the biological activity, classification and health and environmental implications of herbicides. Topics discussed include productive degradation of dichlorprop by transconjugant strains; weeds and their mechanisms of resistance to herbicides; application of 31P-NMR spectroscopy to glyphosate studies in plants; lethal and sublethal glyphosate effects on non-target fish species; chemical properties and genotoxic effects of phenylurea herbicides; action mode of triazines and toxic effects on vertebrates; photolysis experiments on alloxydim herbicide and biological response of its transformation product; glyphosate-resistant weeds in Southern Europe; herbicide paraquat genotoxicity-enhancement by the phenolic antioxidants Dl-A-tocopherl and 2,6-Di-Tert-Butyl-P-Cresol; glyphosate adsorption in a soil depth profile; adsorption-desorption processes and mobility of (4-Chloro-2-Methylophenoxy) acetic acid (MCPA) in Irish grassland soils; pesticides and cancer with a study on the interaction of phenoxy acid herbicides with DNA; herbicidal activity of pyrazole derivatives; behavior of herbicides in paddy water and soil after application; and laboratory lysimeter for pesticide transport studies.

Chapter 1 – Plasmid pMC1 from *Delftia acidovorans* MC1, carrying the genes for productive degradation of racemic 2-phenoxypropionate (dichlorprop, mecoprop) and phenoxyacetate herbicides (2,4-D, MCPA) was transferred to plasmid-free strains of *Cupriavidus necator* and *Comamonas testosteroni*. Cultivation of the transconjugant strains on (*RS*)-2-(2,4-dichlorophenoxy)propionate revealed a limited efficiency of *C. necator* JMP222 (pMC1) to use these enantiomers, which, however, could be improved by a period of adaptation to the racemic compound. Real-time qPCR data revealed that the *rdpA* gene was only present at about 20% in the initial culture at steady state at low dilution rates but increased about 4-fold after prolonged selective growth. As a result, maximum growth rates of 0.24 – 0.27 h^{-1} on (*R*)-2,4-DP were achieved, which slightly exceeded those of the host strain (0.21 h^{-1}). In general, the activity pattern of JMP222 (pMC1) was similar to that of strain MC1 (pMC1). Yet, at low dilution rates the transconjugant strain left a significant amount of the substrate in the initial cultures unutilized whereas MC1 showed effective use throughout. The underlying higher catalytic efficiency with (*R*)-2,4-DP in MC1 went along with the finding that depending on the cellular environment posttranslational modification by reactive oxygen species (ROS) changed RdpA into multiple variants notably differing in *pI* values and individual kinetic characteristics. The presence of these enzyme variants was demonstrated by 2D gel electrophoresis (2D-GE) in accordance with a previous study

(Leibeling et al., 2013). The distribution of the enzyme variants differed in strains MC1 and JMP222 (pMC1) with more oxidative modifications in the latter. In contrast, *C. testosteroni* B337 (pMC1) utilized both enantiomers effectively at relatively low maximum growth rates of about 0.13 h^{-1}, but became apparently poisoned by accumulating toxic intermediates due to imbalances in the activity of the initial degradative oxygenases and the consecutive modified *ortho*-cleavage pathway. The reason might be due to the fact that RdpA and SdpA were very active in this strain with a catalytic efficiency and affinity about 3 to 5-fold higher than that of RdpA in MC1. The enzyme variant pattern of strain B337 (pMC1) was similar to that of MC1 regarding the 2D PAGE position (*pI*) of the main spots, but showed a higher total spot number. Together these results indicate that the strains' genetic background determines the extent to which an introduced catabolic trait takes effect. This knowledge will influence our understanding of selective forces working in contaminated environments.

Chapter 2 – The increase in human population has generated proportional food needs. However, agricultural productivity is constantly affected by factors decreasing land fertility. Weeds cause a constant reduction in the quantity and quality of important crops making their elimination a common necessity.

During the 40's and in subsequent years, the use and constant cycling of selective chemical herbicides considerably improved crop yields. However, overuse of herbicides to minimize the impact of weeds created the development of resistant species.

Since the first case in 1970, confirming the resistance of the common groundsel (*Senecio vulgaris*) against triazine herbicide, the number of resistant weeds against various herbicides has been considerably increasing. To date, more than 200 species have been reported to be resistant to different types of herbicides.

This review aims to address the herbicide resistance mechanisms developed by weeds, which the authors grouped as: Changes in the uptake and translocation of herbicides; herbicide detoxification mechanisms; altered target sites and herbicide cross-resistance.

Chapter 3 – Glyphosate is the world's most widely used herbicide due to favorable attributes of low cost, low toxicity, and high efficacy in controlling a wide range of weed species. The genetic engineering of important crops with the glyphosate-tolerant trait has led to extensive and largely exclusive glyphosate usage in many areas. The resulting selective evolutionary pressure has resulted in the emergence of glyphosate-resistant weed biotypes. Glyphosate contains a phosphonate ($-CPO_3^{2-}$) chemical group, with variable protonation depending upon pH, that provides for convenient detection and monitoring *via* ^{31}P NMR in plant tissue extracts as well as *in vivo*. Insights provided through ^{31}P-NMR studies of weed species exposed to glyphosate have improved our understanding of herbicide uptake and compartmentalization, resulting in key insights regarding resistance mechanisms. For example, the authors have previously reported the discovery and characterization of glyphosate vacuole sequestration as the principal resistance mechanism in horseweed (*Conyza canadensis*) and in ryegrass (*Lolium* spp.) from several continents. Herein, they expand their prior published ^{31}P-NMR studies to include additional weed biotypes, characterizing glyphosate uptake, vacuole sequestration and chloroplast exclusion.

Chapter 4 – Environmental degradation in Latin America has increased in the last three decades. The agricultural frontier of Argentina is spreading out, and this phenomenon has involved a significant increase in the use of biocides. Glyphosate (N-phosphonomethylglycine, PMG) is the most widely used herbicide in this country. Particularly the application of formulated glyphosate increased from 70 million kg in 1999 to more than 210 million in 2011. Although this herbicide is considered as "environmentally friendly" with moderate persistence, the mobility and fate will be conditioned by the chemical and biological profile of the environment, and regarding that, several authors have determined environmentally relevant levels of glyphosate and its metabolites in surface water. The risk associated with the use of PMG, could be assessed by the evaluation of the impact on a non-target species: *Cnesterodon decemmaculatus,* a Neotropical endemic teleost, widely distributed in the region. Experimental results would indicate that glyphosate (active ingredient, analytical grade, acid form) does not present a high level of toxicity (LC_{50}-96h > 100 $mg.L^{-1}$) on *C. decemmaculatus*. However, the toxicity of one of the glyphosate-based herbicide formulation was found to be significantly higher (LC_{50}-96h = 29 $mg.L^{-1}$). These results are consistent with other evidences that indicate a remarkable increase in the toxicity of formulations regarding to the active ingredient in freshwater teleosts.

Several scientists have focused their research on the identification of biomarkers of environmental stress in non-target species. These could be specific and early signals of the toxic-organism interaction. In this regard, the utility of morphological and biochemical parameters as quantitative bioindicators of xenobiotic exposure has been shown. Within this framework, the authors measured the impact of sublethal exposure of *Cnesterodon decemmaculatus* to PMG on antioxidant defenses (catalase activity) and neurotoxicity (acetylcholinesterase activity). After acute exposure, PMG caused changes in the enzyme activities: an increase in the activity of catalase and an inhibition of the acetylcholinesterase activity. These results are the first report indicating that sublethal PMG affect these biomarkers in a native teleost species.

Chapter 5 – A high number of herbicides is used to protect important crops, but an intensive use of these chemical compounds produces a general increasing of environmental pollution, and could be hazardous for human health. Thus, it is very important the identification of their mode of action, and the evaluation of their genotoxic properties, especially to assess the exposition level of people that come into contact via inhalation, skin contact, or ingestion of these compounds, and working in the areas where herbicides are used. Results concerning genotoxic activity of chemical compounds could be obtained using *in vitro* short term tests such as the evaluation of cytogenetic effects on exposed mammalian cell lines. Phenylurea herbicides are selective agents, widely used in agriculture for the control of weeds, belonging to C2 group of the Herbicide Resistance Action Committee (HRAC) classification, and are endowed by a mode of action that determines inhibition of photosynthesis at photosystem II complex. The genotoxic properties of some of these compounds were evaluated by a number of tests, including cytogenetic effects (chromosomal aberrations, and sister chromatid exchanges) on Chinese hamster cell lines. Here the authors present a review of chemical properties of the phenylurea herbicides, and their genotoxic effects evaluated with mutagenic assays.

Chapter 6 – Triazines are the family of herbicides that include atrazine, ametryn and simazine that are widely used in Brazil and can contaminate groundwater. Cattle can accumulate herbicides in their body through ingestion plants infested with these compounds and one of the ways, by which, human beings are exposed to atrazine is through cattle meat and milk consumption. The toxicity of these compounds can be explained mainly by their interaction with microsomal biotransformation processes. The herbicides per se or their metabolites or the secondary products of oxidative stress interact with biomolecules such proteins and DNA affecting a lot of cellular types. Cellular effects of chemicals might involve recruitment or de-repression of cell death mechanisms. Whether a cell survives or dies in the presence of a chemical insult is often determined by proliferative status, repair enzyme capacity, and the ability to induce proteins that either promote or inhibit the cell death process. In this chapter the authors will present an extensive bibliographical review about this herbicide class focusing its effects on vertebrates, looking for defense cellular mechanisms, at morphological and biochemical levels.

Chapter 7 – The photochemical degradation of alloxydim herbicide was investigated in different organic solvents simulating natural constituents of plant cuticles. Methanol and n-hexane were used as surrogates for primary alcohol groups and saturated hydrocarbon chains, respectively. Alloxydim was rapidly photodegraded in both solvents, showing a half-life of 21.94 ± 0.51 min in methanol and 7.5 ± 0.75 min in n-hexane.

During the photolysis, two transformation products were detected and their identification was achieved using liquid chromatography coupled to quadrupole time-of-flight mass spectrometry with electrospray ionization (LC-ESI-Qtof). The main photolysis reaction was the reduction of the oxime group leading to the formation of imine compound, whereas the isomerization of C-N bond was observed to a lesser extent.

Imine by-product was stable and it was isolated in order to study its phytotoxicity in a succeeding cereal crop (wheat) and a grass weed (Bromus). Alloxydim herbicide was more toxic than its main by-product to the tested species, in both bioassays tested, seed germination and plant growth. For instance, in seed germination bioassay, EC50 for alloxydim ranged from 0.33 to 0.88 ppm while for its by-product ranged from 100 to 277 ppm. Roots of all species were more sensitive than coleoptile to both compounds.

Chapter 8 – Glyphosate has been extensively used for more than three decades and it is clearly considered to be the most common herbicide worldwide. However, its use is now in danger because of the evolution and spread of glyphosate-resistant weeds. Adoption of glyphosate-resistant crops has certainly increased the reliance on glyphosate. Unfortunately, repeated use of glyphosate along with the absence of other proactive methods has greatly increased the risk of glyphosate resistance, even in the absence of glyphosate-resistant crops. That is also the case in southern Europe, where many herbicides are no longer registered, making weed control even more difficult and depense on glyphosate more intense. As a result, totally 5 weed species have been reported to have developed resistance to glyphosate in Spain, Greece, Italy, France and Portugal. Even if the number of weed species is rather low (being only about 20% of the global total), the problem is very serious and ongoing since most of these cases from southern Europe have arisen the last decade and the number of glyphosate-resistant biotypes keeps increasing at a worrying rate. This review summarizes the current status of glyphosate resistant weeds reported in these countries, as well as several methods to avoid future spread of glyphosate resistance. Moreover, the importance of diversity and good agricultural practices is highlighted, since they seem to be the only way for the maintenance of sustainability. Increased awareness of weed resistance by farmers, agronomists and other stakeholders is clearly among the first requirements of a proactive approach.

Chapter 9 – This study investigates the enhancement mechanism of toxicity of the herbicide 1,1'-dimetyl-4,4'-bipyridium dichloride (paraquat, PQ) in combination with dl-α-tocopherol (vitamin E, α-TH) or vitamin E synthetic analog 2,6-di-tert-butyl-p-cresol (butylated hydroxytoluene, BHT). Acute structural chromosomal damage induced by PQ + α-TH and PQ + BHT was investigated cytogenetically using *Pelophylax* (*P.*) *nigromaculatus* and *Rana* (*R.*) *ornativentris* leukocytes *in vitro*. Chemical PQ-cation reduction by BHT and sodium nitrite was examined by analysis of the results obtained from PQ cation reduction test as well. In a preliminary study, nitrite (one of nitric oxide degradation products) production test was conducted using the tails of *R. rugosa* tadpoles because of investigation of the possible involvement of nitrite in the chromosomal damage-enhancement-mechanism. From these results obtained so far, the following process is proposed for PQ toxicity-enhancement: an increase in PQ monocation radical formation initiated by vitamin E and BHT → reactive oxygen species (ROS) generation → nitric oxide synthase (NOS) activation → nitric oxide (NO) generation → nitrite production → further enhancement of PQ monocation radical formation by nitrite → ROS generation further increased → an acute increase in structural chromosomal damage.

Chapter 10 – Glyphosate [N-(phosphonomethyl)glycine, PMG] is the most popular, non-selective, post-emergent and widely used herbicide in agriculture, especially in soybean crops. The Ap (0-18 cm), AB (18-50 cm) and BC (105-130 cm) horizons of a Typic Haplustoll soil, located at 62°06´W, 27°24´S, in Province of Santiago del Estero, Argentina, were characterized by chemical analysis, X-ray Diffraction (XRD), Point of Zero Charge (PZC) and surface area by water adsorption (S_w). The adsorption of glyphosate by soil samples of the three horizons was measured. Adsorption isotherms were fitted using the Langmuir and Freundlich model, The affinity constant (K_L) and maximum surface coverage (Γ_{max}) were determined and the results showed that Γ_{max} of the horizons increased with depth Ap<AB<BC and K_L followed the sequence BC< Ap < AB.

Chapter 11 – The aim of this study was to quantify the adsorption-desorption isotherms of (4 –chloro-2-methylophenoxy) acetic acid (MCPA) in selected Irish grassland soils, and to assess its mobility potential to groundwater. The sorption isotherms were determined using a batch equilibrium method and were very well described by the linear and the Freundlich equation ($R^2 \geq 0.99$). The adsorption-desorption process tended towards linearity with the isotherm exponent (1/n) ranging from 0.90 to 0.97 for adsorption and from 0.92 to 1.03 for desorption. The adsorption coefficients calculated with the linear and the Freundlich equation were comparable. The Freundlich adsorption coefficient (K_f) ranged from 2.29 to 5.01 ($mg^{1-1/n}$ kg^{-1}) $(L)^{1/n}$ indicating low adsorption. Such behaviour was probably a result of electrostatic repulsion between negatively charged soil particles and herbicide anionic molecules formed in the soil/water system since the pH values of soils were greater than the dissociation constant of MCPA (pK_a = 3.07). Greater adsorption and consequently lower desorption was observed in the soil with the lowest pH, demonstrating some pH dependence, although this relationship was not significant (K_f, -0.732, P = 0.268). The organic carbon normalized coefficient (K_{oc},) ranged from 48.6 to 107.9 L kg^{-1}, suggesting that MCPA will be highly mobile in soils with similar characteristics to those studied, although this can be limited by environmental conditions in the field.

Chapter 12 – Herbicides are important for the control of weed growth but increasing use causes environmental problems and undesirable side effects in crops. Phenoxy acid herbicides with auxin-like activity have been used against grass and broad leaf weeds in many crops, such as rice, winter wheat and soybean. Carcinogenicity in humans and embryotoxicity in animals have been described as the main hazards of phenoxy acid herbicides.

Thermal denaturation of double stranded DNA has been used as a measure of the effect of harmful chemical substances on the structure and stability of the DNA molecule. Hence, the interaction between several phenoxy acid herbicides (phenoxyacetic acid, 4-

chlorophenoxyacetic acid, MCPA, mecoprop, 2,4-D and dichlorprop) and Calf thymus DNA was assessed by UV–Vis absorption spectroscopy. UV spectra and melting curves have been recorded for solutions at constant DNA concentration and using different concentrations of each of the herbicides under study. The transition temperature values and the thermodynamic parameters of DNA thermal denaturation have been determined. The studies performed so far help to understand herbicide–DNA interactions gathering suitable information for the (re)design of new molecules with lower potential risk to humans.

Chapter 13 – Herbicides are widely used in crop production all over the world. Among them, protoporphyrinogen oxidase (Protox) inhibitors are one of the most important classes of herbicides, which inhibit Protox in the chlorophyll biosynthetic pathway, resulting in light-induced membrane lipid peroxidation. Targeting the porphyrin pathway, these herbicides show high activity and low toxicity, and thus have become a hot-point of novel pesticides research.

To search for novel Protox inhibitors, several pyrazole derivatives were synthesized from 1-(4-chloro-2-fluoro-5-methoxyphenyl)ethanone, *via* a condensation, ring closure reaction, methylation, chlorination, demethylation and alkylation. The single crystal of 4-chloro-3-[4-chloro-2-fluoro-5-(2-methyl)allyloxyphenyl]-1-methyl-5-trifluoromethyl-1*H*-pyrazol was prepared, and its structure was further determined by X-ray analysis. The preliminary bioassay showed some promising results to tested gramineous weeds and latifoliate weeds. The study on the structure-activity relationship suggests that chlorine substituent in 4-position of pyrazole is important for high activity. When hydrogen is located in 4-position of pyrazole, there is no herbicidal activity or low activity, but when chlorine is on pyrazole, most of them show good herbicidal efficiency.

Chapter 14 – Degradation and distribution of herbicides in paddy water and paddy soil after application to paddy fields including experimental paddy plots are reviewed to examine the half-life and the organic carbon normalized soil sorption coefficient (K_{OC}). The target compounds are 10 herbicides: bromobutide, pyrazolynate, mefenacet, simetryn and thiobencarb, and two degradation products, bromobutide-debromo and destosyl pyrazolynate. Their half-lives in paddy water and soil after application are assessed. Their respective K_{OC} in paddy fields (K_{OCP}) are calculated based on their reported concentrations in paddy water and soil. The variations of K_{OCP} values are evaluated and discussed through comparison with the reported K_{OC} using laboratory experiments and calculations.

Chapter 15 – The transport of pesticides through the soil is one of many processes controlling their final distribution in the environment. The presence of pesticides in soil, groundwater, surface water and air may cause considerable adverse effects on ecosystems and human health. Human health may be affected by pesticide residues in food and drinking water, while ecosystems may be affected by loss of biodiversity and decreases in populations of sensitive organisms. A preliminary investigation into the use of a laboratory lysimeter apparatus to study transport of pesticides through undisturbed soil columns was undertaken. Two phenoxyalkanoic acid herbicides (MCPA and mecoprop-p) were applied simultaneously to a soil subject to long-term tillage and grassland management. The amounts leached from the tillage soil accounted for 3.5 mg of MCPA and mecoprop-p, which was approximately 66% and 68% of the initially applied amount, respectively. The amounts leached from the grassland soil accounted for 3.6 mg of MCPA (approximately 69% of initial application) and 3.0 mg of mecoprop-p (approximately 58% of initial application). Even though the quantities leached were comparable, the peak concentration appeared much sooner in the soil with the greatest saturated hydraulic conductivity. This observation may be important for pesticides with a long degradation half-life, as the time they need to degrade to a less harmful form (i.e. CO_2) may not be sufficient before reaching groundwater. The study showed the usefulness of the lysimeter apparatus for studying pesticide transport through undisturbed soil columns under controlled laboratory conditions.

In: Handbook on Herbicides
Editors: Daiki Kobayashi and Eito Watanabe

ISBN: 978-1-62948-053-4

Chapter 1

PRODUCTIVE DEGRADATION OF DICHLORPROP BY TRANSCONJUGANT STRAINS CARRYING PLASMID PMC1 OF *DELFTIA ACIDOVORANS* MC1 WITH SPECIAL FOCUS ON THE ACTIVITY OF RDPA

Sabine Leibeling[1], Karin Glaser[1], Monika Neytschev[1], Hauke Harms[1] and Roland H. Müller[1]

[1]UFZ – Helmholtz Centre for Environmental Research,
Department of Environmental Microbiology,
Leipzig, Germany

ABSTRACT

Plasmid pMC1 from *Delftia acidovorans* MC1, carrying the genes for productive degradation of racemic 2-phenoxypropionate (dichlorprop, mecoprop) and phenoxyacetate herbicides (2,4-D, MCPA) was transferred to plasmid-free strains of *Cupriavidus necator* and *Comamonas testosteroni*. Cultivation of the transconjugant strains on (*RS*)-2-(2,4-dichlorophenoxy)propionate revealed a limited efficiency of *C. necator* JMP222 (pMC1) to use these enantiomers, which, however, could be improved by a period of adaptation to the racemic compound. Real-time qPCR data revealed that the *rdpA* gene was only present at about 20% in the initial culture at steady state at low dilution rates but increased about 4-fold after prolonged selective growth. As a result, maximum growth rates of 0.24 – 0.27 h^{-1} on (*R*)-2,4-DP were achieved, which slightly exceeded those of the host strain (0.21 h^{-1}). In general, the activity pattern of JMP222 (pMC1) was similar to that of strain MC1 (pMC1). Yet, at low dilution rates the transconjugant strain left a significant amount of the substrate in the initial cultures unutilized whereas MC1 showed effective use throughout. The underlying higher catalytic efficiency with (*R*)-2,4-DP in MC1 went along with the finding that depending on the cellular environment posttranslational modification by reactive oxygen species (ROS) changed RdpA into multiple variants notably differing in *pI* values and individual kinetic characteristics. The presence of these enzyme variants was demonstrated by 2D gel electrophoresis (2D-GE) in accordance with a previous study (Leibeling et al., 2013). The distribution of the enzyme variants differed in strains MC1 and JMP222 (pMC1) with more oxidative modifications in the latter. In contrast, *C. testosteroni* B337 (pMC1) utilized both enantiomers effectively at relatively low maximum growth rates of about 0.13 h^{-1}, but became apparently poisoned by accumulating toxic intermediates due to imbalances in the activity of the initial degradative oxygenases and the consecutive modified *ortho*-cleavage pathway. The reason might be due to the fact that RdpA and SdpA were very active in this strain with a catalytic efficiency and affinity about 3 to 5-fold higher than that of RdpA in MC1. The enzyme variant pattern of strain B337 (pMC1) was similar to that of MC1 regarding the 2D PAGE position (*pI*) of the main spots, but showed a higher total spot number. Together these results indicate that the strains' genetic background determines the extent to which an introduced catabolic trait takes effect. This knowledge will influence our understanding of selective forces working in contaminated environments.

INTRODUCTION

Plasmids are of vital importance for microbial evolution as these mobile elements provide their host cells with extraordinary traits like antibiotic resistances or degradation pathways and thus enhance their capabilities of adapting to fluctuating environmental conditions (Eberhard, 1990; Fondi et al., 2010; Frost et al., 2005; Slater et al., 2008; Top and Springeal, 2003). Plasmids permit genetic information to be accumulated in the cell without altering the gene content of the bacterial chromosome and to be spread by homologous recombination and horizontal gene transfer (Bergstrom et al., 2000; Davison, 1999; Janssen et al., 2005). A well-studied plasmid is pJP4 of *Cupriavidus necator* JMP134 (former *Ralstonia eutropha;* Vandamme and Coenye, 2004). pJP4 carries the complete set of genes encoding enzymes involved in degrading the xenobiotic phenoxyalkanoate herbicides 2,4-dichlorophenoxyacetate (2,4-D) and 4-chloro-2-methylphenoxyacetate (MCPA). The various structural genes and regulatory elements for transcriptional control are structurally organized in operons (Hoffmann et al., 2003; You and Ghosal, 1995). On plasmid pJP4 two operons were present encoding one set of corresponding genes, each. The expression of both sets was shown to be essential for effective degradation of the herbicides (Laemmli et al., 2000; Leveau et al., 1999; Pérez-Pantoja et al., 2000, 2003). Composition and arrangement of the various genetic elements have been explained based on the underlying structural and functional units (Trefault et al., 2004). pJP4 is a self-transmissible, broad-host range plasmid. Its dissemination was demonstrated by introducing *C. necator* JMP134(pJP4) into biotopes resulting in conjugative transfer to indigenous strains (Bathe et al., 2004; Dejonghe et al., 2000; DiGiovanni et al., 1996; Kleinsteuber et al., 2001; Newby et al., 2000a, b). Although screening methods showed evidence for a successful transfer of the plasmid, significant expression was verified only for a limited number of species in the biotope (Dejonghe et al., 2000). This was in line with earlier findings revealing that after acquiring pJP4, recipient strains did not always express it in an effective manner (Don and Pemberton, 1981; Ghosal et al., 1985; Kleinsteuber et al., 2001; Pérez-Pantoja et al., 2000). Apparently, the host strain discriminates whether the genetic information carried on this plasmid is employed or not.

Delftia acidovorans MC1 hosts a 120 kb plasmid that carries genes of the *ortho*-cleavage pathway in addition to *tfdB* (coding for chlorophenol hydroxylase), as well as *rdpA* and *sdpA* (Müller et al., 2001; Schleinitz et al, 2004). The latter two genes encode α-ketoglutarate-dependent dioxygenases which catalyze the initial degradation step of phenoxyalkanoate herbicides. RdpA and SdpA have distinct enantio- and substrate-specific properties (Westendorf et al., 2003, 2006). Hence, these two enzymes determine the substrate spectrum of strain MC1, which comprises the racemic compounds (*RS*)-2-(2,4-dichlorophenoxy)propionate (dichlorprop, (*RS*)-2,4-DP) and (*RS*)-2-(2-methyl-4-chlorophenoxy) propionate (mecoprop, (*R*S)-MCPP) as well as the phenoxyacetate derivatives 2,4-dichlorophenoxyacetate (2,4-D) and (2-methyl-4-chlorophenoxy)acetate (MCPA) (Müller et al., 2001; Westendorf et al., 2003, 2006). In addition to *D. acidovorans* MC1, *Rhodoferax* sp. P230 (Ehrig et al., 1997; Müller et al., 2001) and *Sphingobium herbicidovorans* MH (Nickel et al., 1997; Zipper et al., 1996) feature RdpA enzymes identical in their primary amino acid sequence (Müller et al. 2004; Schleinitz et al., 2004). Nevertheless, the kinetic performances were significantly different in these three strains (Westendorf et al., 2006). This, too, indicates that the physiological background of the species modulates the respective degradative performance.

Based on these findings, the present investigation was directed at elucidating the degradative capacities of plasmid pMC1 after conjugative transfer into other gram-negative strains, i.e. in plasmid-free strains of *Comamonas testosteroni* and *Curpriavidus necator*. The effects were followed during transient state cultivation by determining the maximum substrate consumption rates, i.e. the maximum growth rates with (*RS*)-2,4-DP as substrate. Furthermore, at increasing dilution rates, the kinetic characteristics were determined and relative gene expression was evaluated by real-time qPCR. Special attention was paid to the properties of RdpA since previous investigations were able to refer differences in the kinetic properties of RdpA of *D. acidovorans* MC1, *Rhodoferax* sp. P230 and *Sphingobium herbicidovorans* MH to alterations in the enzyme variant pattern of the respective strains (Westendorf et al., 2006). Such differences were also associated with enhanced 2,4-D degradation capabilities of RdpA (Leibeling et al., 2013). It was therefore of interest to obtain the RdpA enzyme variant patterns of the transconjugant strains by 2D-GE (gel electrophoretic separation). Comparison of the transconjugant strains with *D. acidovorans* MC1 as the host of plasmid pMC1 helped to gain more information on strain specific influences regarding the expression and performance of plasmid-encoded catabolic enzymes.

MATERIALS AND METHODS

Strains, Cultivation Conditions and Conjugation

The strains used in this study are *Delftia acidovorans* MC1 (Müller et al., 1999), *Cupriavidus necator* (formerly *Wautersia eutropha* > *Ralstonia eutropha* > *Alcaligenes eutrophus*) JMP222 (provided by Pemberton et al., 1979, UFZ storage no. B619), and *Comamonas testosteroni* B337 (UFZ storage no. B337). *D. acidovorans* MC1 served as donor strain since it possessed plasmid pMC1 which carries both, *rdpA* and *sdpA* as well as the lower pathway genes (Schleinitz et al., 2004). It was stored on 1.5% agar plates (pH 7 – 7.5) with minimal medium (MM stock), 1/1000 (v/v) trace element solution (TES) and 1.7 mM (*RS*)-2,4-DP or (*R*)-2,4-DP as carbon and energy source. The MM stock contained (in mg l^{-1}): NH_4Cl, 760; KH_2PO_4, 340; K_2HPO_4, 485; $CaCl_2 \times 6\ H_2O$, 27; $MgSO_4 \times 7\ H_2O$, 71.2) and the TES contained (in mg l^{-1}): $FeSO_4 \times 7\ H_2O$, 4.98; $CuSO_4 \times 5\ H_2O$, 0.785; $MnSO_4 \times 4\ H_2O$, 0.81;

$ZnSO_4 \times 7\ H_2O$, 0.44; $Na_2MoO_4 \times 2\ H_2O$, 0.25). (*RS*)-2,4-DP was obtained from Sigma-Aldrich (purity 97 – 99%). (*R*)-2,4-DP was prepared from Duplosan (BASF®) in our laboratory. Purification by precipitation and ether extraction resulted in a 95.6% pure product. (*S*)-2,4-DP was synthesized according to Zipper (Zipper, 1998).

For conjugation colonies of *D. acidovorans* MC1 were picked to inoculate liquid rich medium that contained 1 g l^{-1} of each peptone and yeast extract, 3.3 mM fructose, and 0.85 mM (*RS*)-2,4-DP (PYE + FH medium). *C. necator* JMP222 and *C. testosteroni* B337 are the plasmid-free derivative strains of *C. necator* JMP134 (Müller et al., 2001; Top et al., 1996) and *C. testosteroni* (DSM-No. 50244). Plasmid-free strains were used since in case of *C. necator* pJP4 already encodes genes of the 2,4-dichlorophenoxyacetic acid degradation pathway (Top et al., 1996). JMP222 and B337 were stored and grown on rich medium and served as acceptor strains for plasmid pMC1. Plate mating was performed by spreading 200 μl of overnight cultures of each, the donor strain and one of the acceptor strains onto PYE + FH medium. After incubation for one day at 30°C, the cell suspension was stepwise diluted, and plated onto agar plates containing either minimal medium with 1.7 mM (*RS*)-2,4-DP or rich medium. The media were supplemented with nalidixin (10 mg l^{-1}) in the case of strain B337 or with moxalactam (30 mg l^{-1}) in the case of strain JMP222 to suppress growth of the donor strain. Colony formation was followed on minimal media plates with (*RS*)-2,4-DP. Colonies showing strong growth were selected and purified by further serial dilutions and selective growth on herbicide plates. Finally, the colonies were identified using MicroPlate test arrays (Biolog Inc.) and subsequent 16S rDNA analysis. Care was taken to disregard positive antibiotics-resistant clones of strain MC1 which occurred with low frequency in the presence of both 10 mg l^{-1} moxalactam and 10 mg l^{-1} nalidixin, respectively. Clones of the transconjugant strains were tested for their capability to completely degrade (*RS*)-2,4-DP. Therefore, they were inoculated into liquid PYE medium supplemented with 0.43 mM (*RS*)-2,4-DP and incubated overnight. Subsequently, (*RS*)-2,4-DP degrading suspensions were transferred two more times on minimal medium with 0.85 mM (*RS*)-2,4-DP using 10% of the inoculum. The resulting suspension was plated onto agar plates with minimal medium containing 0.85 mM (*RS*)-2,4-DP. Colonies with strong growth were selected and again transferred onto a new agar plate with minimal medium containing 0.85 mM (*RS*)-2,4-DP. These plates were used to prepare stocks of the transconjugant strains *C. necator* JMP222 (pMC1) and *C. testosteroni* B337 (pMC1) after transfer into 50% glycerol to be stored at –80°C under nitrogen. Colonies from the above plates were also used to inoculate an overnight culture in PYE + FH medium. The resulting suspensions served as inoculum for continuous cultivation.

Continuous Cultivation

Continuous cultivation was performed in a lab fermenter (Biostat D, B. Braun, Melsungen, Germany) at a working volume of 0.8 l at 30°C, pH 7.5 ± 0.1; the pH was kept constant by automatic titration of 1 N HCl and 1 N NaOH. The aeration rate was set to 0.1 m^3 h^{-1}, the stirrer speed was 700 rpm, and dissolved oxygen in the suspension was ≥ 50%. Cultivation was initiated from fed-batch mode with minimal medium that contained 0.43 – 0.85 mM (*RS*)-2,4-DP, fed at a rate too low for the toxic intermediate 2,4-dichlorophenol (2,4-DCP) to accumulate. Substrate degradation and intermediate formation were monitored by HPLC. Once the substrate consumption had stabilized, continuous cultivation was started by stepwise increasing the dilution rate to D = 0.02 – 0.055 h^{-1} until a steady state was reached. The medium was fed via two separate streams of MM stock solution and herbicide solution (herbicide stock) contributing 2/3 and 1/3 of the total volumetric flow, respectively. The herbicide stock contained 12.8 mM (*R*)-2,4-DP or (*RS*)-2,4-DP dissolved in 26 mM NaOH. Trace elements were fed separately from a stock solution (TES) of pH 2.0 at 1/1000 (v/v). To analyze the degradation capacities of the transconjugant strains and *D. acidovorans* MC1, transient-state cultivation techniques were applied as described elsewhere (Müller et al., 1995). Starting from steady state this was achieved by increasing the dilution rate in a step-wise manner with ΔD = 0.003 h^{-1} or 0.005 h^{-1} and at time intervals of Δt = 2 h or 4 h. Specifics are given in the respective figure descriptions. Growth, i.e. biomass and herbicide degradation via stationary herbicide concentrations were monitored by OD_{700} and HPLC, respectively (Müller and Babel, 2001). The maximum growth rate (μ_{max}) was estimated based on accumulation properties of residual substrate (Hoffmann and Müller, 2006; Müller et al., 1995). For practical reasons μ_{max} of some cultures was estimated by performing a wash-out experiment, i.e. by raising the dilution rate above a (putative) maximum growth rate (Esener et al., 1981). The μ_{max} value was then obtained from the logarithmically decay of the biomass concentration [$\mu_{max} = D - (\ln x_t - \ln x_o/t)$].

Determination of Enzyme Activity in Cell-Free Extracts, and after Purification

Enzyme activities were determined from cell-free extracts and partially purified enzyme. To obtain cell-free extracts, fermenter effluent was collected on ice, washed, re-suspended at a density of about 15 g dry mass l^{-1} in Tris/HCl buffer and disintegrated by three passage through a French pressure cell (American Instrument Company, Silver Spring, MD, USA) as described elsewhere (Leibeling et al., 2013). RdpA was partially purified by precipitation with protamine sulfate and ammonium sulfate followed by ion exchange chromatography according to Westendorf et al. (Westendorf et al., 2002, 2003, 2006). The fractions containing highest RdpA activity were pooled and concentrated by ultrafiltration using Vivaspin 10 tubes (GE Healthcare). The enzyme was stored in fractions at –20°C until use. RdpA and SdpA activities were assayed colorimetrically by tracing the formation of 2,4-DCP, the product of the cleavage reaction (Westendorf et al., 2003, 2006). Both enzymes were incubated at their optimum temperatures of 30°C and 25°C, respectively. The reaction was followed over a time period of 2 min. Samples were sequentially taken every 30 s starting at 0 min. GraphPad Prism software (www.graphpad.com/prism) was used to determine the catalytic parameters K_m, calculated by nonlinear regression, and the relative catalytic efficiency of cell-free extracts ($V_{max\ rel}/K_m$). $V_{max\ rel}$ was used to determine the catalytic efficiency since the usually measured k_{cat} can only be determined for purified enzymes. Using relative rates, moreover, ensures to exclude individual differences in enzyme activities caused by preparation and storage conditions and facilitates direct comparison of catalytic efficiencies in the various strains.

Chlorocatechol-1,2-dioxygenase (TfdC) and chloromuconate cycloisomerase (TfdD) are the first two enzymes of the chlorocatechol degradation pathway. Their activities were determined in crude extracts by following product formation colorimetrically as reported previously (Müller et al., 2001). The specific enzyme activities were expressed as U (mg protein)$^{-1}$.

2D Gel Electrophoresis, Determination of Protein Spot Positions and Relative Protein Content

For proteomic analyses fermenter effluent of the transconjugant strains was collected at steady state and treated by SDS lysis to obtain cell-free extracts (Leibeling et al., 2010). 2D gel electrophoresis (2D-GE) followed a standard protocol (Benndorf et al., 2004). Specifics regarding sample purification by phenol extraction, protein separation and staining procedures using colloidal Coomassie Brilliant Blue G-250 (Neuhoff et al., 1988) are described in detail by Leibeling et al. (2013). RdpA spot positions were acquired using ImageJ 1.43t software (Wayne Rasband, National Institute of Health, USA, http://rsb.info.nih.gov/ij) as described previously (Leibeling et al., 2013). In brief, RdpA spot positions were deduced from their relative positions to reference proteins of *Delftia acidovorans* MC1100. Strain MC1100 itself is deficient of RdpA and SdpA. Thus, 50 μg purified RdpA of *C. testosteroni* B337 (pMC1) were admixed to 150 μg crude extract of MC1100. *C. necator* JMP222 was treated accordingly. RdpA variants and reference spots were identified by MALDI-MS after trypsin digestion (Hehemann et al., 2008). Delta 2D software (version 3.3, Decodon, Greifswald, Germany) was used for matching 2D gels and determining relative spot intensities after normalization as reported elsewhere (Leibeling et al., 2013). RdpA spot positions and relative spot quantities of the transconjugant strains were compared to *D. acidovorans* strain MC1071 (Leibeling et al., 2013) a derivative strain of the plasmid donor strain MC1 which expresses RdpA but not SdpA (RdpA$^+$, SdpA$^-$) (Leibeling et al., 2013; Müller, 2007; Müller and Hoffmann, 2006).

Real-Time qPCR

For real-time quantitative polymerase chain reaction (real-time qPCR) fermenter effluent was collected at different dilution rates, centrifuged, and the cell pellets stored at –80°C until further use. DNA was isolated from thawed cell pellets using the MasterPure™ DNA Purification Kit (Epicentre Biotechnologies) according to the manufacturer's protocol. Real-time qPCR reaction mixes were prepared in triplicate using 5 μl SsoFast™ EvaGreen® supermix (Bio-Rad) according to the manufacturer's instructions, 1 μl template DNA (dilution 1:100), 0.6 μM of each primer and sterile nano-pure water for PCR as required for a final volume of 10 μl. Real-time qPCR was performed in a CFX96 Real-Time System (Bio-Rad) along with CFX Manager Software (Bio-Rad) for analysis. The cycling program consisted of an initial denaturation step at 98°C for 2 min followed by 40 cycles of 5 s at 98°C, 5 s at 52.7°C, 5 s at 60°C and 5 s at 82°C, where the fluorescence acquisition was performed and the primer dimer had melted but the specific product had not. Subsequently, the melt curve with an initial soak step of 95°C for 10 s was recorded with increments of 0.5°C for 3 s, starting at 65°C up to 95°C.

All primer pairs used in this study are listed in Table 1. Target genes were the plasmid pMC1 encoded *rdpA* and *sdpA* genes coding for the initial phenoxyalkanoate herbicide cleaving enzymes RdpA and SdpA. As reference gene, *trfA* was chosen. *trfA* encodes the replication initiation protein and is a single-copy gene located on the plasmid. To evaluate the expression level of the target genes it is normalized to *trfA* giving the x-fold different expression in relation to plasmid pMC1. Relative expression was calculated using the algorithm published by Pfaffl (2001). For the corrected calculation, the efficiency of each primer pair was determined by measuring a 1:10 dilution series of a representative sample. Presuming that the expression of the target genes is optimal at maximum growth rates we set the expression level at the highest dilution rate to one and assessed the gene expression at all other dilution rates accordingly. Normalizing samples to the plasmid-encoded reference gene *trfA* allowed assessing changes in the expression of *rdpA* and *sdpA* per plasmid. Each real-time qPCR run also included an inter-run calibrator (IRC) in triplicate to correct for run-to-run differences allowing all samples to be treated as if they were analyzed in the same run (Derveaux et al., 2010). Therefore a fresh culture of *D. acidovorans* MC1 (50 ml) was grown in batch on (*RS*)-2,4-DP for two days. DNA was isolated as described above, diluted (1:3000) with sterile nano-pure water for PCR and stored in ready-made aliquots of 15 μl at –20°C.

Analytic Methods

Biomass concentration was measured as optical density at 700 nm (OD_{700}) and as dry mass obtained in quadruple after drying the biomass samples to constant weight at 104°C and cooling them down in a desiccator. Phenoxyalkanoic acids were measured by HPLC according to Oh and Tuovinen (1990). Separation and determination of R- and S-enantiomers was achieved using a Nucleodex-α-PM column (200 by 4 mm) with permethylated α-dextrin as chiral stationary phase (Macherey-Nagel, Düren, Germany).

Table 1. Primer pairs used for real-time qPCR

Target gene	Primer name	Sequence 5' → 3'	Product length (bp)
rdpA	rdpA_f*	CACTTACCAGGTCATCTA	
	rdpA_r*	GTGTACATGCTCAGAAAAC	285
sdpA	sdpA_f*	ATGCCGACAGCACCTACAT	
	sdpA_r1*	GGTCTCGGGATGCACCTT	287
trfA	trf_f1	TTCGCAAGCTGTCGCCCACC	
	trf_r1	GCACCAGGTCGTCAT TGATC	210
rpoB	rpo_f2	GTCCKTACTCKCTGGTGACGC	
	rpo_r2	TCCTTWACCAGCACGTTGAAC	242

*Primers were designed and tested by Paulin et al. 2010.

Results

Growth Patterns

In order to examine the degradation capacity of the transconjugant strain *C. necator* JMP222 (pMC1), cells were grown continuously in a transient-state manner by raising the dilution rate in distinct steps and time intervals. The given dilution rates are in accordance with the specific growth rates (Müller et al., 1995). Although the strain was stored on agar plates with (*RS*)-2,4-DP as selective substrate, restricted herbicide utilization became obvious already at low dilution rates of 0.055 h^{-1}. By then concentrations of about 0.2 and 0.14 mM of the R- and S-enantiomer had accumulated, respectively (Figure 1) whereas they remained below the detection limit under the same conditions with the donor strain *D. acidovorans* MC1. When a dilution rate of 0.1 h^{-1} was reached the residual herbicide concentration amounted to 0.5 – 0.6 mM of each enantiomer (Figure 1). Surprisingly, a further increase of the dilution rate to about 0.15 h^{-1} did not result in a continued accumulation of residual herbicide but rather a stable substrate level (Phase 1). In response to this observation, the transient state cultivation was interrupted, and the dilution rate decreased to D = 0.12 h^{-1} until a new steady state was established (initiating phase 2). This time the culture showed very different substrate utilization characteristics at increasing dilution rates. Unlike in phase 1, the stationary (*R*)-2,4-DP concentration tended to be zero and remained low even after increasing D to about 0.18 h^{-1}. Although the stationary (*S*)-2,4-DP concentration remained lower than in phase 1 it did not fall below 0.29 mM and increased steadily with an increasing dilution rate until a new maximum at D = 0.175 h^{-1} was reached. Growth seems to be mainly supported by (*R*)-2,4-DP since it resided on a low level even at high dilution rates whereas the concentration of the S-enantiomer increased in a progressive manner under these conditions. To estimate the maximum growth rate (μ_{max}) of *C. necator* JMP222 (pMC1) a wash-out experiment was conducted by raising the dilution rate in two steps to D = 0.275 h^{-1} and D = 0.305 h^{-1}. Based on the logarithmical decay in biomass a μ_{max} of 0.24 h^{-1} on (*R*)-2,4-DP was calculated. The presence of residual (*S*)-2,4-DP seemed not to inhibit the utilization of the R-enantiomer significantly as performing a wash-out experiment with (*R*)-2,4-DP as sole substrate revealed a similar μ_{max} of 0.27 h^{-1}. The growth patterns appear to be characteristic for *C. necator* JMP222 (pMC1) as repeated experiments (data not shown) using new inoculums showed similar behavior, i.e. (*R*)- and (*S*)-2,4-DP accumulated to high levels in a first growth phase when applying moderate dilution rates up to 0.1 h^{-1}, while interrupting and restarting the gradient at low dilution rates enabled elevated maximum growth rates in a second phase.

Transconjugant strain *C. testosteroni* B337 (pMC1) behaved completely different from *C. necator* JMP222 (pMC1). Continuous cultivation at D = 0.055 h^{-1} indicated effective use of (*RS*)-2,4-DP. The residual (*S*)-2,4-DP concentration remained low at about 0.032 mM and that of (*R*)-2,4-DP was even below the detection limit over a wide range of dilution rates (Figure 2). This is in general similar to the behavior of *D. acidovorans* MC1, which utilizes both enantiomers when cultivated on (*RS*)-2,4-DP and achieves a μ_{max} of about 0.21 (Hoffmann and Müller, 2006). Step-wise increasing the dilution rate with *C. testosteroni* B337 (pMC1) resulted in the accumulation of both (*R*)- and (*S*)-2,4-DP but only when D approached and exceeded 0.13 h^{-1} thus defining μ_{max} with both enantiomers.

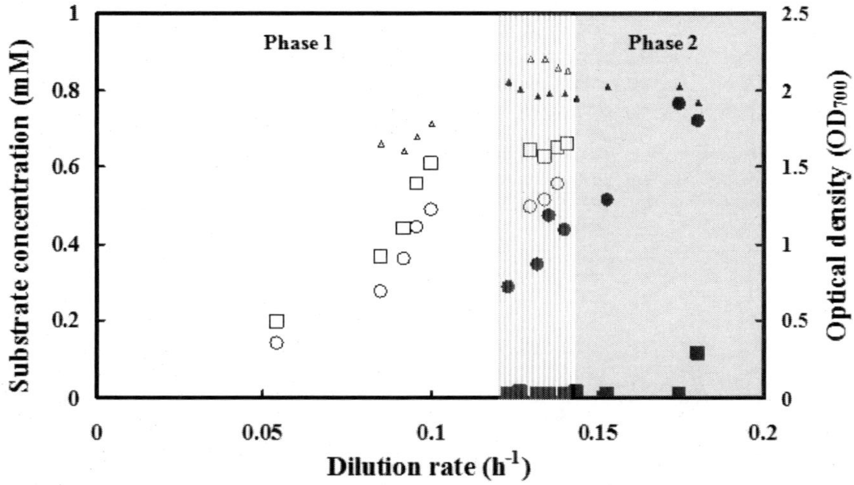

Figure 1. Optical density (triangles) and substrate consumption patterns of *C. necator* JMP222 (pMC1) during transient state cultivation on (*RS*)-2,4-DP at increasing dilution rates with $\Delta D = 0.003\ h^{-1}$ at time intervals of $\Delta t = 2$ h. The strain was first cultivated on (*RS*)-2,4-DP until reaching steady state conditions at $D = 0.055\ h^{-1}$. Then, the dilution rate was step-wise increased in a range of $D = 0.055 – 0.14\ h^{-1}$ (Phase 1, open symbols). Phase 2 was initiated by reducing the dilution rate to $D = 0.12\ h^{-1}$ until a new steady state was reached. Subsequently, the dilution rate gradient was restarted from 0.12 to 0.18 h^{-1} (Phase 2, filled symbols). (*R*)-2,4-DP (squares) and (*S*)-2,4-DP concentrations (circles) were measured using an enantioselective HPLC column. Shadings visualize Phase 1 (white) and 2 (grey) while stripes indicate overlapping dilution rates.

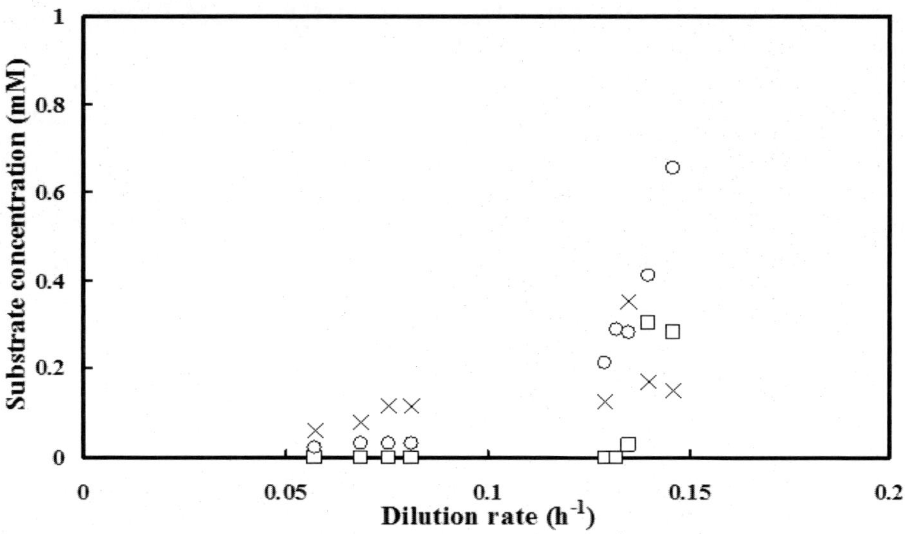

Figure 2. Substrate consumption patterns indicated by residual substrate and formation of 2,4-DCP (crosses) of *C. testosteroni* B337 (pMC1) during transient state cultivation on (*RS*)-2,4-DP at increasing dilution rates with $\Delta D = 0.003\ h^{-1}$ at time intervals of $\Delta t = 2$ h. The strain was first cultivated on (*RS*)-2,4-DP until reaching steady state conditions at $D = 0.055\ h^{-1}$. Then, the dilution rate was step-wise increased in a range of $D = 0.055 – 0.15\ h^{-1}$. (*R*)-2,4-DP (squares) and (*S*)-2,4-DP (circles) concentrations were measured using an enantioselective HPLC column.

Running D-gradient experiments with (*R*)-2,4-DP as sole growth substrate (data not shown) revealed similar μ_{max} values as with (*RS*)-2,4-DP. Interestingly, the toxic intermediate 2,4-dichlorophenol (2,4-DCP) was permanently detectable during cultivation with both substrates (*RS*)-2,4-DP and (*R*)-2,4-DP. Repeated experiments with longer cultivation times and lower gradients did not result in higher threshold rates for both enantiomers and did not avoid the excretion of 2,4-DCP. Approaching dilution rates of D = 0.13 h^{-1} or higher was accompanied by turning the culture purple, caused by the accumulation of further intermediates like 3,5-dichloromuconic acid besides 2,4-DCP. Subsequently, the culture was washed out.

Enzyme Activities and Kinetics

Enzyme activities were measured in crude extracts taken from chemostat runs with racemic substrate. The specific degradation activity of *C. necator* JMP222 (pMC1) with (*R*)-2,4-DP appeared to be correlated with the dilution rate gradient of the chemostat. It increased from 0.1 U (mg protein)$^{-1}$ at initial steady state (D = 0.055 h^{-1}) to 0.27 U (mg protein)$^{-1}$ when almost reaching μ_{max} at D = 0.22 h^{-1}. The latter slightly exceeded the specific activity of *D. acidovorans* MC1 with (*R*)-2,4-DP obtained after chemostatic cultivation on the racemic substrate at pH 8.5 (Hoffmann and Müller, 2006). In contrast, the specific activity of strain JMP222 (pMC1) with (*S*)-2,4-DP remained at its initial low level of 0.007 – 0.023 U (mg protein)$^{-1}$ throughout the course of the experiment (Figure 3). In comparison, strain MC1 exhibited higher specific activities of 0.12 – 0.14 U (mg protein)$^{-1}$ with (*S*)-2,4-DP (Hoffmann and Müller, 2006). Similar activity trends for *C. necator* JMP222 (pMC1) were observed in repeated chemostat runs with (*R*)-2,4-DP as sole substrate (Figure 3). The specific activity of TfdC, the key enzyme which initiates the modified *ortho*-cleavage pathway, was determined by measuring the conversion of 3,5-dichlorocatechol into 2,4-dichloro-cis-cis-muconic acid (Müller et al., 2001). The highest specific TfdC activity in *C. necator* JMP222 (pMC1) was determined at steady state conditions amounting to 0.123 U (mg protein)$^{-1}$. With increasing D it initially dropped to a minimum of 0.05 U (mg protein)$^{-1}$ but then increased again eventually reaching 0.094 U (mg protein)$^{-1}$ (Figure 3). Despite the varying specific enzyme activities the constellation seemed to be appropriate for a balanced metabolism in the peripheral pathway since the 2,4-DCP concentration as a common indicator of imbalances in the metabolite flow remained below the detection limit at all times. The specific activity profile with (*R*)- and (*S*)-2,4-DP was in general coincident with the substrate consumption pattern on both enantiomers allowing distinctly higher rates with (*R*)-2,4-DP (Figure 1, 3).

Regarding *C. testosteroni* B337 (pMC1) maximum specific activities of 0.16 U (mg protein)$^{-1}$ with (*R*)-2,4-DP and 0.22 U (mg protein)$^{-1}$ with (*S*)-2,4-DP were achieved at a dilution rate of about 0.08 h^{-1} after increasing four to five-fold when comparing to activities at steady state (D = 0.05 h^{-1}). However, further raising the dilution rate towards μ_{max} only resulted in declining specific activities (Figure 4). Apparently, (*R*)- and (*S*)-2,4-DP degradation capacities of *C. testosteroni* B337 (pMC1) were similar which is in contrast to observations with *D. acidovorans* MC1 (Hoffmann and Müller, 2006) and *C. necator* JMP222 (pMC1) (Figure 3) where (*S*)-2,4-DP degradation activities were distinctly lower than with the R-enantiomer.

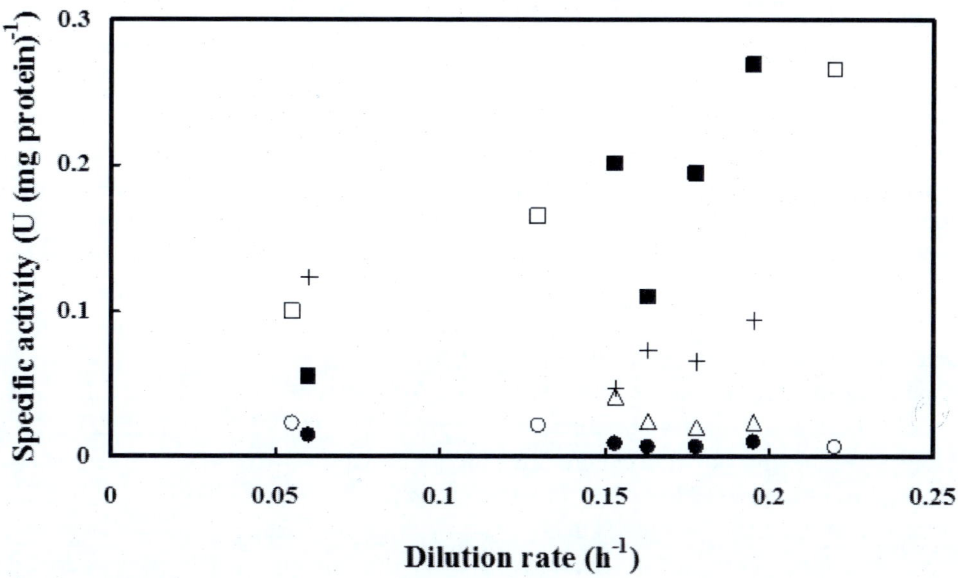

Figure 3. Enzyme activity of *C. necator* JMP222 (pMC1) from crude extracts derived during transient state cultivation on (*RS*)-2,4-DP (open symbols and crosses) and (*R*)-2,4-DP (closed symbols) at increasing dilution rates. Specific activities of α-ketoglutarate-depended dioxygenases (RdpA, SdpA) were determined with (*R*)-2,4-DP (squares), (*S*)-2,4-DP (circles), 2,4-D (triangles) as substrates and of chlorocatechol-dioxygenase (TfdC) with 3,5-dichlorocatechol (crosses) as substrate, respectively.

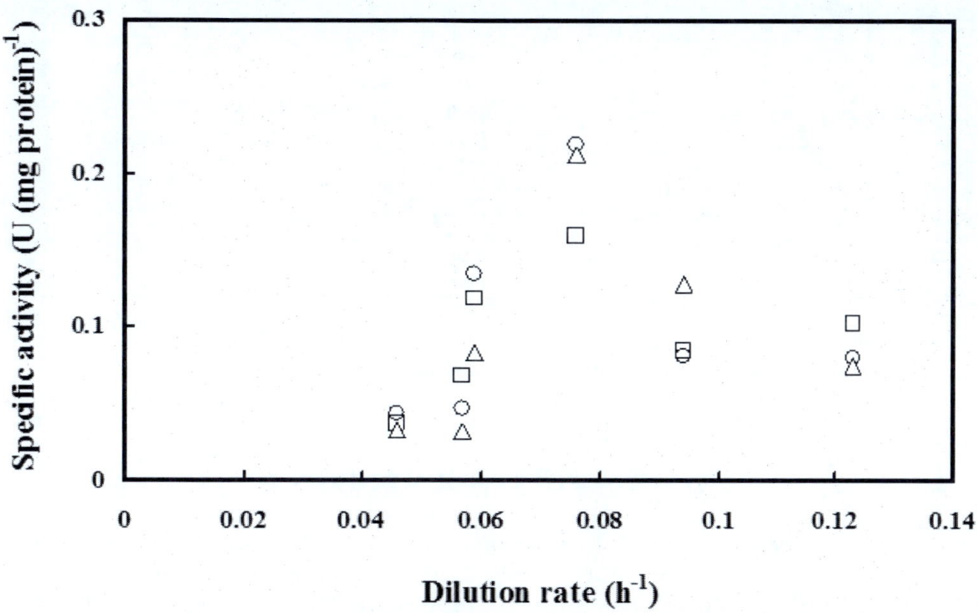

Figure 4. Enzyme activity of *C. testosteroni* B337 (pMC1) from crude extracts derived during transient state cultivation on (*RS*)-2,4-DP at increasing dilution rates. Specific activities of α-ketoglutarate-depended dioxygenases (RdpA, SdpA) were determined with (*R*)-2,4-DP (squares), (*S*)-2,4-DP (circles) and 2,4-D (triangles) as substrates.

Strain B337 (pMC1) even cleaved 2,4-D at rates of 0.22 U (mg protein)$^{-1}$ (Figure 4), i.e. twice as fast as strain MC1 (Hoffmann and Müller, 2006). Yet, the culture turning purple at dilution rates around 0.13 h^{-1} suggested that the catalytic activities of RdpA and SdpA were too high in strain B337 (pMC1) to be balanced by the metabolic capacity of the *ortho*-cleavage pathway (see below). This would also explain our observations of destabilized growth accompanied by the liberation of purple intermediates in some cultures already at dilution rates of about 0.08 h^{-1}, which otherwise enabled maximum specific activities. Measurement of TfdC was on average 0.125 ± 0.011 U (mg protein)$^{-1}$ over dilution rates up to maximum growth. This was in the range of JMP222 (pMC1) but seemed insufficient to eliminate 2,4-DCP in strain B337. The specific activity of the subsequent enzyme TfdD even amounted to only 0.055 U (mg protein)$^{-1}$ in B337.

Kinetic characteristics of the various strains were determined with partially purified RdpA derived at steady state after chemostatic growth with (*RS*)-2,4-DP. Kinetics with (*R*)-2,4-DP as substrate are shown in Figure 5 for the transconjugant strains *C. testosteroni* B337 (pMC1) and *C. necator* JMP222 (pMC1); *D. acidovorans* MC1 as the host of plasmid pMC1 was included for reasons of comparison.

Also for direct comparison relative activities are given, which were derived by normalizing the specific RdpA activity equating it with one at an (*R*)-2,4-DP concentration of 0.5 mM for each strain.

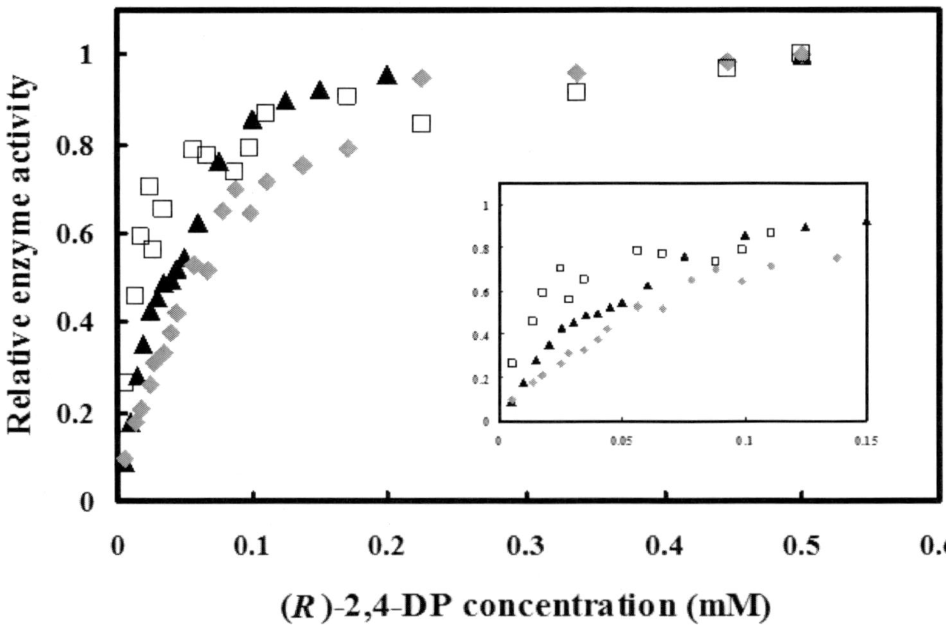

Figure 5. Kinetic characteristics of RdpA derived from partially purified enzyme after growth at steady state on (*R*)-2,4-DP at D = 0.1 h^{-1}. Shown are the data of the transconjugant strains *C. necator* JMP222 (pMC1) (grey diamonds), *C. testosteroni* B337 (pMC1) (white squares) and for reasons of comparison from *D. acidovorans* MC1 (black triangles) as the host of plasmid pMC1. The imbedded window is a magnification at low and intermediate (*R*)-2,4-DP concentrations (0.005 – 0.15 mM) to visualize the discontinuous curve shape and plateau formation which indicates complex kinetics. Relative activities were derived by normalizing the specific RdpA activity equating it with one at the highest (*R*)-2,4-DP concentration of 0.5 mM for each strain.

Table 2. Catalytic characteristics of RdpA from crude extracts derived at steady state after chemostatic growth of the various strains on (*RS*)-2,4-DP. (*R*)-2,4-DP was used as substrate for all catalytic measurements. The half-saturation constant K_m and the relative catalytic efficiency ($V_{max\ rel}/K_m$) were determined using GraphPad Prism software

	K_m (μM)	R^2 (K_m)	$V_{max\ rel}/K_m$ ($\times 10^{-3}\ \mu M^{-1}$)
D. acidovorans MC1	48.8	0.99	4.91
C. necator JMP222 (pMC1)	75.05	0.99	3.60
C. testosteroni B337 (pMC1)	13.93	0.96	11.50

The kinetic characteristics follow a complex pattern which is not purely hyperbolic but discontinuous exhibiting intermediary plateau regions at intermediate (*R*)-2,4-DP concentrations (Figure 5, imbedded window) typical for RdpA from strain MC1 (Westendorf et al., 2003, 2006). Apparently, this holds also true for RdpA of the transconjugant strains. Kinetic parameters were derived as apparent values from the rate data by using non-linear regression based on Michaelis Menten kinetics (Table 2). It is apparent and in accordance with Figure 5, that RdpA of *C. testosteroni* B337 (pMC1) has the highest affinity for its substrate represented by the lowest K_m whereas RdpA of *C. necator* JMP222 (pMC1) has the lowest affinity when compared with *D. acidovorans* MC1. This is also in line with strain B337 (pMC1) exposing a significantly increased relative catalytic efficiency ($V_{max\ rel}/K_m$) which is 2.3- and 3.2-fold higher than found for RdpA from *D. acidovorans* MC1 and *C. necator* JMP222 (pMC1), respectively (Table 2).

RdpA Variant Patterns

Earlier studies (Leibeling et al., 2010, 2013; Westendorf et al., 2006) revealed a correlation between differences in the kinetic parameters of RdpA and the appearance of multiple RdpA forms after separation by 2D gel electrophoresis. Spurred by that, we also examined the protein patterns of the transconjugant strains as described earlier (Leibeling et al., 2013) and summarized the results in Table 3. MALDI-MS was used to identify protein spots. The RdpA distribution pattern of *D. acidovorans* strain MC1071 ($RdpA^+$, $SdpA^-$) (Leibeling et al., 2013; Müller, 2007; Müller and Hoffmann, 2006) at steady state ($D = 0.05\ h^{-1}$) on (*R*)-2,4-DP was used as reference showing a main spot at *pI* 5.91 ± 0.02 and three more acidic spots at *pI* 5.81 ± 0.02, 5.72 ± 0.02, and 5.59 ± 0.05. The distances between the various spots were almost equal amounting to Δ*pI* of 0.11 ± 0.017 (Leibeling et al., 2013). Transconjugant strain *C. testosteroni* B337 (pMC1) showed seven spots spanning a *pI* range from 6.09 ± 0.02 to 5.47 at steady state conditions, grown on (*R*)-2,4-DP. The main spot at *pI* 5.98 ± 0.01 was accompanied by five more acidic spots and another minor alkaline spot at *pI* 6.09 ± 0.02. The spot pattern is similar to MC1071 with respect to the position of the main spot and the distances between spots (Δ*pI* = 0.10 ± 0.023), but comprises a larger total number of spots.

Table 3. Relative distributions of RdpA variants resulting from the transconjugant strains *C. testosteroni* B337 (pMC1) and *C. necator* JMP 222 (pMC1) cultivated with (*RS*)-2,4-DP, and as reference from *D. acidovorans* MC1071 grown with (*R*)-2,4-DP. The relative spot intensities (%) of each variant at the given *pI* in relation to the most intense spot in each lane are listed. The last column (Δ*pI*) indicates the intermediate distances between the various spots for each strain

	Relative spot volume (%) of RdpA variants														
pI	5.09 ±0.01	5.27 ±0.04	5.37 ±0.03	5.47	5.50 ±0.04	5.55	5.59 ±0.05	5.65	5.72 ±0.02	5.81 ±0.02	5.85	5.91 ±0.02	5.98 ±0.01	6.09 ±0.02	Δ*pI*
D. acidovorans MC1071 on (*R*)-2,4-DP							19.2		51.2	84.3		100.0			0.11 ±0.017
C. testosteroni B337 (pMC1) on (*RS*)-2,4-DP				7.6		10.0		24.3	19.9		82.8		100.0	6.8	0.10 ±0.023
C. necator JMP222 (pMC1) on (*RS*)-2,4-DP	10.9	9.6	39.6		90.1				100.0			7.4		1.2	0.17 ±0.04

Like the other transconjugant strain and under the same conditions *C. necator* JMP222 (pMC1) also exhibited seven RdpA spots.

However, the main spot is located at a more acidic *pI* of 5.72 ± 0.02 and the distances between the spots are larger (ΔpI = 0.17 ± 0.04) compared to *C. testosteroni* B337 (pMC1) and *D. acidovorans* MC1071.

Genetic Disposition

To address the possibility that genetic effects were responsible for the strain dependent kinetic characteristics and for determining the maximum growth rate, we checked the presence and content of the respective genes by real-time quantitative polymerase chain reaction (real-time qPCR).

In addition to the genes coding for the initial phenoxypropionate cleaving enzymes RdpA and SdpA we included a protein-coding marker gene in our study in order to evaluate the expression levels in relation to pMC1. Earlier studies (Klumpp, 2011; Nordström and Dasgupta, 2006) found that the number of plasmids per cell is regulated and rather constant even at rising growth rates. We thus considered the gene expression normalized to the plasmid content by referring to the plasmid marker gene *trfA*.

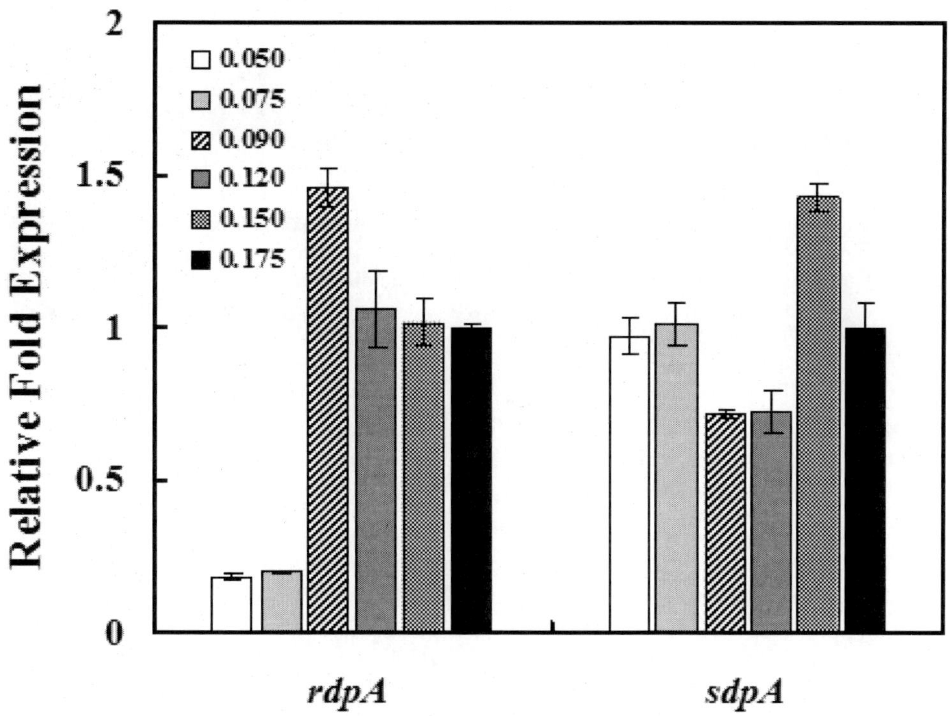

Figure 6. Normalized expression of selected genes from *D. acidovorans* MC1 after real-time qPCR derived during transient state cultivation on (*RS*)-2,4-DP at increasing dilution rates. The dilution rates per hour are given in the figure legend. Expression is normalized to the plasmid marker gene *trfA* and assessed relative to the expression level at the highest dilution rate (black bars). *rdpA* and *sdpA* are coding for the initial phenoxyalkanoate herbicide cleaving enzymes RdpA and SdpA.

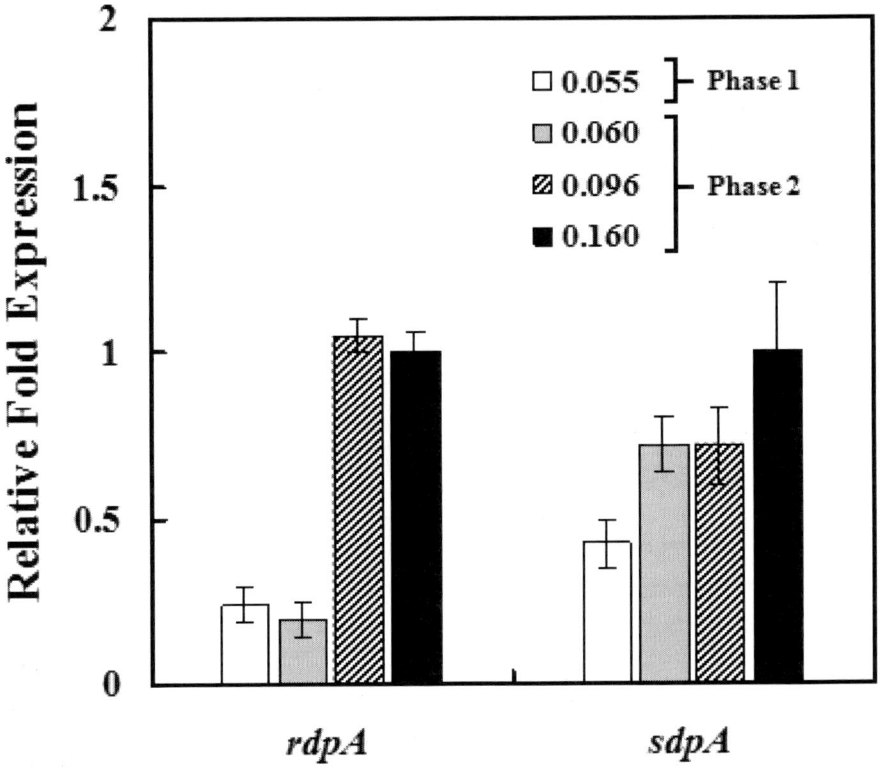

Figure 7. Normalized expression of selected genes from *C. necator* JMP222 (pMC1) after real-time qPCR derived during transient state cultivation on (*RS*)-2,4-DP at increasing dilution rates. The dilution rates per hour are given in the figure legend. Expression is normalized to the plasmid marker gene *trfA* and assessed relative to the expression level at the highest dilution rate (black bars). *rdpA* and *sdpA* are coding for the initial phenoxyalkanoate herbicide cleaving enzymes RdpA and SdpA. The division in two phases is similar to the description in Fig. 1. In phase 1 the dilution rate was step-wise increased in a range of D = 0.025 – 0.1 h^{-1}, then decelerated to D = 0.025 h^{-1}. In phase 2 the dilution rate gradient was restarted in a range of 0.025 – 0.2 h^{-1}.

Expression levels were evaluated by assuming an optimal genetic disposition of *rdpA* and *sdpA* at maximum dilution rates. Thus, the relative expression change was set to one at that state. Relative gene expression at lower dilution rates was assessed accordingly. The qPCR results with *D. acidovorans* MC1 are shown in Figure 6. The relative *sdpA* expression per plasmid remained on average rather constant at about 1.1 ± 0.26-fold disregarding the increasing dilution rate. In contrast, the expression of *rdpA* at low dilution rates is only about 20% of what it appeared to be at high dilution rates. It is further remarkable that the expression steeply increased at an intermediate dilution rate of D = 0.09 h^{-1} and remained elevated at high dilution rates close to μ_{max}.

Real-time qPCR results have already been obtained with *C. necator* JMP222 (pMC1) but have yet to be generated with *C. testosteroni* B337 (pMC1). The growth profile of JMP222 (pMC1) (Figure 1) was characterized by restricted herbicide utilization capabilities in a first phase of transient state cultivation (D = 0.025 – 0.1 h^{-1}) but yet by being able to overcome the limitations in a second phase (D = 0.025 – 0.2 h^{-1}) achieving μ_{max} values typical for *D. acidovorans* MC1. Similarly to MC1 the relative expression of *rdpA* of JMP222 (pMC1) was about 20 – 25% at low dilution rates (D = 0.055 – 0.06 h^{-1}) in phase 1 and 2 of the transient state cultivation. However, it increased significantly at intermediate dilution rates in phase 2 (D = 0.096 – 0.16 h^{-1}) reaching its maximum. The expression of *sdpA* per plasmid also appeared to be affected by the dilution rate changes. While it resided at 42% of its maximum at steady state conditions in phase 1 it continued to increase at low and intermediate dilution rates in phase 2 (Figure 7). Altogether, an increased expression of both *rdpA* and *sdpA* in phase 2 correlated with the respective growth pattern (Figure 1) allowing both enantiomers to be degraded more effectively, although the consumption of (*R*)-2,4-DP was still at advantage.

CONCLUSION

In this study plasmid pMC1 was transferred from its host *Delftia acidovorans* MC1 to plasmid-free strains of *Cupriavidus necator* and *Comamonas testosteroni*. The resulting transconjugant strains apparently dealt differently with the information offered by pMC1 with respect to the newly gained, plasmid-encoded capacity to degrade phenoxyalkanoate herbicides. The presence of the genetic information did not guarantee expression of the respective enzymes in a way that enabled utilization of the herbicides as carbon and energy source as effectively as the plasmid donor strain. Furthermore the activity of the initial cleaving enzymes, RdpA and SdpA, appeared remarkably different in *C. necator* JMP222 (pMC1) and *C. testosteroni* B337 (pMC1). Whereas the former showed similarities to *D. acidovorans* MC1, *C. testosteroni* rather resembled the properties of *Rhodoferax* sp. P230 with respect to an elevated activity with (*S*)-2,4-DP and 2,4-D (Westendorf et al., 2006). Differences became also evident regarding the consecutive metabolization of degradation intermediates via the modified *ortho*-cleavage pathway. Real-time qPCR and 2D-GE analyses indicated that rather adaptive processes were required on both genetic and enzymatic level for the transconjugant strains to handle pMC1 most efficiently.

Continuous cultivation of *D. acidovorans* MC1 under steady state and transient state conditions revealed highest growth rates with (*S*)- and (*R*)-2,4-DP (0.17 and 0.21 h^{-1}) compared with 2,4-D (0.045 h^{-1}) (Hoffmann and Müller, 2006). Determining the cleavage rate in crude extracts with (*R*)-2,4-DP, (*S*)-2,4-DP, and 2,4-D as substrates revealed a ratio of the activities of 1 : 0.48 : 0.26 (Müller et al., 2001). In other words the activity of RdpA dominates over SdpA in the host strain although the latter enzyme carries the main activity not only for (*S*)-2,4-DP but also for the etherolytic cleavage of 2,4-D (Westendorf et al., 2006). The properties of *C. necator* JMP222 (pMC1) proved rather similar regarding the specific enzyme activity with (*R*)-2,4-DP as well as μ_{max} with both enantiomers of the phenoxypropionates. Like MC1 strain JMP222 (pMC1) exhibited major activity with (*R*)-2,4-DP achieved by the RdpA enzyme whereas the activity with (*S*)-2,4-DP was significantly smaller in *C. necator* JMP222 (pMC1) (Figure 3). It remained almost constant but even decreased when increasing the growth rate. The activity with 2,4-D was also low (0.026 U (mg protein)$^{-1}$) in JMP222 (pMC1). In contrast to MC1, however, it exceeded the value for (*S*)-2,4-DP. Moreover, the activity of RdpA of JMP222 (pMC1) steeply increased by a factor of about three after rising the growth rate from 0.055 h^{-1} (0.1 U (mg protein)$^{-1}$) to 0.22 h^{-1} (0.27 U (mg protein)$^{-1}$). This apparently correlated with the high growth rate the strain achieved in a second phase of transient state cultivation (Figure 1, 3). Finally, it slightly exceeded that of the donor strain MC1.

As outlined in the above results the capabilities provided by plasmid pMC1 were differently expressed by the various strains although it was directly transferred from the donor strain to the plasmid-free acceptor strains. Genetic analyses proved both similarities and differences when comparing the plasmid donor strain MC1 and the transconjugant JMP222 (pMC1) on the gene expression level as shown by real-time quantitative PCR. To evaluate expression levels, we assumed an optimal genetic disposition of *rdpA* and *sdpA* i.e. maximal gene content (relative expression, Figure 6, 7) at maximum consumption rates by equating this state with one. The low growth rates of *C. necator* JMP222 (pMC1) in initial cultivations prompted us to consider a decrease in the *rdpA* gene in initial cultivations rather than its multiplication later on in stabilized cultures able to reach maximum growth rates. From this it follows that the *rdpA* content in initial cultures of JMP222 (pMC1) was indeed about 20 – 25% of what it appeared at maximum growth rates after prolonged cultivation in transient state mode. Surprisingly, the same was true for *rdpA* of MC1. In both cases the *rdpA* content increased drastically at intermediate dilution rates. The expression of *sdpA* in MC1, however, was independent from the growth rate and at or close to its maximum throughout our experiment. In contrast, *sdpA* of JMP222 (pMC1) behaved more *rdpA*-like, meaning its initial content was low and increased with increasing dilution rate. Together these results indicated that in both cases establishment of the *rdpA* gene and additionally of the *sdpA* gene in JMP222 (pMC1) was required to enable high growth rates. The results let us further conclude that once the plasmid was established, the transconjugant strain was able to handle pMC1 in a similar manner as *D. acidovorans* MC1.

The observation that *rdpA* and *sdpA* expression apparently differed, has to be considered in view of the fact that the genes are organized in separate transposon structures within the degradative operon (Schleinitz et al., 2004). Furthermore, *rdpA* and *sdpA*, the only genes enabling phenoxyalkanoate degradation, are encoded in one single copy, each, on pMC1 (Schleinitz et al., 2004). This allows for separate loss of the respective genes as has been observed earlier and proven by generating *rdpA* and/or *sdpA* deficient mutants through selective cultivation on opposite enantiomers, respectively (Leibeling et al., 2013; Müller, 2007; Müller and Hoffmann, 2006; Schleinitz, 2005; Müller et al. 2004a,b). Quantification of this effect showed that 99% of the capacity to utilize (*RS*)-2,4-DP is lost after 15 generations under non-selective cultivation conditions (Schleinitz, 2005). A similar effect can be alleged for the present observations with *D. acidovorans* MC1 since growth on a racemic substrate is not entirely selective when separate enzymes or pathways are involved regarding the different enantiomers. It rather represents a mixed substrate situation. Hence chemostatic growth at (very) low dilution rates is not necessarily dependent on the presence of the entire set of initial cleavage enzymes in the whole population as long as maintenance requirements can be satisfied and the input herbicide does not accumulate to a concentration where it becomes toxic. Especially an enzyme like RdpA meets these requirements due to its approved high catabolic efficiency to utilize the offered substrate. To that effect subpopulations with only an *rdpA* gene can coexist with others which are completely equipped or which solely express SdpA. The latter is in coincidence with the initially low *rdpA* expression per plasmid (Figure 6). One also has to keep in mind that plasmids may be subject to structural change. They can lose and gain genes in response to selective pressure (Dutta and Pan, 2002; Frost et al., 2005; Osborn and Boltner, 2002; Rysz et al., 2013). Thus, it was not surprising that increased selective pressure, i.e. increased growth rates, demanded increased RdpA activity, meaning that subpopulations containing or gaining *rdpA* were selectively favored at the expense of the gene-deficient ones. SdpA, however, is always the limiting factor because of its inferior degradation capabilities (Hoffmann and Müller, 2006; Westendorf et al., 2006). Thus, already the initial *sdpA* content is near or at a maximum. Although the formation of subpopulations similar to MC1 can be assumed for JMP222 (pMC1) it does not seem to have equal control mechanisms. *rdpA* and *sdpA* expression per plasmid were equally low in the beginning but required that subpopulations expressing at least either *rdpA* or *sdpA* were present. Cells without initial herbicide degradation activity, however, will not be able to sustain themselves in the bioreactor, and will wash out as they are devoid of any alternative carbon and energy source. After a period of adaptation, enabled by restarting the dilution rate gradient, the expression of both genes was increased, probably by increasing the subpopulation(s) containing the respective genes. However, since the expression pattern of *rdpA* was similar in both strains throughout transient state cultivation, it does not explain the diverging kinetic parameters at steady state. Obviously, the genetic level is not the only influence that has to be taken into account. The fact that growth deficits were more evident in JMP222 might be due to the deviating enzyme affinity and catalytic efficiency determined at steady state (Table 2). In other words, with featuring a 1.5-fold higher affinity towards (*R*)-2,4-DP strain MC1 also converts the substrate 1.5-fold faster during transient cultivation enabling MC1 to readily utilize this enantiomer from the beginning and thus evoking a selective advantage by avoiding high stationary herbicide concentrations.

In our previous work, we found a non-genetic mechanism affecting the herbicide degradation capability of the RdpA enzyme evident through the formation of enzyme variants. We thus performed 2D-GE with MC1 and the transconjugant strains in order to reveal possible differences in the RdpA variant pattern. Comparing the RdpA spot patterns at steady state (Table 3) already indicates a strain specific modification of the RdpA protein. Whereas MC1 displayed four RdpA spots with a main spot at *pI* 5.91 ± 0.02 and distances between the spots of 0.11 ± 0.017 (Leibeling et al., 2013), JMP222 (pMC1) exhibited seven spots with a far more acidic main spot at *pI* 5.72 ± 0.02 and larger distances between spots (ΔpI = 0.17 ± 0.04). Previously (Leibeling et al., 2013) we discussed post translational modification (PTM) via reactive oxygen species (ROS) being involved in an adaptive mechanism leading to altered kinetic characteristics of the enzyme, and above all enhanced 2,4-D degradation of the initial key enzyme RdpA. There, increased numbers of RdpA variants were observed during adaptation while the distance between spots (ΔpI) remained unaffected. However, the mentioned difference between the spot patterns of JMP222 (pMC1) and MC1 in combination with our previous observations (Leibeling et al., 2013) let us hypothesize that RdpA of JMP222 (pMC1) was oxidatively modified in a more drastic manner already at steady state conditions. Thus it explains the impaired activity with (*R*)-2,4-DP which contributes to the accumulation of residual herbicide (Figure 1). Oxidative modifications via ROS are in general considered to impair proteins and affect enzyme activities. We have shown that positive effects are associated with growing cultures in which the extent of modification was associated with increased activity (Leibeling et al., 2013) whereas in static cultures the oxidative effects were accumulated eventually leading to an extinction of enzyme activity (Leibeling et al., 2010).

The presence of enzyme variants as well as their variability in proportion and position is associated with deviating enzyme kinetics. As was shown in Figure 5 both the wild type and the two transconjugant strains showed complex characteristics which apparently seem to be caused by individual kinetics of the various variants of the respective RdpA enzymes. Hints to deviating kinetics are given by Westendorf et al. (2006) although a clear separation and stable isolation of the various variants proved difficult with RdpA of MC1. With respect to the present results it appears that variants with improved affinity for (*R*)-2,4-DP would dominate the RdpA enzyme in case of *C. testosteroni* B337 (pMC1), whereas the opposite would be true for RdpA of *C. necator* JMP222 (pMC1). Confirming this hypothesis is presently beyond our possibilities but proves to be a vital step towards understanding the influence of enzyme variants on the overall enzyme activity. It will thus be a task of further investigations.

A completely different picture was observed with the second transconjugant strain *C. testosteroni* B337 (pMC1). The specific activity of RdpA and SdpA, which was almost identical for both enzymes, showed a strong dependency on the growth rate. Maximum specific enzyme activities were not obtained at the maximum dilution rate (D = 0.13 h^{-1}) but at a dilution rate of about 0.08 h^{-1} (Figure 4). Further increasing the dilution rate only resulted in a significant decline of the enzyme activity. This enzyme activity pattern is in line with the observation that both enantiomers resided on a low level but progressively accumulated at D $\geq$ 0.13 h^{-1} defining the maximum consumption capacity (μ_{max}) for both enantiomers. Apparently, this is in contrast to the growth patterns of MC1 and JMP222 (pMC1), which were characterized by higher μ_{max} values and specific RdpA activities (1.7-fold) but a lower specific SdpA activity (about half). Whereas *C. necator* JMP222 (pMC1) needed two phases of transient state cultivation to improve its (*R*)- and (*S*)-2,4-DP degradation capabilities, the only similarity of B337 (pMC1) to the donor strain was its ability to readily degrade (*R*)- and (*S*)-2,4-DP at low and intermediate dilution rates. While, in case of (*R*)-2,4-DP, this can be ascribed to the high affinity and catalytic efficiency of the RdpA enzyme it does not explain the limitation of μ_{max}. It was, however, remarkable that in cultivations of strain B337 (pMC1) 2,4-DCP was constantly present along the entire range of dilution rates. Noteworthy was furthermore that approaching μ_{max} was accompanied by the accumulation of a purple-colored intermediate which indicated the formation of 3,5-dichloromuconic acid as confirmed by HPLC analyses. Some chemostat cultures did not even achieve the maximum consumption rate (0.13 h^{-1}) stopping to grow at about half that rate while the purple-colored intermediate appeared simultaneously. 2,4-DCP and 3,5-dichloromuconic acid are known toxicants which poison the cell at elevated concentrations (Ledger et al., 2006; Perez-Pantoja et al., 2003; Plumeier et al., 2002). Their accumulation is typically a sign of an impaired modified *ortho*-cleavage pathway. Thus, we determined the specific activity of enzymes catalyzing the subsequent steps, i.e. TfdC and TfdD. TfdC of B337 (pMC1) has a specific activity of around 0.125 ± 0.11 U (mg protein)$^{-1}$ which is lower than the cumulated activity of RdpA and SdpA (0.28 U (mg protein)$^{-1}$). In case of TfdD the enzyme activity was with 0.055 U (mg protein)$^{-1}$ even less. This is distinctly lower than found in the donor strain MC1 where activities for TfdC and TfdD amounted to 0.385 U (mg protein)$^{-1}$ and 0.13 U (mg protein)$^{-1}$, respectively (Müller et al., 2001). This points to an imbalance of the metabolic capacity of the *ortho*-cleavage pathway in combination with an excessive catalytic activity of RdpA and SdpA in B337 (pMC1). Regarding strain MC1 little is known about the genetic disposition of the *ortho*-cleavage pathway. There is, however, evidence from *D. acidovorans* P4a that a duplication of the two *ortho*-cleavage pathway genes, *tfdC* and *tfdE*, appeared necessary for effective degradation of 2,4-D in this strain (Leveau et al., 1999; Laemmli et al., 2000;

Hoffmann et al., 2003). It is evident that in strains MC1 and JMP222 the activity of TfdC is sufficient to manage the flow of 2,4-DCP which results from the initial cleavage and consecutive hydroxylation reactions. This is also apparent from the respective enzyme activities and the maximum specific substrate consumption rate $q_s = \mu_{max} * Y^{-1}$ (mmol h^{-1} (g dry mass)$^{-1}$). Taking into account a $\mu_{max} = 0.22\ h^{-1}$ and Y = 0.07 g dry mass (mmol 2,4-DP)$^{-1}$ then q_s = 3.12 mmol h^{-1} (g dry mass)$^{-1}$ with *C. necator* JMP222 (pMC1). While the capacity of SdpA is negligibly small, the capacity of RdpA in this strain amounts to 9.7 mmol h^{-1} (g dry mass)$^{-1}$ which clearly exceeds the demand. The activity of TdfC seems sufficient as its rate approaches 4.4 mmol h^{-1} (g dry mass)$^{-1}$ and thus, too, exceeds q_s. Similar values were found for strain MC1. With regard to B337 (q_s = 1.86 at $\mu_{max} = 0.13\ h^{-1}$) the initial cleavage reactions greatly exceeded the rates observed for the former two strains. The cleavage activity of RdpA and SdpA together amounted to 13.7 mmol h^{-1} (g dry mass)$^{-1}$ whereas the values of TfdC and TfdD were as low as 4.5 mmol h^{-1} (g dry mass)$^{-1}$ and 2.0 mmol h^{-1} (g dry mass)$^{-1}$, respectively. Hence, the initial cleavage activities might challenge and exhaust the capacity of the succeeding metabolic steps. Answering this question requires elucidation of the genetic structure of the plasmid in both wild type and transconjugant strains and shall be addressed in a future investigation.

According to JMP222 (pMC1), one would expect the different degradation properties of B337 (pMC1) and the plasmid donor strain MC1 to affect the RdpA spot variant pattern and/or the gene expression level. Unfortunately, genetic data still has to be generated for strain B337 (pMC1). However, the spot variant pattern showed that B337 (pMC1) possessed a similar number of RdpA spots like the other transconjugant strain *C. necator* JMP222 (pMC1) but it was more similar to MC1 regarding the difference between spots and the *pI* of the main spot (Table 3). With respect to the increased spot number it was moreover similar to MC1071 and MC1073 obtained after the strains were adapted to 2,4-D consumption. The latter was ascribed to RdpA as the only responsible key enzyme present in the respective strains (Leibeling et al., 2013). Analyzing the transconjugant strains further corroborated our hypothesis that the distance between spots, the number, and the position of spots are indicative of certain enzyme properties.

Together the results are in coincidence with findings in the wild type strains MC1, P230 and MH which express the degradative pathway for phenoxyalkanoate herbicide degradation while employing RdpA and SdpA activity. Whereas SdpA is encoded by genes which differ in part of their sequence, the genes for RdpA are identical in those three strains (Schleinitz et al., 2004). Nevertheless, the enzyme activity of RdpA is different in the various strains with regard to the K_m values for (*R*)-2,4-DP as the main substrate (Westendorf et al., 2006). Therefore, differences in the RdpA variant patterns, in their individual characteristics and proportion were made responsible. Moreover, it was shown that selective forces, which changed the enzymatic properties of RdpA, i.e. substrate specificity, were associated with alterations of the RdpA variant pattern (Leibeling et al., 2013). The present results are in accordance with these observations. A wider spectrum of enzyme variations was revealed giving evidence of a wider spectrum of physiological responses to the herbicides. To which extent changes in the RdpA/SdpA activity will determine the overall substrate flux and thus the capability to utilize the respective herbicides cannot be deduced from the present results. Since RdpA (and SdpA) play a key role in this peripheral metabolism this is expected to be of primary relevance. Due to their flexible reaction mechanism which involves the activation of oxygen α-ketoglutarate-dependent enzymes are likely to be both a potential source and directly susceptible to locally formed ROS (Borowski et al., 2004; Costas et al., 2004; Hegg et al., 1999; Hegg and Que, 1997; Hewitson et al., 2005; Kovaleva and Lipscomp, 2008; Ryle and Hausinger, 2002; Solomon et al., 2000, 2003). ROS derived from this reaction mechanism were made responsible for the generation of enzyme variants. Such microdiversity in enzyme variants seems an excellent tool for regulating the metabolic flux rates and maintaining the maximum growth rate. This would take effect especially when assuming that these variants exhibit different individual kinetic characteristics and that they are serially connected judging from their organization in distinct spots in 2D gels. In response to fluctuations, the prevailing variant changes in order to regain the catalytic optimum which would be of selective advantage. Here, the species-specific background comes into effect as it affects the extent of posttranslational modification. As a result, the enzyme activities which finally determine the maximum consumption rate are deviating in the various strains as they deal with the provided herbicides individually. It can be stated that the strain which expresses a plasmid has because of its genetic and metabolic background a crucial influence on the expression and finally on the activity of the encoded enzymes which is not solely genetically determined but likely due to alterations on the protein level.

In our present investigation, we focused on RdpA but certainly, research has to be extended as several steps relying on oxygenases are included in the primary metabolism of herbicide compounds. An enzyme variant pattern similar as described for RdpA can be expected and was already observed for SdpA (Westendorf et al., 2003) and TfdC (Leibeling et al., 2013) but little is known about associated kinetic characteristics and adaptation capabilities. To which extent TfdC and subsequent enzymes of the *ortho*-cleavage pathway are capable of adaptation or susceptible to oxidative damage has yet to be elucidated. Oxidative damage might be facilitated additionally due to an enhanced ROS-generation by the extended activity of both RdpA and SdpA in *C. testosteroni* B337 (pMC1). The latter is supported by our finding of an optimum in RdpA and SdpA activity at about two-thirds of the maximum growth rate which indicates deleterious effects of too high turnover rates. Beneficial as well as impairing effects during herbicide degradation have been discussed (Leibeling et al., 2010, 2013).

Microbial adaptation is a complex process and merely the acquisition of genetic information, via e.g. transferable plasmids, is not necessarily associated with a successful expression of the respective genes. This has formerly been demonstrated for the canonical plasmid pJP4 by Don and Pemberton (1981) who showed that the expression of pJP4 became effective for 2,4-D degradation in *Alcaligenes eutrophus*, *Variovorax paradoxus*, and *Pseudomonas putida* whereas this was unsuccessful for *Escherichia coli*, *Rhodopseudomonas sphaeroides*, *Agrobacterium tumefaciens*, *Rhizobium* sp., *Pseudomonas fluorescence*, and *Acinetobacter calcoaceticus*. Expression and positive degradation traits after acquisition of pJP4 were verified in various species in consecutive investigations (Bathe et al., 2004; DiGiovanni et al., 1996; Feng et al., 1994; Friedrich et al., 1983; Goris et al., 2002). In one of our studies (Kleinsteuber et al., 2001) we pursued the distribution of this plasmid to alkaliphilic strains. Despite the high conjugation efficiency into strains of a natural consortium, degradative capabilities were rather restricted. About 90% of the strains that showed mercury resistance as the selective marker encoded on pJP4 contained *tfdC* as an indication for the presence of the modified *ortho*-cleavage pathway. Although 88% of these *tfdC*-positive clones contained a plasmid of 80 kB, which corresponds to pJP4, only 4% were capable of degrading 2,4-D (Kleinsteuber et al., 2001; Kiesel et al., 2007). It is another example, where the regulatory disposition of the various strains decided on effective expression of the respective genes. Studies of Newby at al. (2000a,b) and Top et al. (1995) also report a limited degradation activity after transfer of plasmids encoding genes for 2,4-D degradation to soil microorganisms. This is in line with our observations which shed light on alterations on the protein level that could be held accountable for causing deviating enzyme activities in transconjugant strains and thus finally impact the productive degradation of the herbicides.

REFERENCES

Bathe, S., Mohan, T. V., Wuertz, S. and Hausner, M. (2004). Bioaugmentation of a sequencing batch biofilm reactor by horizontal gene transfer. *Water Sci. Technol.,* 49, 337-344.

Benndorf, D., Davidson, I. and Babel, W. (2004). Regulation of catabolic enzymes during long-term exposure of *Delftia acidovorans* MC1 to chlorophenoxy herbicides. *Microbiology,* 150, 1005–1014.

Bergstrom, C. T., Lipsitch, M. and Levin, B. R. (2000). Natural selection, infectious transfer and the existence conditions for bacterial plasmids. *Genetics*, 155, 1505-1519.

Borowski, T., Bassan, A. and Siegbahn, P. E. M. (2004). Mechanism of dioxygen activation in 2-oxoglutarate-dependent enzymes: a hybrid DFT study. *Chem. Eur. J.*, 10, 1031-1041.

Costas, M., Mehn, M. P., Jensen, M. P. and Que Jr, L. (2004). Dioxygen activation at mononuclear nonheme iron active sites: enzymes, models, and intermediates. *Chem. Rev.,* 104, 939–986.

Dejonghe, W., Goris, J., El Fantroussi, S., Höfte, M., De Vos, P., Verstraete, W. and Top, E. M. (2000). Effect of dissemination of 2,4-dichlorophenocyacetic acid (2,4-D) degradation plasmid on 2,4-D degradation and on bacterial community structure in two different soil horizons. *Appl. Environ. Microbiol.,* 66, 3297-3304.

Derveaux, S., Vandesompele, J. and Hellemans, J. (2010). How to do successful gene expression analysis using real-time PCR. *Methods,* 50, 227–230.

Davison, J. (1999). Genetic exchange between bacteria in the environment. *Plasmid,* 42, 73-91.

DiGiovanni, G. D., Neilson, J. W., Pepper, I. L. and Sinclair, N. A. (1996). Gene transfer of *Alcaligenes eutrophus* JMP134 plasmid pJP4 to indigenous soil recipients. *Appl. Environ. Microbiol.*, 62, 2521-2526.

Dutta, C. and Pan, A. (2002). Horizontal gene transfer and bacterial diversity. *J. Biosci.,* 27, 27-33.

Don, R. H. and Pemberton, J. M. (1981). Properties of six pesticide degradation plasmids isolated from Alcaligenes paradoxus and Alcaligenes eutropha. *J. Bacteriol.*, 145, 681-686.

Eberhard, W. (1990). Evolution in bacterial plasmids and levels of selection. *Quarter. Rev. Biol.*, 6, 3-22.

Ehrig, A., Müller, R. H. and Babel, W. (1997). Isolation of phenoxy herbicide-degrading *Rhodoferax* species from contaminated building material. *Acta Biotechnol.*, 17, 351-356.

Esener, A. A., Roels, J. A., Kossen, N. W. F. and Roozenburg, J. W. H. (1981). Description of microbial growth behaviour during wash-out phase; determination of the maximum growth rate. *J. Appl. Microbiol. Biotechnol.*, 1, 141-144.

Feng, L., Van Zwieten, L., Kennedy, I. R., Rolfe, B. G. and Gartner, E. (1994). Expression of the 2,4-D degrading plasmid pJP4 of Alcaligenes eutrophus *in* Rhizobium trifolii. *Acta Biotechnol.*, 14, 119–129.

Friedrich, B., Meyer, M. and Schlegel, H. G. (1983). Transfer and expression of the herbicide-degrading plasmid pJP4 in aerobic autotrophic bacteria. *Arch. Microbiol.*, 134, 92-97.

Fondi, M., Bacci, G., Brilli, M., Papaleo, M. C., Mengoni, A., Vaneechoutte, M., Dijkshoorn, L. and Fani, R. (2010) Exploring the evolutionary dynamics of plasmids: the *Acinetobacter* pan-plasmidome. *BMC Evol. Biol.,* 10, 59.

Frost, L. S., Leplae, R., Summers, A. O. and Toussaint, A. (2005). Mobile genetic elements: the agents of open source evolution. *Nature Rev. Microbiol.*, 3, 722-732.

Ghosal, D. I., You, S., Chaterjee, D. K., and Chakrabarty, A. M. (1985). Genes specifying degradation of 3-chlorobenzoic acid in plasmids pAC27 and pJP4. *Proc. Natl. Acad. Sci. USA,* 82, 1638-1642.

Goris, J., Dejonghe, W., Falsen, E., De Clerck, E., Geeraerts, B., Willems, A., Top, E. M., Vandamme, P. and De Vos, P. (2002). Diversity of transconjugants that acquired plasmid pJP4 or pEMT1 after inoculation of a donor strain in the A- and B-horizon of an agricultural soil and description of *Burkholderia hospita* sp. nov. and Burkholderia terricola sp. nov. *Syst. Appl. Microbiol.*, 25, 340-352.

Hegg, E. L. and Que, L. Jr. (1997). The 2-His-1-carboxylate facial triad – An emerging structural motif in mononuclear non-heme iron(II) enzymes. *Eur. J. Biochem.*, 15, 625-629.

Hegg, E. L., Whiting, A. K., Saari, R. E., McCracken, J., Hausinger, R. P. and Que, L. J. (1999). Herbicide-degrading alpha-keto acid-dependent enzyme TfdA: Metal coordination environment and mechanistic insights. *Biochem.,* 38, 16714-16726.

Hehemann, J. H., Redecke, L., Murugaiyan, J., von Bergen, M., Betzel, C. and Saborowski, R. (2008). Autoproteolytic stability of a trypsin from the marine crab Cancer pagurus. *Biochem. Biophys. Res. Commun.*, 370, 566–571.

Hewitson, K. S., Granatino, N., Welford, R. W. D., McDonough, M. A. and Schofield, C. J. (2005). Oxidation by 2-oxoglutarate oxygenases: non-haem iron systems in catalysis and signalling. *Phil. Trans. R. Soc. A.,* 363, 807-828.

Hoffmann, D., Kleinsteuber, S., Müller, R. H. and Babel, W. (2003). Structural organization and localization of the chlorocatechol pathway in the 2,4-D degrading alkalitolerant strain *Delftia acidovorans* P4a. *Microbiology,* 149, 2545-2556.

Hoffmann, D. and Müller, R. H. (2006). 2,4-Dichlorophenoxyacetic acid (2,4-D) utilization by *Delftia acidovorans* MC1 at alkaline pH and in the presence of dichlorprop is improved by introduction of the *tfdK* gene. *Biodegradation,* 17, 263-73.

Janssen, D. B., Dinkla, I. J. T., Poelarends, G. J. and Terpstra, P. (2005). Bacterial degradation of xenobiotic compounds. *Environ. Microbiol.*, 7, 1868–1882.

Kiesel, B., Müller, R. H. and Kleinsteuber, S. (2007). Adaptive potential of alkaliphilic bacteria towards chloroaromatic substrates assessed by a *gfp*-tagged degradation plasmid. *Engin. Life Sci.,* 7, 361-372.

Kleinsteuber, S., Müller, R. H. and Babel, W. (2001). Expression of the 2,4-D degradative pathway of pJP4 in an alkaliphilic, moderately halophilic soda lake isolate, *Halomonas* sp. EF43. *Extremophiles,* 5, 375-384.

Klumpp S. (2011). Growth-rate dependence reveals design principles of plasmid copy number control. *PloS One,* 6, e20403.

Kovaleva, E. G. and Lipscomb, J. D. (2008). Versatility of biological non-heme Fe(II) centres in oxygen activation reaction. *Nat. Chem. Biol.,* 4, 186-193.

Laemmli, C., Leveau, J. H. J., Zehnder, A. J. B. and van der Meer, J. R. (2000). Characterization of a second *tfd* gene cluster for chlorophenol and chlorocatechol metabolism on plasmid pJP4 in *Ralstonia eutropha* JMP134(pJP4). *J. Bacteriol.,* 182, 4165-4172.

Ledger, T., Pieper, D. H. and González, B. (2006). Chlorophenol hydroxylases encoded by plasmid pJP4 differentially contribute to chlorophenoxyacetic acid degradation. *Appl. Environ. Microbiol.,* 72, 2783–2792.

Leibeling, S., Schmidt, F., Jehmlich, N., von Bergen, M., Müller, R. H. and Harms, H. (2010). Declining capacity of starving *Delftia acidovorans* MC1 to degrade phenoxypropionate herbicides correlates with oxidative modification of the initial enzyme. *Environ. Sci. Technol.,* 44, 3793-3799.

Leibeling, S., Maeß, M. B., Centler, F., Kleinsteuber, S., von Bergen, M., Thullner, M., Harms, H. and Müller, R. H. (2013). Posttranslational oxidative modification of (*R*)-2,4-dichlorophenoxypropionate/α-ketoglutarate-dioxygenase (RdpA) leads to improved degradation of 2,4-dichlorophenoxyacetate (2,4-D). *Engin. Life Sci.,* 13, 278–291.

Leveau, J. H. J., König, F., Füchslin, H., Werlen, C. and van der Meer, J. R. (1999). Dynamics of multigene expression during catabolic adaptation of *Ralstonia eutropha* JMP134(pJP4) to the herbicide 2,4-dichloropheoxyacetate. *Mol. Microbiol.,* 33, 396-406.

Müller, R. H. (2007). Activity and reaction mechanism of the initial enzymatic step specifying the microbial degradation of 2,4-dichlorophenoxyacetate. *Eng. Life Sci.,* 7, 311-321.

Müller, R. H. and Babel, W. (2001). Pseudo-recalcitrance of chlorophenoxyalkanoate herbicides – Correlation to the availability of α-ketoglutarate. *Acta Biotechnol.,* 21, 227-242.

Müller, R. H., Neytschev, M. and Schumann, C. (2004a) Bacterial strains *Delftia acidovorans* MC1-S, production and use thereof for the production of S enantiomers of 2-phenoxypropionic acid derivatives. WO 2004/055172.

Müller, R. H., Neytschev, M. and Schumann, C. (2004b) Bacterial strains *Delftia acidovorans* MC1-R, their production and their use in the production of R enantiomers of 2-phenoxypropionic acid derivatives. WO 2004/055218.

Müller, R. H. and Hoffmann, D. (2006). Uptake kinetics of 2,4-dichlorophenoxyacetate by *Delftia acidovorans* MC1 and derivative strains: complex characteristics in response to pH and growth substrate. *Biosci. Biotechnol. Biochem.,* 70, 1642-1654.

Müller, R. H., Bley, T. and Babel, W. (1995). Transient state cultivation as a means for determining maximum growth rates of microorganisms in inhibition kinetics. *J. Microbiol. Meth.,* 22, 209-219.

Müller, R.H., Jorks, S., Kleinsteuber, S., and Babel, W. (1999). *Comamonas acidovorans* MC1: a new isolate capable of degrading the chiral herbicides dichlorprop and mecoprop and the herbicides 2,4-D and MCPA. *Microbiol. Res.,* 15, 241-246.

Müller, R. H., Kleinsteuber, S., and Babel, W. (2001). Physiological and genetic characteristics of two strains utilizing phenoxypropionate and phenoxyacetate herbicides. *Microbiol. Res.,* 156, 121-131.

Müller, T. A., Byrde, S. M., Werlen, C., van der Meer, J. R. and Kohler, H. P. E . (2004). Genetic analysis of phenoxyalkanoic acid degradation in *Sphingomonas herbicidovorans* MH. *Appl. Environ. Microbiol.,* 70, 6066-6075.

Neuhoff, V., Arold, N., Taube, D. and Ehrhardt, W. (1988). Improved staining of proteins in polyacrylamide gels including isoelectric focusing gels with clear background at nanogram sensitivity using Coomassie Brilliant Blue G-250 and R-250. *Electrophoresis,* 9, 255–262.

Newby, D. T., Josephson, K. L. and Pepper, I. L. (2000a). Detection and characterization of plasmid pJP4 transfer to indigenous soil bacteria. *Appl. Environ. Microbiol.,* 66, 290-296.

Newby, D. T., Gentry, T. and Pepper, I. L. (2000b). Comparison of 2,4-dichlorophenoxyacetic acid degradation and plasmid transfer in soil resulting from bioaugmentation with two different pJP4 donors. *Appl. Environ. Microbiol.* 66, 3399-3407.

Nickel, K., Suter, M. J. and Kohler, H. P. (1997). Involvement of two alpha-ketoglutarate-dependent dioxygenases in enantioselective degradation of (*R*)- and (*S*)-mecoprop by *Sphingomonas herbicidovorans* MH. *J. Bacteriol.,* 179, 6674-6679.

Nordström, K. and Dasgupta S. (2006). Copy-number control of the *Escherichia coli* chromosome: a plasmidologist's view. *EMBO Reports,* 7, 484-489.

Oh, K. H., and Tuovinen, O. H. (1990). Degradation of 2,4-dichlorophenoxyacetic acid by mixed cultures of bacteria. *J. Ind. Microbiol.,* 6, 275-287.

Osborn, A. M. and Boltner, D. (2002). When phage, plasmids, and transposons collide: genomic islands, and conjugative- and mobilizable-transposons as a mosaic continuum. *Plasmid,* 48, 202-212.

Paulin, M. M., Nicolaisen, M. H. and Sørensen, J. (2010). Abundance and expression of enantioselective *rdpA* and *sdpA* dioxygenase genes during degradation of the racemic herbicide (*R,S*)-2-(2,4-dichlorophenoxy)propionate in soil. *Appl. Environ. Microbiol.,* 76, 2873-2883.

Pemberton, J. M., Corney, B., and Don, R. H. Evolution and spread of pesticide degrading ability among soil microoraganisms. In: Timmis, K. N. and Pühler, A. editors. Plasmids of medical, environmental and commercial importance. Amsterdam: Elsevier North Holland Biomedical Press; 1979; 287-299.

Pérez-Pantoja, D., Guzmán, D. L., Manzono, M., Pieper, D. H. and González, B. (2000). Role of $tfdD_ID_IE_IF_I$ and $tfdC_{II}D_{II}E_{II}F_{II}$ gene modules in catabolism of 3-chlorobenzoate by *Ralstonia eutropha* JMP134 (pJP4). *Appl. Environ. Microbiol.,* 66, 1602-1608.

Pérez-Pantoja, D., Ledger, T., Pieper, D. H. and González, B. (2003). Efficient turnover of chlorocatechols is essential for growth of *Ralstonia eutropha* JMP134(pJP4) in 3-chlorobenzoic acid. *J. Bacteriol.,* 185, 1534-1542.

Pfaffl, M. W. (2001). A new mathematical model for relative quantification in real-time RT-PCR. *Nucleic Acids Research*, 29, e45.

Plumeier, I., Pérez-Pantoja, D., Heim, S., González, B. and Pieper, D. H. (2002). Importance of different *tfd*tfd genes for degradation of chloroaromatics by *Ralstonia eutropha* JMP134. *J. Bacteriol.,* 184, 4054-4064.

Ryle, M. J. and Hausinger, R. P. (2002). Non-heme iron oxygenases. *Curr. Opin. Chem. Biol.,* 6, 193-201.

Rysz, M., Mansfield, W. R., Fortner, J. D. and Alvarez, P. J. J. (2013). Tetracycline Resistance Gene Maintenance under Varying Bacterial Growth Rate, Substrate and Oxygen Availability, and Tetracycline Concentration. *Environ. Sci. Technol.*, doi: 10.1021/es3035329.

Schleinitz, K. M., Kleinsteuber, S., Vallaeys, T., and Babel, W. (2004). Localization and characterization of two novel genes encoding stereospecific 2(2,4-dichlorophenoxy)propionate cleavage in *Delftia acidovorans* MC1. *Appl. Environ. Microbiol.*, 70, 5357-5365.

Schleinitz, K. M. (2005) Genetic of the stereospecific cleavage of the chiral herbicide 2-(2,4-dichlorophenoxy)propionate (dichlorprop) in *Delftia acidovorans* MC1. Leipzig, Germany: Ph.D. thesis, University of Leipzig.

Slater, F. R., Bailey, M. J., Tett, A. J. and Turner, S. L. (2008). Progress towards understanding the fate of plasmids in bacterial communities. *FEMS Microbiol. Ecol.,* 66, 3-13.

Solomon, E. I., Brunold, T. C., Davis, M. I., Kemsley, J. N., Lee, S.-K., Lehnert, N., Neese, F., Skulan, A. J., Yang, Y.-S. and Zhou J. (2000). Geometric and Electronic Structure/Function Correlations in Non-Heme Iron Enzymes. *Chem. Rev.,* 100, 235-350.

Solomon, E. I., Decker, A. and Lehnert, N. (2003). Non-heme iron enzymes: Contrasts to heme catalysis. *Proc. Natl. Acad. Sci. USA,* 100, 3589-3594.

Top, E. M., Holben, W. E. and Forney, L. J. (1995). Characterization of diverse 2,4-dichlorophenoxyacetic acid-degrading degradative plasmids isolated from soil by complementation. *Appl. Environ. Microbiol.*, 61, 1691-1698.

Top, E. M., Maltseva, O. V. and Forney, L. J. (1996). Capture of catabolic plasmid that encodes only 2,4-dichlorophenoxyaceticacid: α-ketoglutaric acid dioxygenase (TfdA) by genetic complementation. *Appl. Environ. Microbiol.*, 62, 2470-2476.

Top, E. M. and Springeal, D. (2003). The role of mobile genetic elements in bacterial adaptation to xenobiotic compounds. *Curr. Opin. Biotechnol.*, 14, 262-269.

Trefault, N., de la Iglesia, R., Molina, A. M., Manzano, M., Ledger, T., Pérez-Pantoja, D., Sánchez, M. A., Sguardo, M. and González, B. (2004). Genetic organization of the catabolic plasmid pJP4 from *Ralstonia eutropha* JMP134(pJP4) reveals mechanisms of adaptation to chloroaromatic pollutants and evolution of the specialized chloroaromatic degradation pathways. *Environ. Microbiol.*, 6, 655-668.

Vandamme, P. and Coenye, T. (2004). Taxonomy of the genus *Cupriavidus*: a tale of lost and found. Intern. *J. Syst. Evol. Microbiol.*, 54, 2285-2289.

Westendorf, A., Benndorf, D., Müller, R. H., Babel, W. (2002). The two enantiospecific dichlorprop/alpha-ketoglutarate-dioxygenases from *Delftia acidovorans* MC1 - Protein and sequence data of RdpA and SdpA. *Microbiol. Res.*, 157, 317-322.

Westendorf, A., Müller, R. H. and Babel, W. (2003). The two enantioselective dichlorprop/α-ketoglutarate-dioxygenases from *Delftia acidovorans* MC1 - protein data of RdpA and SdpA. *Microbiol. Res.,* 157, 317-322.

Westendorf, A., Benndorf, D., Pribyl, T., Harms, H. and Müller, R. H. (2006). Kinetic traits and enzyme form patterns of (*R*)-2-(2,4-dichlorophenoxy) propionate/alpha-ketoglutarate dioxygenase (RdpA) after expression in different bacterial strains. *Eng. Life Sci.,* 6, 552-559.

You, I. S. and Ghosal, D. (1995). Genetic and molecular analysis of the regulatory region of the herbicide 2,4-dichlorophenoxyacetate catabolic plasmid pJP4. *Mol. Microbiol.,* 16, 321-331.

Zipper, C., Nickel, K., Angst, W. and Kohler, H. P. E. (1996). Complete microbial degradation of both enantiomers of the chiral herbicide Mecoprop [(*RS*)-2-(4-chloro-2-methylphenoxy)propionic acid] in an enantioselective manner by *Sphingomonas herbicidovorans* sp. nov. *Appl. Environ. Microbiol.*, 62, 4318-4322.

Zipper, C. (1998). Microbial degradation and environmental fate of chiral phenoxyalkanoic acid herbicides. Zurich, Switzerland: Ph.D. thesis. ETH Zurich.

In: Handbook on Herbicides
Editors: Daiki Kobayashi and Eito Watanabe
ISBN: 978-1-62948-053-4

Chapter 2

WEEDS AND THEIR MECHANISMS OF RESISTANCE TO HERBICIDES

***Félix Morales-Flores*[1], *Beatriz King-Díaz*[1], *María Isabel Aguilar*[2] *and Blas Lotina-Hennsen*[1*]**
[1]Departamento de Bioquímica
[2]Departamento de Farmacia, Facultad de Química,
Universidad Nacional Autónoma de México, Ciudad Universitaria,
Delegación Coyoacan, México, México

ABSTRACT

The increase in human population has generated proportional food needs. However, agricultural productivity is constantly affected by factors decreasing land fertility. Weeds cause a constant reduction in the quantity and quality of important crops making their elimination a common necessity.

During the 40's and in subsequent years, the use and constant cycling of selective chemical herbicides considerably improved crop yields. However, overuse of herbicides to minimize the impact of weeds created the development of resistant species.

Since the first case in 1970, confirming the resistance of the common groundsel (*Senecio vulgaris*) against triazine herbicide, the number of resistant weeds against various herbicides has been considerably increasing. To date, more than 200 species have been reported to be resistant to different types of herbicides.

This review aims to address the herbicide resistance mechanisms developed by weeds, which we grouped as: Changes in the uptake and translocation of herbicides; herbicide detoxification mechanisms; altered target sites and herbicide cross-resistance.

Keywords: Acetolactate synthase; acetyl coenzyme A carboxylase; cross-resistance; cytochrome P450; herbicides detoxification; glutathione S-transferase; herbicide resistance, resistance mechanisms

*Author to whom correspondence should be addressed: E-Mail: blas@unam.mx; Tel.: +55-5622-5294.

1. Introduction

Pesticides, chemicals such as herbicides, fungicides and various other substances are used to prevent, destroy, or repel pests, and are the major technological tools successfully used worldwide. However, adverse consequences of their persistent application are the emergence of resistant pest populations [1], along with herbicide resistance, both an increasing problem in world agriculture [2]. The first confirmed report of an herbicide resistant weed was the common groundsel (*Senecio vulgaris*) tolerant to triazine herbicide. Thereafter the number of resistant species against various herbicides continued to rise. In a recently reported survey by the International Survey of Herbicide Resistant Weeds, 400 biotypes belonging to 217 species (129 dicots and 88 monocots) with resistance against various herbicides were reported [3].

The Weed Science Society of America (WSSA) defines herbicide resistance as the inherent ability of a plant to survive and reproduce following exposure to a dose of an herbicide normally lethal to the wild type [3]. However, not all "herbicide resistant plants" are herbicide-resistant weeds; they may be herbicide resistant crops or laboratory creations. With respect to weed resistance to herbicides, this is defined as the inherited ability of a weed population to survive an herbicide application that is normally lethal to the vast majority of that weed species [4, 5]. Resistance is not acquired due to mutations caused by the herbicide; rather, it arises from the selection of natural mutations or small pre-existing populations of resistant plants, by the selection pressure exerted by herbicides [6].

Resistance to herbicides may be conferred by changes in uptake, translocation, metabolism, transport or target sites, gene amplification and vacuolar sequestration [7-11].

Two other important characteristics of a weed in terms of herbicide-resistance evolution are the size and viability of the soil seed bank and the weed fitness. The soil seed bank may act as a buffer delaying the development of resistance [12]. This is because over the years the soil seed bank has been enriched by the predominant susceptible individual's seeds shed. In some cases, individuals carrying a mutation that confers resistance to herbicides are penalized by being less adapted or less fit in absence of the herbicide. It is difficult to measure reduced fitness but it can be related to impaired efficiency of key physiological processes such as photosynthesis or whole plant characteristics such as decreased seed production or reduced competitive ability. Often resistant biotypes, however, are not less fit than the normal, susceptible ones [12].

The majority of weed resistant biotypes (R) produce less above-ground biomass than sensitive (S) phenotypes during the vegetative stage; resistant biotypes are generally less productive than susceptible biotypes [13] due to a reduced relative growth rate, a lower rate of carbon dioxide fixation and reduced rates of PS II electron transport [14,15,16].

Several explanations are considered to account for the existence of a physiological cost for herbicide resistance [17]: (1) a mutation associated with target site resistance may in some way compromise normal functioning of the target enzyme, resulting in pleiotropic changes in cellular metabolism [18-21]; (2) where the resistance mechanism is non-target-site based it may be associated with the diversion of resources for increasing production of detoxification enzymes [22,23]; (3) resistance loci may be tightly linked and co-segregate with other loci affecting plant growth attributes [24,25].

2. Criteria for Confirmation of Herbicide-Resistant Weeds

2.1. Data Confirmation of Resistance

Resistance must be confirmed through comparison of resistant and susceptible plants of the same species by a scientifically validated test. There is a big difference between testing for the initial characterization of a putative resistant weed biotype and the routine testing of hundreds of samples that may follow after the initial discovery. The most herbicide resistance assays are: Petri dish shoot bioassays, target enzyme assays and fluorescence tests [26]. However, for the initial characterization of a putative resistant weed biotype, the most preferred test is a dose response experiment under controlled conditions (growth cabinet, glasshouse, etc.) using whole plants. Six to eight herbicide doses are often selected for evaluation of putative R weeds.

Doses may range from sub-lethal to 10- to 1,000-fold greater than the lethal dose for S plants. The dose required to reduce shoot weight or plant number by 50% relative to untreated plants (GR_{50} and LD_{50}, respectively) of R and S biotypes is calculated from the regression equations. The relative level of resistance is often expressed as the ratio of these values (R/S). A weed is considered resistant if the R/S ratio is higher than 10 [26].

2.2. The Resistance Must Be Inheritable

Testing procedures usually require the collection of seeds from resistant and susceptible populations. For the purpose of listing a putative resistant weed as a real resistant weed, assays with collected seeds are generally required for sexually propagated species, sometimes a second generation of seeds from greenhouse grown plants of R and S populations are collected and tested for resistance, and his descendent should have inherited the resistance.

2.3. Demonstration of Practical Field Impact

If there is no detectable difference in control of the weed under field conditions at the recommended rate, then it will not be classified as an herbicide resistant-weed. Although valid scientific arguments may be made for the occurrence of low-level resistance even when the weed is controlled by the field rate under field conditions, the weed must present a problem to the farmer when using the herbicide at the recommended rate.

To be classified as an herbicide resistant-weed, the plant in question must be a weed and identified down to the species level. Cases of deliberate selection for herbicide resistance, including herbicide resistant-crops as volunteers, are not included in the classification of herbicide-resistant [27].

3. Cross-Resistance

An herbicide-cross-resistant weed is defined as a biotype that has developed resistance from one herbicide and then exhibits resistance to other herbicides that differ chemically and/or have different modes of action [2]. For example, an infestation of rigid ryegrass resistant to diclofop-methyl was reported in the Border town region of South Australia in 1982 [27]. Further studies with these biotypes revealed cross-resistance to other aryloxyphenoxy propionate herbicides, to unrelated sulfonylurea herbicides and to the dinitroaniline herbicide trifluralin [28]. This was the first report of a weed species exhibiting extensive cross-resistance to chemically dissimilar herbicides and to herbicides with different modes of action. There are two broad cross-resistance categories: target site cross-resistant and non-target cross-resistant.

3.1. Target Site Cross-Resistance

Target site cross-resistance occurs when a change at the biochemical site of action of one herbicide also confers resistance to herbicides from a different chemical class that inhibits the same site of action in the plant and does not necessarily result in resistance to all herbicide classes with similar modes of action or indeed all herbicides within a given herbicide class. In some cases the resistance mechanism of weeds to sulfonylurea herbicides is a change in the target site enzyme acetolactate synthase (ALS) [29-39]. In most cases, the sulfonylurea-resistant biotypes with a resistant ALS enzyme exhibit varying levels of target site cross-resistance to the chemically dissimilar ALS-inhibiting, imidazolinone and/or triazolopyrimidine herbicides [40-44].

In the 1970's and 1980's, two chemically different herbicide groups, the aryloxyphenoxypropionic acid (APP) and cyclohexanedione (CHD) herbicides, which target the enzyme acetylcoenzyme A carboxylase (ACCase), were commercially developed and widely adopted. Two species of *Lolium rigidum* resistant to APP [45] and CHD herbicides [46] were selected, both species developed target site cross-resistance to both the APP and CHD herbicides; however, the level of resistance to APPs was greater than to CHDs [47].

Preston and Powles [48] determined the frequency of individuals resistant to two acetolactate synthase (ALS)-inhibiting herbicides in previously untreated populations of L. rigidum. The frequency of individuals resistant to sulfometuron-methyl and imazapyr varied from 2.2×10^{-5} to 1.2×10^{-4} and from 1×10^{-5} to 5.8×10^{-5}, respectively, depending on the population. The high initial frequency of individuals resistant to ALS-inhibiting herbicides helps to explain the rapid evolution of resistance in this species once these herbicides were used. Triazine and substituted urea herbicide groups are toxic to plants; they are potent and specific inhibitors of photosynthesis at the photosystem II (PS2) level.

These herbicides bind to the quinone B (Q_B) binding site on the D1 protein within the PS2 reaction center. Triazine herbicides have been continually used for weed control in maize production in many parts of the world and this practice has led to widespread resistance in target weeds. Plants containing triazine-resistant PS2 are also resistant to other PS2-inhibiting herbicide chemistries including triazinones, uracils, and pyridazinones [49-52].

3.2. Non-Target Site Cross-Resistance

Non-target site cross-resistance is defined as cross-resistance to different herbicide classes conferred by a mechanism(s) other than resistant target enzymes. The study of Heap and Knight [28] and the vast farmer experience in Australia, has demonstrated that many *L. rigidum* populations that developed resistance following selection with the ACCase-inhibiting herbicide diclofop-methyl displayed resistance to cereal-selective ALS herbicides without previous exposure to ALS herbicide. These biotypes of *L. rigidum* are resistant as a result of an increase in the rate that the herbicide is metabolized, the increased metabolism of *L. rigidum* biotypes is catalyzed by Cyt P450 enzyme-based mechanisms operating in wheat [41,53].

3.3. Changes in the Uptake and Translocation of Herbicides

Resistance to herbicides occurs as a result of the molecule that cannot reach its site of toxic action. In other words, it is the inability of the herbicide molecule to concentrate in lethal amounts at the point of action within the weed plant. Such exclusion of the herbicide from its action site can be due to several reasons.

3.3.1. Morphological Barriers

Differential herbicide uptake is due to morphological barriers on the leaves such as extraordinarily increased waxy coating on the cuticle, hairy epidermis, low foliage number and size, etc. [54]

Herbicide glyphosate inhibits 5-enolpyruvylshikimate-3-phosphate synthase (EPSPS), which is essential in the biosynthesis of the aromatic amino acids as well as many secondary plant products [55]. There are glyphosate-resistant populations of giant ragweed (*Ambrosia trifida* L.) that have an unusual phenotype of rapid desiccation of treated leaves which then fall off the plant, and the translocation of glyphosate to the growing points is reduced. The mechanism of this resistance is unknown [56].

3.3.2. Differential Translocation

Differential translocation whereby apoplastic (Xylem tubes) or symplastic path cells (Phloem cells) may restrict or delay movement of the right concentration of herbicide to the site of action [57]. It is likely that the herbicide undergoes a sequence of locations before reaching the site of action e.g. some lipophilic herbicides may become immobilized by partitioning into lipid rich glands or oil bodies [58]. For example, one biotype of annual grass weed is *Hordeum glaucum* Steud. Resistant to the bipyridyl herbicide, paraquat [59] its resistance is not due to any change in the active site at photosystem I [60]. Moreover, Bishop et al. [61] found that the resistance to paraquat is not due to a change in cuticular penetration, however it does not enter the cytoplasm of the resistant plants because the resistant biotype has the capacity to prevent this, probably by some mechanisms in the vascular tissue and/or cell walls allowing paraquat sequestration before it could enter the cytoplasm.

Glyphosate is a potent herbicide because of its ability to translocate in the plant to the apical meristems, root meristems and underground reproductive organs of perennial plants.

There are at least two mechanisms of glyphosate uptake, an active system that operates at low concentrations of the herbicide, which may involve a phosphate transporter, and a passive mass flow system [62]. The herbicide has to enter the phloem lumen, probably through the cell symplast, either by mass diffusion into the mesophyll cells followed by movement to the phloem through the plasmodesmata or, it may be actively taken into the mesophyll or into companion cells via a phosphate-transporter, as previously described [62]. Once glyphosate enters the sieve element, it is trapped and transported to the sink tissues due to its hydrophilic properties [62]. The amount of translocated glyphosate from source leaves to sink tissues is self-limiting owing to its mechanism of action that inhibits assimilation-translocation. Glyphosate resistance (GR) identified in horseweed [63,64], hairy fleabane [65], rigid ryegrass [66] and Italian ryegrass [67,68] species, appears to be limiting glyphosate translocation to the meristematic sinks. The pattern of glyphosate movement in these GR biotypes differs from the glyphosate sensible (GS) biotypes; these resulted in a 3 to 10-fold difference in the G_{50} values (concentration that inhibits 50% growth) [62]. The mechanism for this reduced absorption has not been elucidated. However, Shaner [62] suggested the following potential mechanisms by which the cellular absorption and subsequent translocation of glyphosate could be reduced: (1) alteration in a putative phosphate transporter responsible for the active cellular uptake of glyphosate, if the transporter is no longer present or no longer recognizes glyphosate; (2) evolution of a new transporter that pumps glyphosate into the vacuole, thus sequestering the herbicide and preventing it from reaching either the chloroplast or the phloem as has been demonstrated in horseweed-resistant glyphosate [10]; (3) evolution of a new transporter that actively pumps glyphosate out of the cell into the apoplast; or (4) evolution of a transporter at the chloroplast envelope that pumps glyphosate out of the chloroplast, preventing the herbicide from reaching its target site. [62].

3.4. Changes in the Detoxification of Herbicides

Higher plants are equipped with a remarkably versatile system that protects them from the potentially phytotoxic action of xenobiotics like herbicides. During the last decades, research has revealed that plants contain a large number of enzymes that metabolize herbicides and other xenobiotics to non-phytotoxic products [68]. Herbicides can be metabolized by oxidation, hydrolysis or conjugation reactions to an endogenous moiety such as glutathione, glucose or aminoacids [69] generating: a) molecules incapable of interacting with the target site, b) more hydrophobic molecules and thus less mobile in the plant than the parent herbicide and c) molecules susceptible to further processing, which may include secondary conjugation, degradation and compartmentalization [68].

3.4.1. Cytochrome P_{450}

Cytochrome P_{450} is a family of enzymes that are involved in the biosynthesis of numerous secondary products such as phenylpropanoids, terpenes, fatty acids, cyanogenic glucosides and others [69]. It has been established that Cyt P_{450}-dependent mono oxygenases play an essential role in the oxidation of herbicides in plants [60,61]. The most common Cyt P_{450}-mediated reactions with herbicides are hydroxylation of aromatic rings or alkyl groups and heteroatom release (O, N-dealkylation).

It has been widely documented that the basis for selectivity of atrazine in tolerant weeds such as *Digitaria* [70], *Panicum* [70,71], *Setaria* [71,72] and *Shorgum* [72], is their biochemical ability to degrade the herbicide to non-toxic metabolites; atrazine is detoxified by hydroxylation, N-dealkylation and conjugation with glutathione.

Burnet's group (1993) [7] studied the mechanism of resistance of two biotypes of *L. rigidum* to triazine herbicides. The resistant biotypes showed different behaviors when they were exposed to the herbicide atrazine; however both strains developed major resistance to simazine, a herbicide structurally similar to atrazine. Since a similar concentration of simazine was required for a 50% reduction on thylakoid electron transport isolated from both resistant and susceptible biotypes, the authors conclude that simazine resistance was not due to a change at the target site. Uptake of simazine from a nutrient solution and its distribution between the roots and the shoots is similar in R and S biotypes. However, R biotypes metabolized [^{14}C]-simazine at a greater rate than S plants when this was supplied as either a 12 h pulse or continuously over 7 days. The primary metabolite is the de-ethyl derivative with subsequent removal of the second ethyl residue [7] (Figure 1C). The chemical structure of the metabolites and the inhibition of the metabolism by 1-aminobenzotriazole (ABT) suggest the action of oxidative enzymes in the mechanism of resistance to this herbicide [7].

De Prado and his workgroup [72] detected two resistant biotypes of *Setaria: S. faberi* and *S. viridis* in fields that had been continuously treated with triazine herbicides. The R biotypes of *S. faberi* and *S viridis* were 10 and 6.96 times less sensitive to atrazine than the S biotype. Absorption and translocation of [^{14}C]-atrazine from a nutrient solution via the roots to the shoot were similar for R and S in both *Setaria* species, which implies that the resistance is not given by changes in the absorption and translocation of the herbicide; however, R biotypes metabolized the atrazine to its conjugates faster than S biotypes. The non-phytotoxic atrazine conjugate was the main product formed in the shoots of *S. faberis* and *S. viridis* biotypes ($R \geq 75 \leq S$) after 24 h of incubation with the herbicide.

An analysis of atrazine metabolism in five different species of *Setaria* (*S. adherens, S. faberi, S. glauca, S. verticilata* and *S. viridis*) that had never been treated with this herbicide or any other PS II-inhibiting herbicide showed that the more tolerant species, *S. adherens* and *S. verticilata*, had a highly efficient atrazine-conjugation system (34.1 and 97.7% after incubation for 3 and 72 h, respectively), even though it is still slower than in maize conjugation [70]. Although *S. faberi* and *S. viridis* have a less efficient conjugation system, a de-alkylation system was present, so that the levels of the non-toxic metabolite reached 9.4 and 78.1% after 3 and 72 hours of incubation, respectively. In agreement with these results, De Prado et al [73] suggest that dealkylation is the most important mechanism for atrazine detoxification.

3.4.2. Glutathione S-Transferase (GST)

Multiple-herbicide resistance (MHR) in black-grass (*Alopecurus myosuroides*) and annual rye-grass (*L. rigidum*) is a global problem leading to a loss of chemical weed control in cereal crops [74]. In both annual rye-grass and black-grass, MHR was associated with the overexpression of a plant (phi) F class glutathione transferase (GSTF1) that had a restricted ability to detoxify herbicides. When the black-grass *A. myosuroides* (Am) AmGSTF1 was expressed in *Arabidopsis thaliana*, the transgenic plants acquired resistance to multiple herbicides and showed similar changes, in their secondary, xenobiotic, and antioxidant metabolism, to those determined in MHR weeds. Transcriptome array experiments showed that these changes in biochemistry were not due to changes in gene expression. Rather, AmGSTF1 exerted a direct regulatory control on metabolism that led to an accumulation of protective flavonoids. Further evidence for a key role for this protein in MHR was obtained by showing that the human glutathione transferase GSTP1- and multiple-drug resistance (MDR) in tumors, MDR-inhibiting pharmacophore 4-chloro-7-nitro-benzoxadiazole was also active toward AmGSTF1 and helped to restore herbicide control in MHR black-grass. These studies demonstrated a central role for specific GSTFs in MHR in weeds that have parallels with similar roles for unrelated GSTs in MDR in humans and showed their potential as targets for chemical intervention in resistant weed management [74]. Before these findings, it was already known that the evolution of herbicide resistance in Glutathione S-Transferases (GSTs) was due to multifunctional dimeric proteins that catalyze the nucleophilic attack of the tiolate anion of glutathione (GSH) to electrophilic groups on various substrates [75].

Atrazine-resistant velvetleaf (*Abutilon theophrasti* Medic) biotype discovered in a field in Maryland in 1986 had a 10-fold greater tolerance to atrazine than the susceptible biotype from Minnesota [76]. Andersen and collaborators [76] found that GST-activity measured with atrazine in the R biotype as a substrate, was 4.4 and 3.5 fold greater in leaf and stem extracts respectively, than the S biotype. Kinetic analyses of GST activity in leaf extracts from resistant and susceptible biotypes showed similar apparent Km values for atrazine and GSH; however, the V_{max} were approximately 3 and 3.5-fold greater for atrazine and GSH, respectively, in the R biotype compared to the S one. Therefore, the authors concluded that the R biotype was tolerant to atrazine due to its capacity to rapidly metabolize the herbicide via GSH conjugation. In plants, GSTs catalyze the detoxification of the s-triazine [77-83], chloroacetanilide [77,81,84,85] and tiocarbamate herbicides [81,86].

3.5. Altered Target Site

Many biotypes resistance to herbicides is conferred by an altered target site, i.e. a modified target protein with reduced affinity to the herbicide(s) tested, where their action sites are better known, e.g. the inhibition of photosynthetic electron transfer at Photosystem II, acetyl-CoA carboxylase and acetolactate synthase [87]. In a *Latuca sativa* Rbiotype, the action site of sulfonylurea in the ALS enzyme is modified in such a way that the herbicide cannot bind [88].

3.5.1. Acetolactate Synthase (ALS)

Acetolactate synthase (ALS) is a plastid enzyme [89], which catalyzes the first common step in the biosynthesis of branched chain aminoacids such as leucine, isoleucine and valine [90,91]. Inhibition of ALS leads to a rapid growth cessation in susceptible species. In a classification based on its mode of action, ALS inhibitors are the more prone herbicide group to cause resistance having the highest number of resistance biotypes in the last 10 years [89].

Altered target sites are the main mechanism of resistance to ALS inhibitors for different weeds: Stellaria media, Setaria arvensis, Descorainia shopia, Alisma plantago-aquatica, Papaver rhoeas, Scirpus mucronatus, Monochoria vaginalis, Cyperus difformis and Amaranthus [30-40].The most frequent mechanism that confers resistance to ALS-inhibitors is a single point mutation in the gene [29,30,32,34-39,92, 93].

Table 1. Examples of ALS point mutations from herbicide resistant weeds modified from Heap [93]

Amino Acid Residue and Number(1)	Substitution conferring resistance	Weed Species	SU(2)	IMI(2)	PTB(2)	TP(2)	SCT(2)	Year(3)
Ala 122	Thr	*Amaranthus powellii*	S	R	ND	ND	ND	2005
	Val	*Apera spica-venti*	R	ND	ND	ND	S	2011
	Tyr	*Raphanus raphanistrum*	R	R	ND	R	ND	2012
Pro 197	Ile	*Sisymbrium orientale*	R	r	ND	R	ND	1999
	Arg	*Papaver rhoeas*	R	r	R	r	ND	2009
	Asn	*Apera spica-venti*	R	ND	ND	r	r	2011
	Gln	*Anthemis cotula*	R	r	ND	R	ND	2011
	Ala	*Conyza canadensis*	R	S	R	R	ND	2011
	Ser	*Conyza canadensis*	R	S	R	R	ND	2011
	Thr	*Capsella bursa-pastoris*	R	ND	ND	ND	ND	2012
	His	*Capsella bursa-pastoris*	R	ND	ND	ND	ND	2012
	Leu	*Capsella bursa-pastoris*	R	ND	ND	ND	ND	2012
Ala 205	Val	*Solanum ptycanthum*	r	R	ND	S	ND	2007
ASP 376	Glu	*Galium spurium*	ND	ND	ND	R	ND	2012
Arg 377	His	*Apera spica-venti*	R	ND	ND	R	R	2011
Trp 574	Leu	*Galium spurium*	ND	R	ND	R	ND	2012
	Gly	*Galium aparine*	R	ND	ND	ND	ND	2011
Ser 653	Ile	*Setaria viridis*	r	R	R	ND	r	2009
	Thr	*Avena fatua*	ND	R	ND	ND	ND	2012
	Asn	*Avena fatua*	ND	R	ND	ND	ND	2012
Gly 654	Asp	*Setaria viridis*	r	R	S	ND	r	2009

1. Amino acid number is standardized to the *Arabidopsis thaliana* sequence.

2. S = Susceptible biotype, r = Moderate resistance (<10-fold relative to sensitive biotype), R = High Resistance (> 10-fold relative to sensitive biotype), ND = Not determined.

3. Year when the mutation was first reported. Visit the web site of reference 93.

SU= Sulfonylureas, IMI= Imidazolines, PTB = Pyrimidinylthiobenzoates, TP = Triazolopyrimidines, SCT = Sulfonylaminocarbonyltriazolinone.

To date, only 8 aminoacid substitutions have been reported from field-selected weed biotypes: Ala_{122}, Pro $_{197}$, Ala_{205}, Trp_{574}, Ser_{653} [94-96], Asp_{376} [97-99], Arg_{377} [100] and Gly_{654} [101]. Table 1 shows some examples of mutations for different resistant species. The mutation, which confers major high levels of resistance to sulfonylurea (SU) herbicides, but low or no cross-resistance to imidazolinone (IMI), pirimidinylthiobenzoates (PTB), triazolopyrimidines (TP) and sulfonylaminocarbonyltriazolinone (SCT) herbicides occurs at the Pro_{197} (Table 1). Mutations in other domains confer different patterns of resistance, i.e. the Asp_{376} to Glu and Trp_{591} to Leu mutations confer high levels of resistance to all ALS inhibitors, whereas the Ser_{653} to Thr, Ala_{122} to Thr and Ala_{205} to Val mutations confer a high level of resistance to IMI but little change in sensitivity to other ALS Inhibitors. Mutations in ALS vary depending on the weed. In this way, weeds present resistance to different molecule structure herbicides with the same action target.

3.5.2. Acetyl Coenzyme aCarboxylase (Accase)

Acetyl Coenzyme A carboxylase (ACCase) is a biotinylated enzyme; it catalyzes the ATP-dependent carboxylation of acetyl-CoA to form malonyl-CoA, the first step of fatty acid biosynthesis in eukaryotes and prokaryotes [102]. ACCase also catalyzes the malonylation required for the synthesis of long-chain fatty acids and flavonoids found in the cytosol of all plants [103,104]. ACCase is the target of widely used herbicides that include three chemically different groups: aryloxyphenoxypropionate (APP), cyclohexanedione (CDH) and recently phenylpyrazoline (DEN), which contains pinoxaden as an active ingredient and has a chemical structure that differs from APPs and CDHs. [105,106].

Altered Chloroplast ACCase causes herbicide resistance [107]. The 400 aminoacid regions, encompassing the carboxyl terminal (CT) domain of wheat chloroplast ACCase, are involved in insensitivity to both CDH and APP herbicides [107]. The results summarized in Table 2 enable a rough separation of three groups of cross-resistance patterns conferred by mutant plastid ACC isoforms. The first ACC isoforms group confers a very high resistance to DEN Pinoxaden, but less resistance to CHDs and to APPs. In a *Lolium spp* species, this cross-resistance pattern has been associated with the substitution of isoleucine (Ile) to Valine (Val) at position 2041 [105]; A second group of mutant ACC isoforms that only confers high resistance to APPs but low or no resistance to CHDs;

The cross-resistance pattern in blackgrass is conferred by substitution of tryptophan (Trp) to cysteine (Cys), Ile to asparagine (Asn), glycine (Gly) to alanine (Ala) at positions 2027, 2041, or 2096, respectively. It should be noted that the Ile to Asn substitution at the position 2041conferred a high resistance to the APP clodinafop (Table 2). In a third group, mutant ACC isoforms confer high resistance to all assayed APPs and CHDs (Table 2). In blackgrass, this cross-resistance pattern is associated with an aspartic acid (Asp) to Gly substitution at position 2078. The resistance to ACC-inhibiting herbicides (Ile_{1781}, Trp_{2027}, Ile_{2041}, Asp_{2078} and Gly_{2096}) is located within the active site cavity of the ACC CT domain associated with five amino acid residues [108]. From tridimensional models derived from yeast homomeric cytosolic ACC, only the substitution at position 2041 interfered directly with the herbicide binding; it has been proposed that the other four mutations cause resistance by hampering herbicide access to its binding site by altering the spatial shape of the herbicide binding site via small allosteric changes [108]. The fact that such a subtle change as the Ile to Leu substitution at position 1781 has such a drastic effect on the sensitivity to sethoxydim, giving the tight fit of the herbicide within the CT active site cavity (Table 2) [109].

Table 2. Amino acid substitutions within plastidic, homomeric ACC and associated cross-resistance patterns observed at the whole plant level[a] (Modified from Délye an Dijon 2011)[169]

Amino acid residue[b].			Resistance										Ref
			APPs[c]					CDHs[d]				DEN[e]	
Wild type	Resistant	Weed species	Cd	Dc	Fx	Fz	Hx	Ct	Cx	Sx	TK	Px	
Ile_{1781}	Leu	*A. myosuroides*	S	R	R	R	S	S	R	R	R	ND	107
	Leu	*A. fatua*	ND	R	ND	ND	ND	ND	ND	R	ND	ND	107
	Leu	*S. viridis*	ND	R	R	ND	ND	ND	ND	R	R	ND	107
	Leu	*Lolium* sp.	R	R	R	ND	R	R	R	R	ND	R	105,107
Trp_{2027}	Cys	*A. myosuroides*	R	ND	R	ND	R	S	S	ND	ND	ND	107
Ile_{2041}	Asn	*A. myosuroides*	R	ND	R	ND	R	S	S	ND	ND	ND	107
	Asn	*Lolium* sp.	R	R	ND	ND	R	S	S	S	ND	S	105,107
	Val	*Lolium* sp	S	ND	ND	ND	R	S	S	S	ND	R	105,107
Asp_{2078}	Gly	*A. myosuroides*	R	ND	R	ND	R	R	R	ND	ND	ND	107
Cys_{2088}	Arg	*Lolium* sp	R	ND	ND	ND	ND	S	ND	R	ND	R	105
Gly_{2096}	Ala	*A. myosuroides*	R	ND	R	ND	R	S	S	ND	ND	ND	107

[a] S, R and ND indicate sensitive, resistant or not determined biotype, respectively.

[b] Amino acid number is standardized to *A. myosuroides* plastidic, homomeric ACC (EMBL accession AJ310767).

[c] Cd, clodinafop; Dc, diclofop; Fx, fenoxaprop; Fz, fluazifop; Hx, haloxyfop.

[d] Ct, clethodim; Cx, cycloxydim; Sx; sethoxydim; Tk, tralkoxydim.

[e]Px Pinoxaden.

3.5.3. Phytoene Desaturase

The carotenoid biosynthetic pathway is essential for plant development while it is absent in animals. Several chemicals that act as phytoene desaturase (PDS) have been developed, including pyridazinones, pyridinecarboxamides and phenoxybutanamides [110]. However, only a few inhibitors of this pathway have been commercialized because most of these compounds lack crop selectivity. Herbicide fluridone was discovered in the mid-1970s for use in aquatic systems and it was an excellent herbicide against hydrilla [*Hydrilla verticillata* (Lf. Royle)], one of the most serious aquatic weed problems in the southern and western USA [111]. Phytoene desaturase is the molecular target site of fluridone; however, in 2004 the decrease in the efficacy of fluridone to control the aquatic weed was reported [112]. At least three dioecious fluridone-resitant biotypes of *Hydrilla* with two- to five-fold higher resistances to herbicides than to the wild-type were identified. *In vitro* assays with resistant biotypes of *Hydrilla* have shown mutations at the arginine 304 codon of the phytoene desaturase (*pds*) gene of *Hydrilla* rendered this enzyme less sensitive to fluridone [112].

3.5.4. p-Hydroxyphenylpyruvate Dioxygenase (HPPD)

The enzyme *p*-hydroxyphenylpyruvate dioxygenase (HPPD) catalyzes the formation of homogentistic acid, the aromatic precursor of plastoquinone and vitamin E. HPPD is the specific target of several herbicide families: isoxazoles, triketones and pyroxazoles. Its inhibition results in the depletion of the plant plastoquinone and vitamin E pools leading to bleaching symptoms, these herbicides have been used for weed control in important crops like maize and rice [113]

Waterhemp [*Amaranthus tuberculatus* (Moq.) Sauer] is a summer annual weed species that can reduce the yield of maize (*Zea mays* L.) [114], soybean (*Glycine max* L.) [115] and sorghum (*Sorghum bicolor* L.) [116]; waterhemp is ideally suited for evolving herbicide resistance by sharing resistance genes among populations and biotypes [117], showing resistance to herbicides that inhibit acetolactate synthase (ALS) [118], photosystem II (PSII) [119], protoporphyrinogen oxidase (PPO) [120], glyphosate [121] and HPPD herbicides [122].

3.5.5. Protoporphyrinogen Oxidase (PPO)

Protoporphyrinogen oxidase (PPO or Protox in some literature), a key enzyme in the synthesis of tetrapyrroles, catalyzes the oxidation of protoporphyrinogen IX (Protogen) to protoporphyrin IX (Proto). Plant PPO1 is compartmentalized in the thylakoid and in the envelope membranes of chloroplasts, whereas the mitochondrial isoform PPO2 is localized on the outer surface of the inner mitochondrial membrane [123,124]. In the chloroplast the porphyrin pathway biosynthesizes both chlorophyll and heme, whereas in the mitochondrion it leads exclusively to the heme group [125]. The three-dimensional (3D) structure of plant mitochondrial PPO revealed that this homodimer folds into a compact structure that includes an FAD-binding, a substrate binding, and a membrane-binding domain [125]. The PPO FAD-binding domain has structural homologies to other flavoenzymes [126] and it is known to be near the binding site of xenobiotic inhibitors [127,128]. A number of different herbicides (e.g., diphenyl ethers, oxadiazoles, phenylphthalimides and triazolinones) act as competitive inhibitors of PPO [124,129], resulting in the cytosolic accumulation of Proto [130-136].

Crystal structures of PPO, complexed with various inhibitors, have revealed that their binding within the catalytic domain is located near the C-terminus capping region of the α-8 helix [131,133]. Gly_{210} plays an important role in the α-8 helix cap. The first weed that developed resistance to PPO-inhibiting herbicides was *A. tuberculatus*, a problematic weed in the Midwestern United States [137]. PPO inhibitor-resistant *A. tuberculatus* biotype revealed that resistance was conferred by a single, nuclear gene. Three genes predicted to encode PPO were identified in *A. tuberculatus*. One gene from the resistant biotype, designated *PPX2L*, contained a codon deletion that was shown to confer resistance by complementation of a *hemG* mutant strain of *Escherichia coli* grown in the presence and absence of the PPO inhibitor lactofen. *PPX2L* is predicted to encode both plastid- and mitochondria-targeted PPO isoforms. [138]. Thinglum et al. [139] observing waterhemp resistance to herbicides that inhibit PPO, determined that if geographically separated, populations from Illinois, Kansas, and Missouri possessed the same mechanism for resistance. A dose–response study with lactofen indicated that the resistant populations had different levels of resistance. These differences, however, could be accounted for by different frequencies of resistant individuals within populations and, therefore, the dose–response data were consistent with the hypothesis that the populations contained the same resistance mechanism. This hypothesis was provided by DNA sequencing, which showed that nearly all resistant evaluated plants contained the DG210 mutation. A variable region of the PPX2 gene was sequenced and resulting sequences were aligned and organized into a phylogenetic tree. The phylogenetic tree did not reveal a clear clustering by either geography or phenotype (resistant vs. sensitive). Possibly recombination within the PPX2 gene has masked its evolutionary history. Resistance to protoporphyrinogen oxidase (PPO)-inhibiting herbicides in waterhemp was previously shown to be the result of a unique mechanism: specifically, a three-base-pair (3-bp) deletion in the PPX2L gene.

3.5.6. 5-Enolpyruvylshikimate-3-Phosphate Synthase (EPSPS)

The first glyphosate-resistant weed population containing an altered target site was found in Malaysia in 2002 and was reported by Lee and Ngim [140]. Baerson et al. [141,142] showed that the resistant Malaysian goosegrass populations exhibited a reduced sensitivity of EPSPS to glyphosate, based on the finding that the glyphosate concentration required to inhibit EPSPS activity by 50% (I_{50}) in resistant populations was 3 times higher than in susceptible populations. The molecular basis for this resistance is a mutation in the EPSPS gene that caused a substitution at amino acid Pro101 to Ser (P101S) [143]. A proline to threonine (P101T) substitution was also found to confer similar levels of glyphosate resistance in Malaysian goosegrass populations [144,145]. The result of these substitutions is a decrease in the affinity of EPSPS for glyphosate binding. These target-site mutations were subsequently linked to Australian [146] and Chilean [67] populations of glyphosate-resistant ryegrass (*Lolium* spp.). The inheritance of the EPSPS P106S target-site mutation is through a single, nuclear encoded gene that is incompletely dominant [144].

3.5.7. Changes in Chloroplast Structure

Triazine resistance in weeds is due to a mutation of the PsbA chloroplast gene [147-149] induced by a Ser-Gly substitution in the 264 position of the PS II reaction center D1 protein, causing less affinity to atrazine and Q_B [150]. On another hand, a comparison of the chemical composition and physical states of the R and S plants showed that thylakoids of R plants contained less polar lipids than S plants (expressed on chlorophyll basis). Also, chloroplasts of R plants have a higher content of monogalactosyl-diacylglycerol and a lower content of digalactosyl-diacylglycerol and phosphatidylglycerol than those of S plants. Furthermore, the chloroplast total lipids in the R biotype exhibited a higher degree of unsaturation making the lipid matrix of the thylakoid membranes more fluid compared to the S biotype, measured by the fluorescence polarization technique [151]. Some researchers [152-154] have suggested that the lipid micro-environment is directly or indirectly involved in the herbicide-binding to D1 protein, by affecting the physical state of the thylakoid membrane, and by an allosteric change that affects the herbicide-binding site in different plants [151,155].

Another example is the soybean cell line STR7, an atrazine resistant mutant with a single mutation in the chloroplast Psb A that implies the substitution of serine 268 by a proline in the D1 protein. The STR7 strain showed some important differences compared to the wild type (WT): slower growth, reduced oxygen-evolving activity, reduced electron transfer rate between the secondary acceptors Q_A and Q_B, presence of more non-Q_B-reducing PSII centers, a larger antenna size [156] and its unusual tolerance to high temperatures and its increased sensitivity to light stress [157,158]; both of the last two properties seem to be related to the alteration of the lipid composition found in STR7. This mutant strain showed an unusually high content of saturated fatty acids compared to the WT and consequently a more rigid thylakoid membrane matrix [157,159]. Higher levels of unsaturated lipids induced other atrazine-resistant mutants [151,155] contrary to what was observed in the SRT7 cell line.

3.6. Gene Amplification

Another mechanism of glyphosate resistance was discovered in the laboratory using tissue cultures. Cultures of carrot (*Daucus carota* L.) were selected for glyphosate resistance by a stepwise increase in glyphosate level, until they could grow in 24–35 mM of glyphosate. Unselected cells died at 1 mM of glyphosate [160]. The resistant cell line had a 12-fold increase in EPSPS protein levels, and the genome contained a 4-25-fold increase in EPSPS copy number [160].

Subsequently, glyphosate resistance was selected in cell cultures of petunia, tobacco, carrot, soybeans, chicory and alfalfa [161]. In all of these cases there was at least a 20-fold increase in the copies of EPSPS genes in the genome. The stability of the increased EPSPS copies, particularly in the absence of glyphosate selection pressure, was tested in many cases. The results were varied, depending on the cell culture. Stable resistance was found in cultures of chicory, tomato and tobacco [161].

However, resistance was slowly decreased or was lost completely in other lines owing to the loss of the multiple copies of the gene [161]. Attempts were made to regenerate plants with EPSPS gene amplification in tobacco, but in some cases, regenerated plants maintained glyphosate resistance, but in others either the regenerated plants were not resistant or they lost their resistance over time [161].

On these results, researchers concluded that gene amplification could not be selected in the field because the ability of selection for resistant lines in the laboratory was low and the selection pressure would not be duplicated in the field [162]. However, it was reported in 2012 [9] that an Italian ryegrass population from Arkansas was resistant to glyphosate, primarily due to increased EPSPS enzyme activity associated with amplification of the *EPSPS* gene copy number.

3.7. Negative Cross-Resistance (*Collateral Sensitivity*)

Individuals resistant to one chemical or chemical family of herbicides have a higher sensitivity to other herbicides [163]. Mixtures of herbicides can be powerful tools in weed control with important advantages; the admixed herbicides can often be applied at much lower rates, especially when interacting synergistically [164,165]. Application of mixtures can be environmentally safer, less expensive and offer control of a broader spectrum of weed species than single compounds. Moreover, resistant individuals are effectively controlled and resistant populations rarely evolve when herbicides with different mechanisms of action are used simultaneously. The use of mixtures requires the adoption of certain guidelines: herbicides should have different target domains or sites of action, the components should have similar half lives, and the herbicidal components should have similar spectra of weed control, etc [166,167].

Beckie and colls. [168] examined the response of homozygous susceptible and acetolactate synthase (ALS) inhibitor–resistant plants from six *Canadian kochia* (common and economically important weeds in crops and rural areas in the southern Canadian prairies and Great Plains of the United States) accessions with the Pro197 or Trp574 mutation to six alternative herbicides of different sites of action. Resistant plants were 80, 60, and 50% more sensitive than susceptible plants to pyrasulfotole, mesotrione (hydroxyphenylpyruvate dioxygenase [HPPD] inhibitors), and carfentrazone (protoporphyrinogen oxidase [PPO] inhibitor), respectively. However, no differential dose response between resistant and susceptible plants of *kochia* to bromoxynil, fluroxypyr, or glyphosate was observed. Negative cross-resistance exhibited by resistant plants to PPO and HPPD inhibitors in this experiment may be a pleiotropic effect related to the Trp574 mutation.

CONCLUSION

Monoculture, with repeated use of a single herbicide, is responsible for the evolution of herbicide resistance. Resistance is not acquired due to mutations caused by the herbicide; rather, it arises from the selection of natural mutations or small pre-existing populations of resistant plants.

In this revision different mechanisms of resistance have been presented; however, when the herbicide target is one enzyme, this enzyme can present resistance to different herbicides in the cross-resistance. The most frequent mechanism is a single point mutation, as the resistance to herbicides that have their action site in ALS, ACC, PDS, PPO or EPSPS enzymes. There are also biotypes resistant to different herbicides because they have different mechanisms of resistance, moreover, one herbicide as glyphosate can be metabolized by different biotypes resistant to it by different mechanisms which could be: changes in uptake by sequestration, altered target and gene amplification. Resistance to glyphosate up to this moment has been the most studied, maybe because glyphosate is the most used herbicide in agriculture.

No matter the mechanism by which the herbicide acts in a weed population, there will always be some individuals that acquire resistance to it. Nature provides different mechanisms of detoxification to these individuals. Presently, there are missing studies that could explain and describe these mechanisms. Rapid evolution of herbicide resistance in weed species is an important problem for farmers; furthermore, herbicide cross-resistance aggravates the problem. There are strategies for herbicide resistance management, such as the use of herbicide mixtures and herbicides with new targets and mechanisms of action. Natural biologically active products could represent a new strategy to solve the resistance to herbicides.

Acknowledgments

The authors thank the financial support of the DGAPA-UNAM IT102012-3 and F. Morales-Flores acknowledges the fellowship awarded by CONACyT México.

References

[1] Powles SB, Howat PD. (1990) Herbicide-resistant weeds in Australia. *Weed Tech.*4: 178-185.

[2] Powles SB, Yu Q. (2010) Evolution in action: plants resistant to herbicides. *Annu. Rev. Plant Biol.*61: 317–347.

[3] Heap I. International Survey of herbicide resistant weeds. [Online]. Weed Science Society of America (2013). Available: http://www.weedscience.org/In.asp [5 June 2013].

[4] Powles SB, Preston C, Bryant IB, Jotsum AR. (1997) Herbicide resistance: Impact and management, *Advances in Agronomy* 58: 57-93.

[5] Tharayil-Santhakumar N. Mechanism of herbicide resistance in weeds. [Online]. Weed Science Society of America (2003). Available: http://www.weedscience.org/In.asp. [15 August 2011].

[6] Duke SO, Christy AL, Hess FD, Holt JS. Herbicide-Resistant Crops, Comments from CAST 1991-1. Council of agricultural Science and Technology Ames Iowa, USA, (1991). pp. 32

[7] Burnet MWM, Lovey SBR, Holtum HAM, Powles SB. (1993) Increased detoxification is a mechanism of simazine resistance in *Lolium rigidum. Pest Biochem. Physiol.* 46: 207-218.

[8] Gaines TA, Zhang W, Wang D, Bukun B, Chisholm ST, Shaner DL, Nissen SJ, Patzoldt WL, Tranel PJ, Culpepper AS, Grey TL, Webster TM, Vencill WK, Sammons RD, Jiang J, Preston C, Leach JE, Westra P. (2010) Gene amplification confers glyphosate resistance in Amaranrthus palmeri. *PNAS* 107: 1029-1034.

[9] Salas RA, Dayan FE, Pan Z, Watson SB, Dickson JW, Scott RC, Burgos NR. (2012) EPSPS gene amplification in glyphosate-resistant Italian ryegrass (*Lolium perene* ssp. *multiflorum*) from Arkansas. *Pest. Manag. Sci.* 68:1223-1230.

[10] Ge X, d'Avignon DA, Ackerman JJH and Sammons RD. (2010) Rapid vacuolar sequestration: the horseweed glyphosate resistance mechanism. *Pest. Manag. Sci* 66:345–348.

[11] Ge X, d'Avignon DA, Ackerman JJH, Collavo A, Sattin M, Ostrander EL, Hall EL, Sammons DR, Preston C. (2012) Vacuolar glyphosate-sequestration correlates with glyphosate resistance in Ryegrass (Lolim spp.) from Australia, South America, and Europe: A ^{31}P NMR investigation. *J. Agric. Food Chem.* 60:1243-1250.

[12] Valverde BE. (2004) Herbicide resistance management in developing countries, in: Labrada R. (Ed), Weed Management for developing countries Addendum 1,*Food and Agriculture Org.Rome* pp. 223-244.

[13] Ort DR, Ahren WH, Martin B and Stoller EW. (1983) Comparison of photosynthetic performance in atrazine-resistant and susceptible biotypes of *Amaranthus hibridus,Plant Physiol* 72:925-930

[14] Vila-Aiub MM, Neve P and Powles SB. (2005) Resistance cost of a cytochrome P450 herbicide metabolism mechanism but not an ACCase target site mutation in a multiple resistant *Lolium rigidum* population. *New Phytologist* 167: 787-796.

[15] Holt JJ, Stemer AJ, Rodoservich SR. (1981) Differential light responses of photosynthesis by triazine-resistant and triazine-susceptible *Senecio vulgaris. Plant. Physiol.* 67:744-748.

[16] Ireland CR, Telfer A, Covello PS, Baker NR, Barber J. (1988) Studies on the limitations to photosynthesis in leaves to the atrazine-resistant mutant of *Senecio vulgaris* L. *Planta* 173: 459-467.

[17] Strauss SY, Rudgers JA, Lau JA and Irwin RE. (2002) Direct and ecological costs of resistance to herbivory. *Trends in ecology and evolution.* 17: 278-285.

[18] Gronwald JW. (1994) Resistance to photosystem II inhibiting herbicides, in: Powles SB, Holtum JAM (Eds), Herbicide Resistance in Plants. Biology and Biochemistry, Lewis Publishers, Florida, pp. 27-60.

[19] Purrington CB, Bergelson J. (1999) Exploring the physiological basis of cost of herbicide resistance in *Arabipdosis thaliana. American Naturalist*154: S82-S91.

[20] Berticat C, Boquien G, Raymond M, Chevillon C. (2002) Insecticide resistance genes induce a mating competition cost in *Culex pipiens* mosquitoes. *Genetical research* 79:4 1-47.

[21] Li M, Yu Q, Han H, Vila-Aiub M, Powles SB. (2013) ALS herbicide resistancemutations in *Raphanus raphanistrum*: evaluation of pleiotropic effects on vegetative growth and ALS activity, *Pest Manag Sci*69: 689-695.

[22] Coley PD, Bryant JP. (1985) F.S.-III Chapin, Resource availability and plant anti-herbivore defense. *Science* 230: 895-899.

[23] Herms DA, Mattson WJ. (1992) The dilemma of plants: to grow or defend. Quarterly. *Review of Biology* 67: 283-335.

[24] Bergelson J, Purrington CB. (1996) Surveying patterns in the cost of resistance in plants. *American Naturalist* 148: 536-558.

[25] Purrington CB, Bergelson J. (1997) Fitness consequences of genetically engineered herbicide and antibiotic resistance in *Arabidopsis thaliana. Genetics* 145: 807-814.

[26] Beckie HJ, Heap IM, Smeda A, Hall LM. (2000) Screening for herbicide resistance in weeds. *Weed Tech* 14: 428-445.

[27] Heap IM, Knight R. (1982) A population of ryegrass tolerant to the herbicide diclofop-methyl. *J. Aust. Inst. Agric. Sci.*48: 56-157.

[28] Heap IM, Knight R. (1986). The occurrence of herbicide cross-resistance in a population of annual ryegrass, *Lolium rigidum*, resistant to diclofop-methyl. *Aust. J. Agric. Res.* 37: 149-156.

[29] Christoffers JM, Nandula VK, Howat KA, Wehking TR. (2006) Target-site resistance to acetolactate synthase inhibitors in wild mustard (*Sinapis arvensis*). *Weed Science* 54:191-197.

[30] Cruz-Hipolto H, Osuna MD, Vidal RA, De Prado R. (2009) Resistant Mechanism to bensulfuron-methyl in biotypes of *Scirpus mucronatus* L collected in Chilean rice fields. *J. Agric. Food. Chem.* 57: 4273-4278.

[31] Duran-Prado M, Osuna MD, De Prado R, Franco AR. (2004) Molecular basis of resistance to sulfonylueras in *Papaver rhoeas.Pest. Biochem. Physiol.* 79: 10-17.

[32] Kudsk P, Mathiassen SK, Catterman JC. (1995) Sulfonylurea resistance in *Stellaria media* (L) Vill. *Weed Research* 35: 19-24.

[33] Kuk VI, Jung HI, Kwon OD, Lee DJ, Burgos NR, Guh JO. (2003) Sulfonylureas herbicide-resistant *Monochoria vaginalis* in Korean rice culture. *Pest. Manage Sci.*59:949-961.

[34] Kuk VI, Kim KH, Kwon OD, Lee DJ, Burgos NR, Jung HI, Guh JO. (2003) Cross-resistance pattern and alternative herbicides for *Cyperus difformis* resistant to sulfonylurea herbicides in Korea. *Pest. Manage Sci.* 60:85-94.

[35] Merotto A, Jasieniuk M, Fischer AJ. (2010) Distribution and cross-resistance patterns of ALS-inhibiting herbicide resistance in smallflower umbrella sedge (*Cyperus difformis*). *Weed Sci.* 58: 22-29.

[36] Scarabel L, Varoltus S, Sattin M. (2010) An Italian population of *Amaranthus retoflexus* resistant to ALS inhibiting herbicides: resistance pattern and molecular basis. 13th EWRS Symposium, Abstr. 202.

[37] Tabacchi M, Scarabel L, Sattin M. (2004) Herbicide resistance in Italian rice crops: a late to developing but fast-evolving story, in Proceedings of the MED-Rice Conference, Torino, Italy (February 13-15)

[38] Warwick SI, Sauder C, Beckie HJ. (2005) Resistance in Canadian biotypes of wild mustard (*Sinapis arvensis*) to acetolactate Synthase inhibiting herbicides. *Weed Sci.*54: 191-197.

[39] Xu X, Wang GQ, Chen SL, Fan CQ, Li BH. (2010) Confirmation of flixweed (*Descurainia Sophia*) resistance to tribenuron-methyl using three different assay methods. *Weed Sci.* 58: 56-60.

[40] Hall LM, Devine MD. (1990) Cross resistance of a chlorsulfuron-resistant biotype of *Stellaria media* to a triazolopyrimidine herbicide. *Plant. Physiol.* 93: 962-966.

[41] Christopher JT, Powles SB, Holtum JAM. (1992) Resistance to acetolactate synthase-inhibiting herbicides in annual ryegrass (*Lolium rigidum*) involves at least two mechanisms. *Plant. Physiol.* 100:1909-1913.

[42] Saari LL, Cotterman JC, Primiani MM. (1990) Mechanism of sulfonuylurea herbicide resistance in the broadleaf weed, *Kochia scoparia. Plant. Physiol.* 93: 55-61.

[43] Saari LL, Cotterman JC, Primiani MM. (1992) Sulfonylurea herbicide resistance in common chickweed, perennial ryegrass and Russian thistle. *Pestic Biochem.Physiol.* 42: 110-118.
[44] Saari LL, Cotterman JC, Thill DC. Resistance to acetolactate synthase inhibiting herbicides, in: Powles SB, Holtum JAM (Eds) Herbicide Resistance in Plants: Biology and Biochemistry, Lewis Publishers, Boca Raton, Fl, (1994) pp. 83-139.
[45] Holtum JAM, Powles SB. (1991) Annual ryegrass: an abundance of resistance, a plethora of mechanisms. Brighton Crop Protection Conference – Weedspp. 1071-1078.
[46] Tardif FJ, Holtum JAM, Powles SB. (1993) Occurrence of a herbicide-resistant acetyl-coenzyme A carboxylase mutant in annual ryegrass (*Lolium rigidum*) selected by sethoxydim. *Planta* 190: 176-181.
[47] Powles SB, Preston C. [Online]. Herbicide Resistance action. Available: http://www.hracglobal.com/publications/HerbicideCrossResistanceandMultipleResistance.aspx [15 August 2011].
[48] Preston C, Powles SB. (2002) Evolution of herbicide 596 resistance in weeds: initial frequency of target site-based resistance to acetolactate synthase-inhibiting herbicides in *Lollium rigidum. Heredity* 88: 8-13.
[49] Fuerst EP, Arntzen CJ, Pfister K, Penner D. (1986) Herbicide cross-resistance in triazine-resistant biotypes of four species. *Weed Sci.* 34: 1986: 344-353.
[50] Ducruet JM, De Prado R. (1982) Comparison of inhibitory activity of amide derivatives in triazine-resistant and -susceptible chloroplasts from *Chenopodium album* and *Brassica campestris. Pestic. Biochem. Physiol.* 18: 253-261.
[51] Oettmeier W, Masson K, Fedtke C, Konze J, Schmidt RR. (1982) Effect of Photosystem II inhibitors on chloroplasts from species either susceptible or resistant towards s-triazine herbicides. *Pestic. Biochem. Physiol.* 18: 357-367.
[52] De Prado R, Domínguez C, Tena M. (1989) Characterization of triazine-resistant biotypes of common lambsquarters (*Chenopodium album*), hairy fleabane (*Conyza bonariensis*) and yellow foxtail (*Setaria glauca*) found in Spain. *Weed Sci.* 37: 1-4.
[53] Christopher JT, Powles SB, Liljegren DR, Holtum JAM. (1991) Cross resistance to herbicides in annual ryegrass (*Lolium rigidum*). *Plant. Physiol.* 95: 1036-1043.
[54] Dekker J, Duke SO. (1995) Herbicide-resistant field crops. Advances in Agronomy 54: 69-116.
[55] Amrhein N, Deus B, Gehrke P, Steinrucken HC. (1980) The site of inhibition of the shikimate pathway by glyphosate, II: interference of glyphosate with chorismate formation *in vivo* and *in vitro*. *Plant. Physiol.* 66: 830-834.
[56] Shaner DL, Lindenmeyer RB, Ostlie MH. (2012) What have the mechanisms of resistance to glyphosate taught us? *Pest. Manag. Sci.*68: 3-9.
[57] Ozair CA, Moshier LJ, Werner GM. (1987) Absorption, translocation, and metabolism of foliage-applied chloramben in velvetleaf (*Abutilion theophrasti*) and soybean
[58] Stegink SJ, Vaughn KC. (1988) Norflurazon SAN-9789 reduced abscisic acid levels in cotton seedlings: A glandless isoline is more sensitive than it's glanded counterpart. *Pestic. Biochem. Physiol.* 31: 269-275.
[59] Powles SB. (1986) The appearance of a biotype of the weed *Hordeum glaucum* resistant to the herbicide paraquat. *Weed Res.* 26: 167-172.
[60] Powles SB, Cornic G. (1987) Mechanism of paraquat resistance in *Hordeum glaucum*. I. Studies with isolated organelles and enzymes. *Aust. J. plant. Physiol.* 14: 81-89.

[61] Bishop T, Powles SB, Cornic G. (1987) Mechanism of paraquat resistance in *Hordeum glaucum*. Paraquat Uptake and Translocation. Australian Journal of Plant Physiology 14(5): 539-547.

[62] Shaner D.L. (2009) Role of translocation as a mechanism of resistance to glyphosate. *Weed Sci.* 57: 118-123.

[63] Feng PC, Tran M, Chiu T. Sammons RD, Heck GR, Jacob CA. (2004) Investigation into GR horseweed (*Conyza canadensis*): retention, uptake, translocation and metabolism. *Weed Sci.* 52: 498-505.

[64] Koger CH, Reddy KN. (2005) Role of absorption and translocation in the mechanism of glyphosate resistance in horseweed (*Conyza Canadensis*). *Weed Sci.*53: 438-445.

[65] Dinelli G, Marotti I, Catizone P, Bonetti A, Urbano JM, Barnes J. (2008) Physiological and molecular basis of glyphosate resistance in *C. bonariensis* (L.) Cronq. biotypes from Spain. *Weed Res.* 48: 257–265.

[66] Wakelin AM, Lorraine-Colwill DF, Preston C. (2004) Glyphosate resistance in four different populations of *Lolium rigidum* as associated with reduced translocation of glyphosate to meristematic zones. *Weed Res.* 44: 453-459.

[67] Perez A, Alister C, Kogan M. (2004) Absorption, translocation and allocation of glyphosate in resistant and susceptible Chilean biotypes of *Lolium multiflorum. Weed Biol. Manag.* 4: 56–58.

[68] Perez-Jones A, Park KW, Colquhoun J, Mallory-Smith C, Shaner D.L. (2005) Identification of glyphosate-resistant Italian ryegrass (*Lolium multiflorum*) in Oregon. *Weed Sci* 53: 775–779.

[69] Kreuz K, Tommasini R, Martinoia E. (1996) Old enzymes for a new job: herbicide detoxification in plants. *Plant Physiol* 111: 349-353.

[70] Durst F, O'keef DP. (1995) Plant Cytochrome P450: an overview. *Drug Metab. Drug Interac.* 12: 171-186.

[71] Thompson LJr, Houghton JM, Slife FW, But HS. Metabolism of atrazine 767 by fall panicum and large crop grass. *Weed Sci.* 19 (1971) 409-412.

[72] Thompson LJr. (1972) Metabolism of chloro-s-triazine herbicide by *Panicum* and *Setaria. Weed Sci* 20: 584-587.

[73] De Prado R, Lopez-Martinez N, Gonzalez-Gutierrez J. (2000) Identification of two mechanisms of atrazine resistance in *Setaria faberi* and *Setaria viridis* biotypes. *Pest. Biochem. Physiol.* 67: 114-124.

[74] Cummins I, Wortley DJ, Sabbadin F, He Z, Coxon CR, Straker HE, Sellars JD, Knight K, Edwards L, Hughes D, Kaundun SS, Hutchings S-J, Steel PG, Edwards R. (2013) Key role for a glutathione transferase in multiple-herbicide resistance in grass weeds. *PNAS* 110: 5812-5817.

[75] Mannervik B, Danielson UH. (1988) Glutathione transferases: structure and catalytic activity. *Crit. Rev. Biochem.* 23: 283-337.

[76] Andersen MP, Gronwald JW. (1991) Atrazine resistance in a velvetleaf (*Abulition theophrasti*) biotype due to enhanced glutathione S-transferase activity. *Plant. Physiol.* 96: 104-109.

[77] Edwars R, Owen WJ. (1986) Comparison of glutathione S-transferases of *Zea mays* responsible for herbicide detoxification in plants and suspension-cultured cells. *Planta* 169: 208-215.

[78] Ezra G, Stephenson GR. (1985) Comparative metabolism of atrazine and EPTC in proso millet (*Penicum miliaceum* L) a corn. *Pest. Biochem. Physiol.* 24: 207-212.
[79] Gronwald JW, Andersen RN, Yee C. (1989) Atrazine resistance in velvetleaf (*Abutilon theophrasti*) due to enhanced atrazine detoxification. *Pest Biochem Physiol* 34: 149-163.
[80] Jachetta JJ, Radosevich SR. (1981) Enhanced degradation of atrazine by corn (*Zea mays*). *Weed Sci* 29: 37-44.
[81] Shimabukuro RH. (1985) Detoxification of herbicides, in: Duke SO (Ed) *Weed Physiology, Vol II: Herbicide Physiology*, CRC, Press, Florida, pp. 215-240.
[82] Hatton PJ, Cole DJ, Edwards R. (1996) Influence of plant age of glutathione levels and glutathione transferases involved in herbicide detoxification in corn and giant foxtail. *Pest Biochem Physiol* 54: 199-209.
[83] Hatton PJ, Cummins I, Cole DJ, Edwars R. (1999) Glutathione transferases involved in herbicide detoxification in the leaves of *Setaria faberi* (giant foxtail). *Physiol. Plant.* 105: 9-16.
[84] Dean JV, Gronwald JW, Eberlein CV. (1990) Induction of glutathione-S-transferase isozymes in sorghum by herbicide antidotes. *Plant. Physiol.* 92: 467-473.
[85] Mozer TJ, Tiemeier DC, Jaworski EG. (1983) Purification and characterization of corn glutathione-S-transferase. *Biochemistry* 22: 1068-1072.
[86] Lay MM, Casida JE. (1976) Dichloroacetamide antidotes enhance thiocarbamate sulfoxide detoxification by elevated corn root glutathione content and glutathione-S-transferase activity. *Pest. Biochem. Physiol.* 6: 442-456.
[87] Devine MD, Shukla A. (2000) Altered target sites as a mechanism of herbicide resistance. *Crop. Protection.* 19: 881-889.
[88] Eberlin CV, Guittieri MJ, Berger PH, Fleming JK, Mallory-Smit CA, Thill DC, Baerg RJ, Belkmap CAST. (1991) Physiological consequences of mutation for ALS-inhibitor resistance. *Weed Sci.* 47: 383-392.
[89] Rosario JM, Cruz-Hipolito H, Smeda RJ, De Prado R. (2011) White mustard (*Sinapis alba*) resistance to ALS-inhibiting herbicides and alternative herbicides for control in Spain. *Europ. J. Agronomy* 35: 57-62.
[90] Bryan JK. Synthesis of the aspartate family and branched-chain aminoacids, in B.F. Miflin (Ed), The Biochemistry of Plants. *A Compressive Treatise Aminoacids and Derivatives* Vol. 5. Academic Press, New York, (1991). pp, 403-452.
[91] McCourt JA, Duggleby RG. (2006) Acetohydroxyacid synthase and its role in the biosynthetic pathways for branched-chain aminoacids. *Amino Acids* 31: 173-210.
[92] Duggleby RG, McCourt JA, Guddat L. (2008) Structure and mechanism of inhibition of plant acetoohydroxyacid synthase. *Plant. Physiol. Biochem.* 46: 309-324.
[93] Tranel PJ, Wright TR, Heap IM. ALS Mutations from herbicide-resistant weed [Online]. (2008). Available: www.weedscience.com [26 December 2008]. *Proc. Northeast Weed Sci. Soc.* 40 (1986) 50-52
[94] Boutsalis P, Karotam J, Powles SB. (1999) Molecular basis of resistance to acetolactate synthase-inhibiting herbicides in *Sisymbrium orientale* and *Brassica tournefortii. Pestic. Sci.* 55: 507-516.
[95] Tranel PJ, Wright TR. Resistance of weeds to ALS-inhibiting herbicides: what have we learned? *Weed Sci.* 50: 700-712.

[96] Han H, Yu Q, Purba E, Li M, Walsh M, Friesen S, Powles SB. (2012) A novel amino acid substitution Ala-122-Tyr in ALS confers high-level and broad resistance across ALS-inhibiting herbicides, *Pest. Manag. Sci.* 68: 1164-1170.

[97] Ashigh J, Corbett CL, Smith PJ, Laplante J, Tardif FJ. Characterization and diagnostic test of resistance to acetohydroxyacid synthase inhibitors due to an $Asp_{376}Glu$ substitution in *Amaranthus powelli. Pestic. Biochem. Physiol.* 95: 38-46.

[98] Warwick SI, Xu R, Sauder C, Beckie HJ. (2008) Acetolactate synthase target-site mutations and a single nucleotide polymorphism genotyping in ALS-resistant kochia (*Kochia scoparia*). *Weed Sci.* 56 (2008) 797-806.

[99] Whaley CM, Wilson HP, Westwood JH. (2007) A new mutation in plant ALS confers resistance to five classes to ALS inhibitors. *Weed Sci.* 55: 83-90.

[100] Massa D, Krenz B, Gerhards R. (2011) Target-site resistance to ALS inhibiting herbicides in *Apica spica-venti* populations is conferred by documented and previously unknown mutations. *Weed Res.* 51: 294-303.

[101] Laplante J, Rajcan I, Tardif J. (2009) Multiple allelic forms of acetohydroxyacid synthase are responsible for herbicide resistance in *Setaria viridis. Theor. Appl. Genet.* 119 (2009) 577-585

[102] Harwood JL. (1988) Fatty acid metabolism. *Annu. Rev.Plant Physiol.* 39: 101-138.

[103] Konishi T, Shinohara K, Yamada K, Sasaki Y. (1996) Acetyl-CoA carboxylase in higher plants: most plants other than Grammineae have both the prokaryotic and the eukaryotic forms of this enzyme, *Plant. Cell. Physiol.* 37: 117-122.

[104] Focke M, Gieringer E, Schwan S, Jänsch L, Binder Sand Braun H-P. (2003) Fatty acid biosynthesis in mitochondria of grasses: malonyl-coenzyme A is generated by a mitochondrial-localized acetyl-coenzyme A carboxylase. *Plant. Physiol.* 133: 875-884.

[105] Scarabel L. Panozzo S, Varotto, Sattin. (2011) Allellic variation of the ACCase gene and response to ACCase. Inhibiting herbicides in pinoxaden-resistant *Lolium spp.Pest. Manag. Sci.* 8: 932-941.

[106] Jang SR, Marjanovic J, Gornicki P. (2013) Resistance to herbicides caused by single amino acid mutations in acetyl-CoA carboxylase in resistant populations of grassy weeds. *New Phytologist* doi: 10.1111/nph.12117

[107] Délye C, Wang T, Darmeney H. (2002) An isoleucine-leucine substitution in chloroplastic acetyl-CoA carboxylase from green foxtail (*Setaria viridis* L. Beauv.) is responsible for resistance to the cyclohexanedione herbicide sethoxydim. *Planta* 214: 421-427.

[108] Délye C, Zhang XQ, Chalopin C, Michel S, Matéjicek A, Powles SB. Molecular bases for sensitivity to Acetyl-coenzyme A carboxylase inhibitors in black-grass, *Plant. Physiol.* 137: 794-806.

[109] Délye C. (2005) Weed resistance to acetyl coenzyme A carboxylase inhibitors: an update. *Weed Sci.* 53: 728-746.

[110] Dayan FE, Duke SO. Herbicides: carotenoid biosynthesis inhibitors, Encyclopedia of agrochemicals, Vol. 2, Ed. by Plimmer JR, Gammon DW, Ragsdale NN and Roberts T. John Wiley & Sons, New York, NY, pp 744-749, 2003

[111] Doong RL, MacDonald GE, Shilling DG. (1993) Effect of fluridone on chlorophyll, carotenoids and anthocyanin content of hydrilla. *J. Aquat. Plant.Manag.* 31: 55-59.

[112] Michel A, Scheffler BE, Arias RS, Duke SO, Netherland M and Dayan FE. (2004) Somatic mutation-mediated evolution of herbicide resistance in the non-indigenous invasive plant hydrilla (Hydrilla verticillata). *Mol. Ecol.* 13: 3229-3237.

[113] Matringe M, Sailland A, Pelissier B, Rolland A, Zink O. (2005) *p*-Hydroxyphenylpyruvate dioxygenase inhibitor-resistant plants. *Pest. Manag. Sci.* 61: 269-276.

[114] Steckel LE and Sprague CL. (2004) Commonwaterhemp interference in corn. *Weed Sci.*52:359-364.

[115] Hager AG, Wax LM, Stoller EW and Bollero GA. (2002) Common waterhemp (*Amaranthus rudis*) interference in soybean. *Weed Sci.*50:607–610.

[116] Feltner KC, Hyrst HR, Anderson LE. (1968)Tall waterhemp competition in grain sorghum. *Weed Sci* 16:214–216.

[117] Steckel LE. (2007) The dioecious *Amaranthus* spp.: here to stay. *Weed Technol.*21: 567–570.

[118] Horak MJ, Peterson DE, Biotypes of Palmer amaranth (*Amaranthus palmeri*) and common waterhemp (*Amaranthus rudis*) are resistant to imazethapyr and thifensulfuron. *WeedTechnol.*9:192–195, 1995

[119] Anderson DD, Roeth FW and Martin AR. (1996) Occurrence and control of triazine-resistant common waterhemp (*Amaranthus rudis*) in field corn (*Zeamays*). *Weed Technol.*10:570–575.

[120] Shoup DE, Al-Khatib K and Peterson DE. (2003) Common waterhemp (*Amaranthus rudis*) resistance to protoporphyrinogen oxidaseinhibiting herbicides. *Weed Sci.*51:145–150.

[121] Legleiter TR and Bradley KW, (2008) Glyphosate and multiple herbicide resistance in waterhemp (*Amaranthus rudis*) populations from Missouri. *Weed Sci.*56:582–587

[122] Nicholas E Hausman, Sukhvinder Singh, Patrick J Tranel, Dean E Riechers, Shiv S Kaundun, Nicholas D Polge, David A Thomas and Aaron G Hagera. (2011) Resistance to HPPD-inhibiting herbicides in a population of waterhemp (*Amaranthus tuberculatus*) from Illinois, United States. *Pest. Manag. Sci.*; 67: 258–261.

[123] Lermontova I, Kruse E, Mock HP, Grimm B, (1997) Cloning and characterization of a plastidal and a mitochondrial isoform of tobacco protoporphyrinogen IX oxidase, *Proc. Nat. Acad. Sci.* U. S. A. 94:8895–8900.

[124] Ferreira GC, Andrew TL, Karr SW, Dailey HA, (1988) Organization of the terminal two enzymes of the heme biosynthetic pathway, *J. Biol. Chem.* 264:835–3839]

[125] Dayan FE, Duke SO. Phytotoxicity of protoporphyrinogen oxidase inhibitors: phenomenology, mode of action and mechanisms of resistance, in: Roe RM, Burton JD, Kuhr RJ (Eds.). Herbicide Activity: Toxicology, Biochemistry and Molecular Biology, IOS Press, Washington, DC, 1997, pp. 11–35.

[126] Koch M, Breithaupt C, Kiefersauer R, Freigang J, Huber R, Messerschmidt A. (2004) Crystal structure of protoporphyrinogen IX oxidase: a key enzyme in haem and chlorophyll biosynthesis. *EMBO J.* 23: 1720–1728.

[127] Ghisla S, Massey V. Mechanisms of flavoprotein-catalyzed reactions. *Eur. J.Biochem.* 191: 1–17.

[128] Heinemann IU, Diekmann N, Masoumi A, Koch M, Messerschmidt A, Jahn M, Jahn D.(2007) Functional definition of the tobacco protoporphyrinogen IX oxidase substrate-binding site. *Biochem J.* 402: 575–580.

[129] Corradi HR, Corrigall AV, Boix E, Mohan CG, Sturrock ED, Meissner PN, Acharya KR. (2006) Crystal structure of protoporphyrinogen oxidase from Myxococcus xanthus and its complex with the inhibitor acifluorfen. *J. Biol. Chem.* 281: 38625–38633.

[130] Dayan FE, Romagni JG, Duke SO. Protoporphyrinogen oxidase inhibitors, in: R. Krieger, J. Doull, D. Ecobichon, D. Gammon, E. Hodgson, L. Reiter, J. Ross (Eds.), Handbook of Pesticide Toxicology, Academic Press, San Diego, CA, 2001, pp. 1529–1542; U.B. Nandihalli, S.O. Duke, The porphyrin pathway as a herbicide target site, *Am. Chem. Soc. Symp. Ser.* 524 (1993) 62–78.

[131] Grossmann K, Niggeweg R, Christiansen N, Looser R, Ehrhardt T. (2010) The herbicide saflufenacil (KixorTM) is a new inhibitor of protoporphyrinogen IX oxidase activity. *Weed Sci.* 58: 1–9.

[132] Matringe M, Scalla R. (1988) Effects of acifluorfen-methyl on cucumber cotyledons: porphyrin accumulation. *Pestic. Biochem. Physiol.*32: 164-172.

[133] Becerril JM, Duke SO. (1989) Acifluorfen effects on intermediates of chlorophyll synthesis in green cucumber cotyledon tissues, *Pestic. Biochem. Physiol.* 35: 119-126.

[134] Becerril JM, Duke SO. (1989) Protoporphyrin IX content correlates with activity of photobleaching herbicides. *Plant. Physiol.* 90: 1175–1181.

[135] Sherman TD, Becerril JM, Matsumoto H, Duke MV, Jacobs JM, Jacobs NJ, Duke SO. (1991) Physiological basis for differential sensitivities of plant species to protoporphyrinogen oxidase-inhibiting herbicides. *Plant. Physiol.* 97: 280–287.

[136] Lee HJ, Duke MV, Duke SO. (1993) Cellular localization of protoporphyrino genoxidizing activities of etiolated barley (Hordeum vulgare L.) leaves. *Plant Physiol* 102: 881–889.

[137] Lee HJ, Duke SO. (1994) Protoporphyrinogen IX-oxidizing activities involved in the mode of action of peroxidizing herbicides. *J. Agric. Food Chem.* 42: 2610-2618.

[138] William L, Hager PAG, McCormick JS, Tranel PJ. (2006) Codon deletion confers resistance to herbicides inhibiting protoporphyrinogen oxidase. *PNAS* 103: 12329-12334.

[139] Thinglum KA, Riggins CW, Davis AS, Bradley KW, Al-Khatib K, Tranel PJ. (2011) Wide Distribution of the Waterhemp (Amaranthus tuberculatus) DG210 PPX2 Mutation, which Confers Resistance to PPO-Inhibiting Herbicides. *WeedScience* 59: 22–27.

[140] [140] Lee LJ, Ngim J. (2000) A first report of glyphosate-resistant goosegrass [*Eleusine indica* (L) Gaertn.] in Malaysia. *Pest. Manag. Sci.*56: 336–339.

[141] Baerson SR, Rodriguez DJ, Tran M, Feng Y, Best NA, Dill GM. (2002) Glyphosate-resistant goosegrass: identification of a mutation in the target enzyme 5-enolpyruvylshikimate-3-phosphate synthase. *Plant. Physiol.*129: 1265–1275.

[142] Baerson SR, Rodriguez DJ, Biest NA, Tran M, You J, Kreuger RW, Dill GM, Pratley JE, Gruys KJ. (2002) Investigating the mechanism of glyphosate resistance in rigid ryegrass (*Lolium rigidum*). *Weed Sci.*50: 721-730.

[143] Ng CH, Wickenswary R, Salmijah S, Teng YT and Ismail BS. (2003) Gene polymorphisms in glyphosate-resistant and -susceptible biotypes of *Eleusine indica* from Malaysia. *Weed Res.*43: 108-115.

[144] Ng CH, Wickenswary R, Surif S, Ismail BS. (2004) Inheritance of glyphosate resistance in goosegrass (*Eleusine indica*). *Weed Sci* 52: 564–570.

[145] Ng CH, Wickenswary R, Salmijah S, Teng YT and Ismail BS. (2005) Glyphosate resistance in *Eleusine indica* from different origins and polymerase chain reaction amplification of specific alleles. *Aust. J. Agric. Res.*55: 407-414.

[146] Wakelin AM, Preston C. (2005) Target-site glyphosate resistance in rigid ryegrass (*Lolium rigidum* Gaudin). *Weed Sci. Soc. Am.Abstr.* 417. Lawrence, KS: Weed Sci Soc Am.

[147] Devine MD, Duke SO, Fedtke C. Herbicidal inhibition of photosynthetic electron transport, in M.O. Devine, S.O. Duke and C. Fedtke (eds) Physiology of Herbicide Action, Prentice Hall, Englewood, Cliffs, (1993) pp. 113-140.

[148] Hirscherberg J, McIntosh L. (1983) Molecular basis of herbicide resistance in *Amaranthus hibridus*. *Science* 224: 1346-1348.

[149] Gressel J. Herbicide tolerance and resistance, alteration of site of activity, in S.O. Duke (ed) *Weed Physiology* Vol II, CRC Press, Florida, (1985) pp. 159-189.

[150] Pfister K, Arntzn CJ. (1979) The mode of action of photosystem II-specific inhibitors in herbicide resistant weed biotypes. *Z Naturforsch* 34c: 996-1009.

[151] Szigeti Z, Lehoczki E. (2003) A review of physiological and biochemical aspects of resistance to atrazine and paraquat in Hungarian weeds. *Pest Manag Sci* 59: 451-458.

[152] Pölös E, Laskay G, Szigeti Z, Pataki S, Lehoczki E. (1987) Photosynthetic properties and cross-resistance to some urea herbicides of triazine-resistant *Conyza Canadensis* Cronq. (L.). *Z Naturforsch* 42c: 783-793.

[153] Siegenthaler PA, Mayor JP. (1992) Changes in the binding and inhibitory properties of urea/triazine-type herbicides upon phospholipid and galactoplipid depletion in the outer monolayer of thylakoid membranes. *Photosynth Res.* 31: 57-68.

[154] St John JB. (1976) Manipulation of galactolipid fatty acid composition with substituted pyridazinones, *Plant.Physiol.* 57: 257-259.

[155] Schwenger-Erger C, Barz W, Weber N. (2001) Fatty acid alteration of plastidic and extra-plastidic membrane lipids in metribuzin-resistant photoautotrophic *Chemopodium rubrum* cells as compared to wild type cells, *Z Naturforsch* 56: 1047-1056.

[156] Roncel M, Yruela I, Kirilovsky D, Guerrero F, Alfonso M, Picorel R, Ortega JM. (2007) Changes in photosynthetic electron transfer and state transitions in an herbicide-resistant D1 mutant from soybean cell cultures, *Biochem. Biophys.Acta* 1767: 694-702.

[157] M. Alfonso, I. Yruela, S. Almárcegui, E. Torrado, M.A. Pérez and R. Picorel, Unusual tolerance to high temperatures in a new herbicide-resistant D1 mutant from *Glicine max* (L) Merr. cell cultures deficient in fatty acid desaturation. *Planta* 212 (2001) 573-582

[158] M. Alfonso, R. Collados, I. Yruela and R. Picorel, Photoinhibition and recovery in a herbicide-resistant mutant from Glycine max (L.) Merr. cell cultures deficient in fatty acid unsaturation, *Planta* 219 (2004) 428–439.

[159] Yruela, M. Alfonso, I. García-Rubio, J.I. Martinez, R. Picorel, P.J. Alonso, Spin label electron paramagnetic resonance study in thylakoid membranes from a new herbicide-resistant D1 mutant from soybean cell cultures deficient in fatty acid desaturation. *Biochim. Biophys. Acta* 1515 (2001) 55–63.

[160] Nafziger Ed, Widholm JM, Steinrucken HC, Killmer JL. (1984) Selection and characterization of a carrot cell line tolerant to glyphosate. *Plant Physiol* 76: 571-574.

[161] Pline-Srnic W. (2006) Physiological mechanisms of glyphosate resistance. *Weed Technol.*20: 290–300.

[162] Bradshaw LD, Padgette SR, Kimball SL, Wells BH. (1997) Perspectives on glyphosate resistance. *Weed Technol.*11:189-198.

[163] Gadamski G, Ciarka D, Gressel J, Gawronski ISW. (2000) Negative cross-resistance in triazine-resistant biotypes of *Echinochloa crus-galli* and *Conyza Canadensis. Weed Science* 48: 176–180.

[164] Gressel J. (1990) Synergizing herbicides. *Rev Weed Sci.* 5: 49–82.

[165] Streibig, J. C. 1987. Joint action of root-absorbed mixtures of auxin herbicides in *Sinapis alba*L. and barley (*Hordeum vulgare* L.). *Weed Res.* 27:337–347.

[166] Owen MKE, Gressel J. (2000) Non-traditional concepts of synergy for evaluating integrated weed management. *In* Hall JC, Hoagland R, Zablotowicz R. (Eds) Pesticide Biotransformations in Plants and Microorganisms: Similarities and Divergences. Clarendon Hills, IL: American Chemical Society Publications. Vol. 777. Pp. 376-393.

[167] Wrubel RP, Gressel J. (1994) Are herbicide mixtures useful for delaying the rapid evolution of resistance? A case study. *Weed Technol.* 8:635–648.

[168] Beckie HJ, Warwick SI, Sauder CA, Lozinski C, Shirriff S. (2011) Occurrence and molecular characterization of acetolactate synthase (ALS) inhibitor-resistant kochia (Kochia scoparia) in western Canada. *Weed Technol.* 25: 170–175.

[169] C. Délye and C. Dijon, ACCase mutation table. [Online]. (2011). Available: http://weedscience.org/In.asp [15 August 2011].

In: Handbook on Herbicides
Editors: Daiki Kobayashi and Eito Watanabe

ISBN: 978-1-62948-053-4

Chapter 3

APPLICATION OF ^{31}P-NMR SPECTROSCOPY TO GLYPHOSATE STUDIES IN PLANTS: INSIGHTS INTO CELLULAR UPTAKE AND VACUOLE SEQUESTRATION CORRELATED TO HERBICIDE RESISTANCE

Xia Ge[1], D. Andre 'd'Avignon[1], Joseph J. H. Ackerman[1,2,3], Elizabeth Ostrander[4] and R. Douglas Sammons[4]

[1]Departments of Chemistry
[2]Radiology, and [3] Internal Medicine,
Washington University, Saint Louis, MO, US
[4]Monsanto Company, Saint Louis, MO, US

ABSTRACT

Glyphosate is the world's most widely used herbicide due to favorable attributes of low cost, low toxicity, and high efficacy in controlling a wide range of weed species. The genetic engineering of important crops with the glyphosate-tolerant trait has led to extensive and largely exclusive glyphosate usage in many areas. The resulting selective evolutionary pressure has resulted in the emergence of glyphosate-resistant weed biotypes. Glyphosate contains a phosphonate ($-CPO_3^{2-}$) chemical group, with variable protonation depending upon pH, that provides for convenient detection and monitoring *via* ^{31}P NMR in plant tissue extracts as well as *in vivo*. Insights provided through ^{31}P-NMR studies of weed species exposed to glyphosate have improved our understanding of herbicide uptake and compartmentalization, resulting in key insights regarding resistance mechanisms. For example, we have previously reported the discovery and characterization of glyphosate vacuole sequestration as the principal resistance mechanism in horseweed (*Conyza canadensis*) and in ryegrass (*Lolium* spp.) from several continents. Herein, we expand our prior published ^{31}P-NMR studies to include additional weed biotypes, characterizing glyphosate uptake, vacuole sequestration and chloroplast exclusion.

INTRODUCTION

Modern agriculture plays a critical role in providing food, fiber, and fuel to the 7 billion people on the Earth. [1] This represents a significant challenge given emergence of food-needy developing countries as well as changes and variability in global weather patterns. Since the beginning of agriculture, the need to control weeds, plant species that compete for crop space, sunlight, water, and nutrients has been a major yield determinant. In the last 50 years, the use of herbicides coupled with new agricultural practices has largely replaced human labor as the primary weed control method and consequently modern farming has seen markedly increased crop productivity. [2, 3]

The herbicide N-(phosphonomethyl)glycine (common name: glyphosate) stands alone as the most popular herbicide used in agriculture. [4, 5] There are a number of reasons for this including low toxicity, bacterial degradation in soil, low-volatility, ease of production, and inherent ability to control a very wide range of weedy species. This molecule targets 5-enolpyruvate shikimate-3-phosphate synthase (EPSPS), a critical enzyme in the shikimate pathway (Scheme 1), located in the chloroplast of the cell. Glyphosate acts as a transition-state inhibitor of EPSPS, interrupting this essential pathway, hence leading to a shortage of aromatic amino acids and other required substrates and eventually to plant death. [6-8] Glyphosate is a foliage-applied herbicide. To reach the target EPSPS enzyme it - first must gain entry to the plant cells of source leaves. From this source tissue glyphosate is then translocated plant-wide by the phloem, following the sucrose gradient pathway to sink tissues (young leaves, tap root, meristem and the root system). [9, 10] Any obstacle preventing glyphosate from reaching and overwhelming the target EPSPS enzyme (poor translocation, poor cellular or chloroplast entry, mutations in, or overexpression of, the EPSPS enzyme, etc.) represents an impediment to the action of this herbicide. [11-15]

Genetically modified (GM) biotech crops carrying the trait for glyphosate resistance dominate the GM crops available worldwide. [16] This revolution in agricultural practice coupled with the urgent need to address hunger in many developing countries has spawned significant societal questions related to food production, including security, cost, environmental considerations and consequences of climate change. [1]

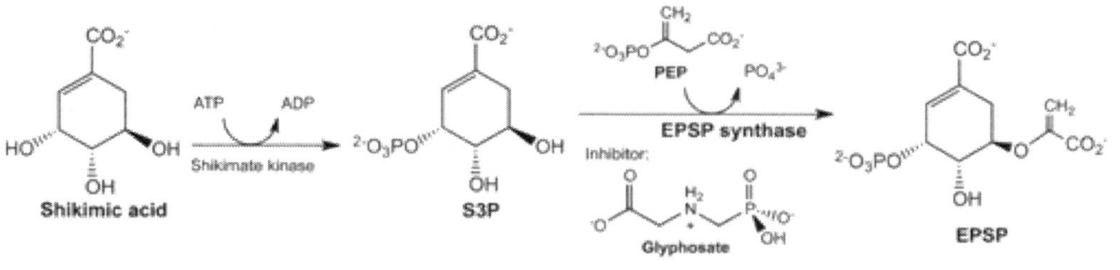

Scheme 1. Key reactions in the chloroplast's shikimate biosynthetic pathway. The enzyme 5-enolpyruvate shikimate-3-phosphate synthase (EPSPS) facilitates coupling of shikimate-3-phosphate (S3P) with phosphoenolpyruvate (PEP) to form EPSP. Glyphosate binds to EPSPS in competition with PEP forming a dead-end ternary complex leading to accumulation of S3P and shikimate (in significant quantities).

For the first 25 years of its usage, glyphosate was unique among all herbicides in that no resistant weed species were reported. However, with the persistent near exclusive use of glyphosate over time, resistant species were selected, with the first detected in 1997 in rigid ryegrass in Australia by scientists at Charles Stuart University. [17, 18] Today, 24 glyphosate resistant (GR) weeds have developed under the intensive selective pressure from the primary usage of this single herbicide. With widespread adaption of GM crops, most bearing the glyphosate tolerant trait, weed resistance has emerged as a significant agricultural problem and drawn increased attention. [19-22]

Identification of GR weeds and associated studies began in 1998. [13, 23, 24]. Since these early observations, a number of glyphosate resistance modes-of-action (MOA) have been identified, including: (i) multiple copies of genes producing EPSPS in pigweed (*Amaranthus palmeri*) and Italian ryegrass (*Lolium multiflorum*), [15, 25] (ii) mutations in genes producing site specific alterations in EPSPS in goosegrass (*Eleusine indica),* [13] ryegrass (*Lolium* spp.), [14, 26-28] and now horseweed and common ragweed (unpublished data), and (iii) vacuolar sequestration of glyphosate in horseweed and ryegrass. [29-31] GR mechanisms (i) and (ii) are categorized as target-site (EPSPS) resistant mechanisms while (iii) is exclusion. Metabolism has not been directly correlated to a glyphosate resistance although several plants (legumes) have been found to metabolize glyphosate suggesting it may yet be a supporting mechanism [32, 33]. Currently 14 glyphosate resistant weeds occur in the United States. GR horseweed, as a case in point, has now infested 21 states in the United States of America and is found on five continents and two related species are also found resistant, *C sumatrensis and C. bonariensis.* [34]

^{31}P NMR is a non-destructive radio-frequency spectroscopy and has been widely adopted to study plant tissue and bacteria since the first report by Roberts on the use of inorganic phosphate (Pi) ^{31}P chemical shifts to measure pH environments inside living cells. [35, 36] Glyphosate has an ionizable phosphono group (-CPO^{3-}) making the application of ^{31}P NMR to investigate glyphosate in plants an attractive and powerful method. The glyphosate ^{31}P chemical shift, like that of Pi, is pH sensitive – an NMR pH meter – in a range that includes subcellular environments. [29, 37, 38] ^{31}P NMR can also monitor cell health by quantifying the presence of ATP, UDPG and preservation of a shift difference (pH difference) in Pi resonances from the cell vacuole and cytosol. Although their pKa's are different, the ^{31}P chemical shifts of Pi and glyphosate are consistent with the finding of a vacuole pH considerably more acidic than that of the cytosol, pH ~5 *vs.* ~7). The three ^{31}P resonances of nucleoside triphosphates (dominated by ATP) are a marker for energy stores and phosphorylation potential as necessary for maintenance of normal cell metabolism. ATP also is a substrate in the normal functioning of the shikimate pathway; so is phosphoenolpyruvate (PEP), which is occasionally visible by ^{31}P NMR. Finally, ^{31}P NMR easily detects increased levels of shikimate-3-phosphate (S3P), which accumulates when EPSPS is inhibited by glyphosate. Overall, ^{31}P NMR is a powerful tool to monitor plant metabolic flux, [39] such as amino-assimilation, [40-43] and glyphosate metabolism in microbes. [44-46] (Because glyphosate metabolism in higher plants is rare and minimal and has not been shown to contribute to resistance in weeds it is only briefly covered in this chapter. [47])

The *in vivo* ^{31}P resonance linewidth depends on tissue homogeneity, cellular viscosity, substrate molecular weight, the presence or lack of paramagnetic metals, and, for ionizable phosphorus groups, the presence or lack of pH gradients within a given compartment or cell type. In perfused plant tissue, natural ^{31}P resonance linewidths vary from ~20-80 Hz (at 12 tesla magnetic field strength) with chemical shifts dispersed over a range of roughly 40 ppm. Plant ^{31}P resonances are detectable if the substrate they originate from is present in moderate concentrations (>0.5 mM). Several important phosphorous containing compounds related to glyphosate exposure as well as plant metabolism in general can be monitored by *in vivo* ^{31}P NMR, including glyphosate, phosphocholine, nucleotide triphosphates (commonly ATP), nucleotide diphosphoglucose (commonly the uridine analog UDPG), S3P [37] and 2-C-methyl-D-erythritol-2,4-cyclopyrophosphate (MEcPP). [48-50]

A representative use of ^{31}P NMR for elucidation of the glyphosate MOA *in vivo* involves comparing GR and glyphosate sensitive (GS) biotypes and monitoring for the accumulation of S3P. In GS biotypes, inhibition of EPSPS by glyphosate should lead to an accumulation of S3P and detection of its ^{31}P-NMR signal (Scheme 1). [7, 51] In GR biotypes (with EPSPS overproduction or point mutations, or any useful resistant mechanism) continued EPSPS activity should lead to less, possibly negligible levels of S3P as quantified by ^{31}P NMR.

Historically, glyphosate metabolism in higher plants was thought to be rare, [47] however Reddy *et al.* recently reported glyphosate conversion to AMPA in herbicide tolerant crops and weeds using HPLC methods. [33] AMPA has also been detected by *in vivo* ^{31}P NMR in glyphosate incubated sycamore cells, arising presumably from metabolism. [37] We have reported the *in vivo* ^{31}P-NMR observation of MEcPP, a cyclic phosphono-containing substrate found in response to glyphosate treatment (toxicity) in higher plants. [48] The discovery of this compound cannot be singularly ascribed to the use of glyphosate as we have not examined other herbicides, many of which also have targets in the chloroplast. For example, bacteria exposed to paraquat also produce MEcPP. [53, 54]

Our initial efforts using *in vivo* ^{31}P NMR to investigate glyphosate resistance were in studies of GR and GS *Conyza canadensis*, [29] a common weed native to North America also known as horseweed or mare's tail. [55] In the first of a series of publications, we reported results with GS and GR horseweed exposed to two glyphosate-treatment methods. [29] The first involved spray-application to whole plants, mimicking the field use of glyphosate, followed by ^{31}P-NMR analysis of perfused (*in vivo*) leaf tissue harvested at different times. The advantage of this method is that the dynamics of glyphosate movement from source leaves to sink tissues by phloem translocation occurs under normal greenhouse growth conditions (i.e., photosynthesis and photorespiration). Further, the subcellular compartmental distributions of glyphosate in sink and source tissues can be quantified by ^{31}P NMR. In another strategy, referred to as the pulse-chase protocol, glyphosate is infused into detached leaf tissue under perfusion conditions (the pulse phase) in a controlled fashion (glyphosate concentration, duration of exposure, pH, nutrients, and temperature). ^{31}P NMR is used to monitor the tissue response to glyphosate over time (typically during the glyphosate-free chase phase), including relative quantification of variables such as the amount of glyphosate that has entered the cell, the amount (proportion) in the cell vacuole, cell health (ATP levels and maintenance of pH gradients), and generation of other phosphorus containing metabolites. Initial work highlighted the utility of the ^{31}P-NMR platform for investigation of a number of dynamic responses to glyphosate exposure in GR and GS biotypes. [29-31]

For glyphosate to be effective in weed control, it first must enter the cells of source leaves and establish a local concentration gradient, then load the phloem and be distributed to sink tissues *via* the assimilate translocation pathway. [9, 56] Complicating this series of events, it has been shown that translocation, intracellular compartmentalization, and toxicity in horseweed are also dose-dependent and growth stage-dependent. [57-61] In understanding weed resistance, all of these factors (cell entry, chloroplast entry, restrictions to translocation, mutations in and over-expression of EPSPS, growth-stage, and availability of ATP) have to be taken into account.

In this chapter, we review our ^{31}P-NMR research into rapid vacuole sequestration of glyphosate as the resistance mechanism for the invasive horseweed. We include previously unreported aspects of this work, including studies of plant tissue other than leaves and evidence for transmembrane substrate pumps. A second objective in this contribution is to present unpublished work on a second resistance mechanism, the restriction of glyphosate entry into the plant cell (an exclusion mechanism). Finally, we survey a number of weedy species that have historically demonstrated difficult-to-control behavior under glyphosate treatment. This survey highlights the role of vacuole sequestration as well as restricted entry in contributing to glyphosate resistance.

A hypothesis consistent with data collected to date is that transporters (energy dependent pumps) play a key role in the selective movement of glyphosate across the tonoplast (vacuole sequestration) or the cell membrane (restricted entry). The GR biotypes have evolved with different expression of these transporters compared to their GS biotypes. Transporters conferring herbicide resistance in higher plants have not yet been widely explored. However, Conte *et al.* highlighted plant ATP binding cassette (ABC) transporters that are responsible for multidrug resistance (MDR) [62]. In that work they identified MDR pumps for glutathione that were associated with the tonoplast, the chloroplast, and the plasma membrane, all with variations and unique genes. The data present here are suggestive of the presence of similar pumps that control the flow of glyphosate into and within the cell, and are intimately linked to the observed response of GR species.

METHODS

Detailed methods are included in published work [29-31] but will be briefly summarized here, with special focus on unpublished studies.

Perfusion Buffer: The standard media employed to infiltrate and perfuse plant tissue is a buffer consisted of 2-[N-morpholino]ethanesulfonic acid (MES) hydrate (12 mM) and 50 mM sucrose in 10% D_2O. The pH is corrected (pH of 5 common for most trials) with 2 mM potassium hydroxide. All compounds are obtained from commercial suppliers unless otherwise noted. Glyphosate is obtained from Monsanto Co.

NMR Considerations: All studies are carried out on a Varian/Agilent DirectDrive™ 11.74-tesla NMR spectrometer equipped with a 10-mm broad-band probe (Nalorac Corp.; now part of Agilent Technologies). A laboratory constructed perfusion apparatus designed for use with 10-mm glass NMR tubes and attached to a peristaltic Mini-Pump (VWR International) recirculated oxygenated perfusion media through plant tissue arranged within the 10-mm NMR tube at a flow rate of 4-6 ml per minute. The sample is placed within the NMR radio-frequency probe and is maintained at 20 °C with an FTS Thermal Products system (SP Industries). Buffer pH is monitored (Mettler-Toledo) throughout the duration of each NMR experiment. The buffer media (perfusate) is aerated with oxygen *via* a fritted stone immersed in the solution, which is stirred.

Methylene diphosphonic acid (MDP), nominally 20 mM in D_2O, is contained in a capillary tube placed with the plant tissue within the NMR tube. MDP serves as an external

^{31}P resonance amplitude (concentration) and frequency (chemical shift) reference. Chemical shifts are formally assigned relative to the chemical shift of 85% orthophosphoric acid, which is taken as 0.00 ppm. ^{31}P-NMR data are collected under the following conditions: 68° flip angle, 1.1 s interpulse recycle time, 3,200 complex time-domain data points, 16,000 Hz spectral width, and two-hour data averaging time blocks unless specified otherwise. ^{31}P resonances are quantified (amplitudes and frequencies) by Bayesian-probability-theory-based modeling of the time-domain data using signal analysis algorithms [63] written by G. Larry Bretthorst and available for free download at uniform resource locator http://bayesiananalysis.wustl.edu/index.html. [31] These tools are especially useful for discrimination of overlapping resonances in crowded spectral regions.

^{31}P-NMR analysis as carried out in the studies reported herein is appropriate for small to modest molecular weight, highly mobile, "solution-state" molecules. These species result in narrow frequency-domain resonance linewidths and are readily detected with standard "pulse-and-detect" NMR signal averaging techniques. Less mobile, high molecular weight species, such as glyphosate complexed to EPSPS or bound within a membrane, are not generally detected with solution-state NMR methods. A wide range of "solid-state" NMR protocols can be used to detect immobile, high molecular weight species but such approaches generally preclude *in vivo* analysis.

Plant Growth Conditions and Treatment: Plants are cultivated from seeds using Readi-Earth® (Osmocote 14-14-14) in 10.2 cm pots under normal greenhouse conditions (30/20 °C day/night temperature, 14/10 h photoperiod, 700-900 μE m^{-2} s^{-1}). Plants are watered daily. Resistant horseweed plants used for this study were originally collected in Delaware and described by VanGessel. [24] The seed stock is maintained by seed collection from 5–10 plants annually from the original plants. These lines have responded faithfully in repetitive glyphosate titrations since 2002. The LD_{50} values of GR (6.4 kg a.e. ha^{-1} or 7.6 times the standard use rate of glyphosate) and GS (0.09 kg a.e. ha^{-1}) lines were determined with at least 20 plants in each glyphosate treatment group and with ten untreated plants. Seeds for Johnsongrass biotypes were generously collected from Creighton County, Arkansas, USA by Bob Scott and Jason Norsworthy of the University of Arkansas. Seeds for the difficult-to-control weed species were obtained from Azlin Seed Service, Leland, Mississippi, USA. Treatment with glyphosate is carried out only after seedlings are well established and less than 15 cm in height. For horseweed studies, plants are investigated at two growth stages. Upper stem tissue is harvested after the "bolt" phase has initiated (18 cm plant height). All other tissue types (source leaf tissue, sink leaves, apical meristem, taproot, root tips) are harvested at the rosette growth phase. Spray application of glyphosate (WeatherMAX®, Monsanto) is carried out either by hand spraying to wet (nominal 3.36 kg a.e. ha^{-1} or 4X field use rate) or by track sprayer (4X field use rate at 10 gallons/acre) and returned to the greenhouse.

Tissue Preparation for Perfusion and NMR Monitoring: For studies described herein, all plant tissues including source and sink leaves, upper shoot stem, apical meristem, tap root and root tips are harvested 24 hours following spraying for separate ^{31}P-NMR studies. Plant tissue directly exposed to glyphosate is washed with deionized (DI) water repeatedly to remove any non-specific (adventitious) surface-adhering glyphosate. Following surface washing, tissues are swirled with perfusate buffer under vacuum to remove inner air pockets. This process markedly reduces magnetic susceptibility inhomogeneities, thus minimizing NMR linewidths and improving spectral resolution, and is especially critical for leaf tissue. [64] Leaf and stem tissues are arranged vertically in the NMR tube and this orientation is maintained throughout NMR data collection. Apical meristem and taproot are cut into smaller pieces to fit into the NMR tube and are secured vertically by glass wool plugs to avoid tissue displacement during perfusion. Root tips are arranged loosely as a bundle and wrapped with thread at the top and bottom of the bundle length to avoid collapse in the NMR tube and subsequent blockage of circulating buffer. Once packed with plant tissues, the 10-mm NMR tube is interfaced to the perfusion apparatus for circulation of the oxygenated buffer solution. Care is taken to not pack tissue too tightly as this inhibits perfusate flow throughout the sample and, in some cases, leads to increased NMR linewidths. Intact, whole horseweed leaves can be used directly. Broader leaves from other weed species are cut into strips prior to loading into the NMR tube. For monocot species, the top and bottom of the leaf is discarded and tissue cuttings representing an age mixture of the remainder of the leaf are placed in the NMR tube to avoid investigation of only the oldest and youngest tissue sections.

Glyphosate Pulse-Chase Assay of GR and GS Johnsongrass: Mature leaves from GR and GS Johnsongrass plants are harvested and the midvein of the leaves removed. Individual strips of the leaf are swirled under vacuum with 10-mM glyphosate-containing perfusate buffer and oriented vertically within the NMR tube fitted with the perfusion apparatus. Glyphosate is delivered by infusion from the perfusion buffer for 10 hours (pulse phase) and then treated with glyphosate-free perfusate (chase portion) for 14 hours. There is a washing period (within the NMR tube during data collection) during the first two hours of the chase period whereby adventitious glyphosate is washed from the inner tissue surface and the perfusion stream and discarded. Otherwise, the perfusate stream is continually aerated with oxygen and recirculated. ^{31}P-NMR data are collected throughout the entire experiment in two-hour data-averaging time blocks.

RESULTS

^{31}P-NMR Studies of GR and GS Horseweed

Horseweed glyphosate uptake is pH independent: A common theory to explain both cellular delivery and retention of ionic herbicides is acid-trapping. This hypothesis is based on the idea that charged molecules do not passively diffuse across membranes efficiently whereas neutral molecules do. Thus, there can be a pH-dependent cellular delivery on either side of the herbicide pKa favoring membrane transport of the neutral or less-charged molecule. Once inside the cell, assuming the herbicide pKa is below the near neutral cellular pH, the herbicide would deprotonate and become "trapped" in a charged state less likely to passively diffuse back out. A good example is the greater than 10-fold improved uptake of the herbicide Bentazon at pH 4.6 *vs.* 6.6. [65]

For glyphosate, if cellular entry is passive and acid-trapping is important, an improvement in delivery to the plant cell is anticipated when the delivery-media (herbicide application solution) is at a pH below the second pKa of the phosphonate group (e.g., pH ~ 6). At this low pH, glyphosate protonation will be favored, resulting in a less charged molecule that is more amenable to passive membrane translocation (diffusion across the plant cell membrane). Once glyphosate is inside the cell, the higher pH of the cytosol (~ 6.8) will favor deprotonation of the phosphonate group, resulting in a more highly charged molecule that is less amenable to passive membrane translocation. However, glyphosate uptake is virtually identical at pH 5 and 8 as shown in Figure 1 indicating that glyphosate delivery to the cell does not follow a passive transport mechanism that would favor a less-charged molecule. An alternate hypothesis consistent with this finding is that a glyphosate transporter facilitates cellular entry.

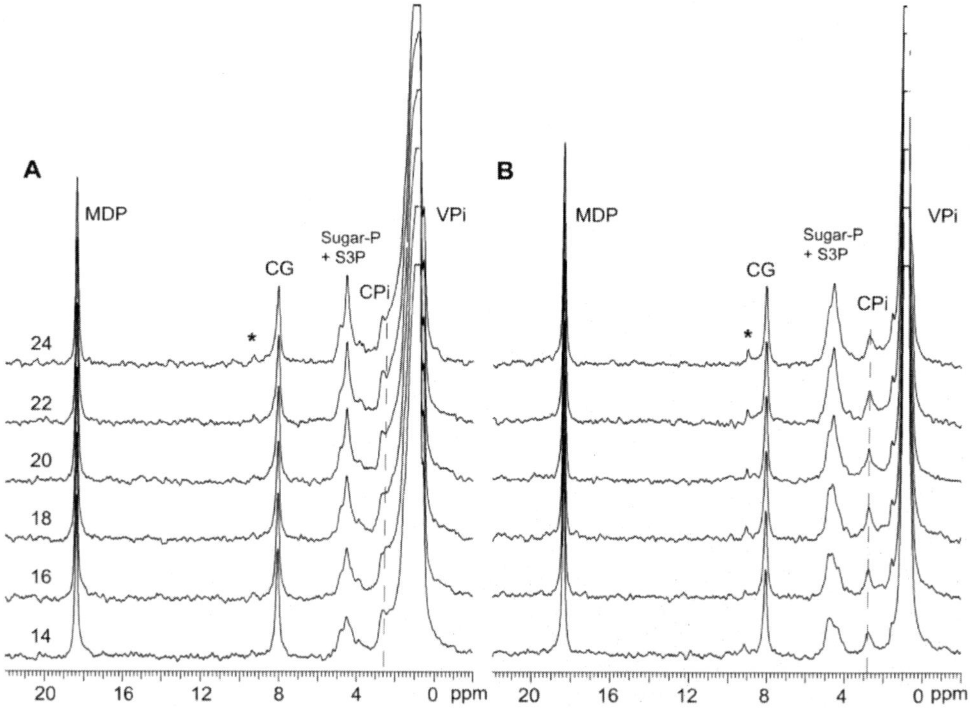

Figure 1. *In vivo* ^{31}P-NMR chase-phase spectra of GS horseweed following glyphosate pulse-phase delivery at pH 5.0 (A) and pH 8.0 (B). The chase phase data are collected in two-hour signal-averaging time blocks during hours 14-24 for mature leaf tissue following a 10-hour, 10-mM glyphosate-containing oxygenated buffer infusion at 20 °C. Abbreviations regarding resonance assignments are: MDP, methylene diphosphonate, an external chemical-shift and resonance-amplitude reference; CG, cytosolic glyphosate; Sugar-P, sugar phosphates (note, other substrates occur in this region and not all sugar phosphates appear at this shift); S3P, shikimate-3-phosphate; CPi, cytoplasmic inorganic phosphate; VPi, vacuolar inorganic phosphate. The star (*) indicates residual glyphosate remaining in the perfusate buffer. MDP is assigned a ^{31}P chemical shift of 18.2 ppm relative to the ^{31}P chemical shift of 85% phosphoric acid, which is taken as 0.00 ppm.

Source to Sink Tissue Translocation: Horseweed source leaves are fully expanded and photosynthetically active when receiving glyphosate during spray treatment. Sink tissues are protected so they do not receive the herbicide directly. These tissues receive glyphosate through phloem-enabled translocation, thus facilitating study of this process. Translocation takes place while photosynthesis is still functioning and can continue up to at least 72 hours post-treatment in horseweed. For healthy plants, at least 50% of total glyphosate translocation occurs within the first 24 hours post-treatment. [57] However, only a portion of the applied herbicide taken up by source leaf cells is translocated to sink tissues. ^{14}C-labeling studies on

GS horseweed have shown that only ~10-15% of the glyphosate taken up by the source leaves is observed in sink tissue over a 72-hour period post-treatment. [57] Translocation rates are plant specific and depend on many environmental factors and plant development stages. Glyphosate is successful, in part, because it kills the plant slowly allowing for translocation to all sink tissues over a period of days, thus enabling a continuous inhibition of EPSPS and subsequent disruption of the shikimate phosphate biosynthetic pathway throughout all plant tissues.

Source Leaves and Vacuole Sequestration: GR and GS horseweed biotypes have an abundance of source leaf tissue, and the biennial habit allows them to be studied over a range of temperatures (0 °C to 35 °C) making horseweed an ideal candidate to receive herbicide by spray treatment and for subsequent study by ^{31}P NMR. Glyphosate is reproducibly and readily observed in high concentrations in both GR and GS horseweed source leaves as reported previously. [29] Once taken up by source leaves, the fate of glyphosate differs considerably for the GS and GR biotypes. Glyphosate is rapidly sequestered (< 24 hours) into the vacuole for the GR biotype and, thus, is unavailable for translocation. Vacuole sequestration occurs more slowly and to a much lesser extent for the GS biotype. Figure 2 shows a representative example of this difference in vacuole sequestration with data collection beginning 8 hours following spraying for the horseweed biotypes. Because the ^{31}P-NMR chemical shift for glyphosate is pH dependent, it is straightforward to distinguish the amount of glyphosate residing in the more acidic vacuole compared to more basic environs such as the cytosol and/or chloroplast.

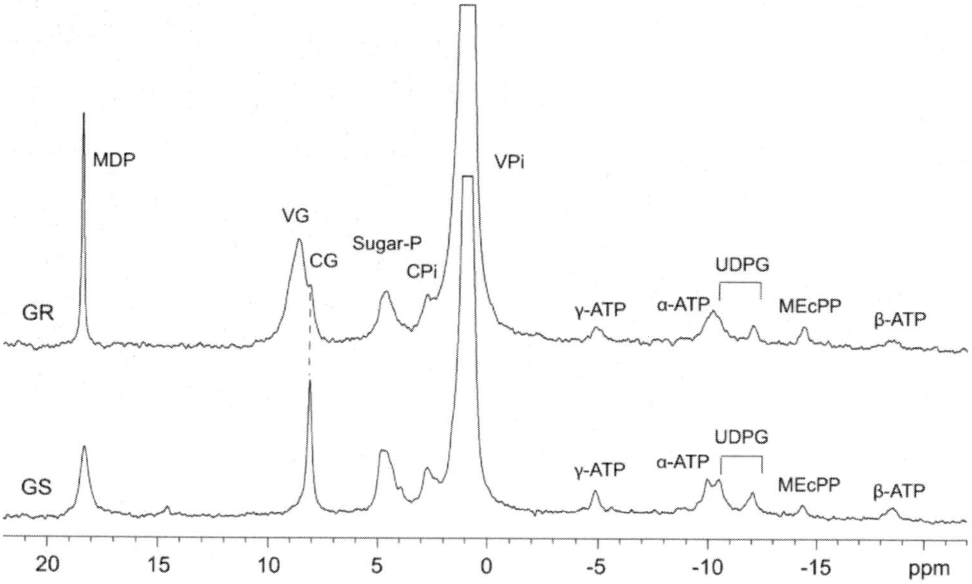

Figure 2. *In vivo* ^{31}P-NMR spectra of GR (top) and GS (bottom) horseweed biotypes obtained from mature source leaves eight hours following spray treatment. Spectra result from two hours of data averaging with oxygenated perfused leaf tissue at 20 °C. Abbreviations regarding resonance assignments are the same as in Figure 1 with the addition of α-, β-, γ-ATP (nucleoside triphosphates, of which adenosine triphosphate is the major component); UDPG, uridine 5-diphosphoglucose; MEcPP, 2-C-methyl-D-erythritol-2,4-cyclopyrophosphate.

The GS horseweed ^{31}P-NMR spectrum in Figure 2 shows a narrow glyphosate resonance near 8.0 ppm assigned to the cytosol, while the ^{31}P-NMR spectrum of the GR biotype reveals a broader resonance near 8.8 ppm assigned to the vacuole, with a slight residual cytosolic resonance (appearing as a lower frequency shoulder on the major vacuole phosphate resonance) present as well. The inherent ability of the GR biotype to move the herbicide to the vacuole compartment over a period of a few hours is termed rapid vacuole sequestration (RVS). This process is irreversible, at least on a multi-day timescale, and, thus, effectively removes a significant portion of the glyphosate delivered to the source leaves from the pool that would normally translocate to sink tissue (and to the chloroplast). Vacuole sequestration accounts for roughly 90% of the glyphosate in source leaves during the 24 hours following treatment when GR horseweed is maintained under greenhouse conditions. Under lower light or at room temperature, 75-80% sequestration is typical for GR horseweed in 24 hours.

It is also common for ^{31}P NMR to detect the presence of S3P in GS horseweed source leaves 24 hours after treatment but the amplitude of the S3P resonance is variable among different plant "subjects" indicating differential shikimate biosynthetic flux. Further barriers to glyphosate inhibition of the shikimate pathway and, thus, the build-up of S3P may be involved in other plant species. The example of chloroplast exclusion is discussed later in the section regarding difficult-to-control weeds. The S3P signal occurs within the sugar phosphate region (3.5-5 ppm) of the ^{31}P-NMR spectrum and is difficult to visualize or quantify unless it accumulates in large quantity (greater than the quiescent levels of sugar phosphates). The concentration of S3P in untreated plants is anticipated to be below normal ^{31}P-NMR detection limits.

RVS in GR horseweed explains earlier observations of translocation deficiencies observed by ^{14}C radiography in this biotype compared to the GS biotype. [57] We further confirmed that RVS is temperature sensitive and that vacuole sequestration can be markedly slowed at cooler temperatures. This reintroduces glyphosate sensitivity and allows GR horseweed to be controlled at normal field usage rates under cool spring conditions. [30]

Sink Leaves: ^{31}P-NMR analysis of rapidly growing sink leaf tissues (that were covered during spray treatment and can receive the herbicide only *via* translocation) reveals expected differences between GR and GS biotypes: the GR biotype shows vacuole sequestration, the GS biotypes shows greater S3P accumulation than GR. [29] There is considerable variability in the amounts of glyphosate and S3P measured by ^{31}P NMR depending on the age of the sink leaf tissue. For example, with the very youngest unexpanded leaves from the apical meristem 24 hours post-treatment (with the apical meristem covered during treatment), the glyphosate signal is not always detected. However, the S3P ^{31}P resonance is commonly observed with significant amplitude (concentration) in sink leaves of both GR and GS biotypes (much greater than observed in source tissue), providing evidence of significant EPSPS inhibition. This observation is consistent with (i) a high binding affinity of glyphosate for EPSPS as only a fraction is translocated, especially in the GR biotype, and (ii) a high shikimate pathway flux in rapidly growing leaves (compared to mature source leaves) leading to a higher glyphosate sensitivity of sink tissue. [66] The unexpanded leaf tissue from the apical meristem is located physically close in the rosette structure to multiple source leaves and it is not unreasonable to expect that both biotypes would receive sufficient glyphosate to interrupt the EPSPS enzyme in spite of vacuole sequestration, especially with the anticipated poorly developed vacuole organelle associated with very young tissue. This ^{31}P-NMR observation is consistent with separate observations that shikimate levels following treatment vary by the tissue's stage of development. [67] The young growing tissues are highly metabolically active and the shikimate pathway flux is presumably high to support growth.

Because of signal-to-noise sensitivity limitations inherent to the method, ^{31}P NMR is challenged to detect and quantify translocated glyphosate (or other phosphorus containing species) present at low concentrations. The ability of solution-state ^{31}P NMR to report reliably on the presence and concentration of any substrate or metabolite is impacted by the molecule's resonance linewidth, a broad linewidth leading to a decrease in frequency-domain signal-to-noise. The ^{31}P-NMR linewidth is generally expected to decrease as the phosphorus-containing molecule's local environment becomes more homogeneous (e.g., reduced intra-compartmental pH and ion gradients) and its molecular motion increases (becomes more solution-state like). The horseweed leaf tissue subcellular compartmental environment is relatively inhomogeneous, compared to grasses, and ^{31}P-NMR linewidths are generally broad (> 60 Hz full width at half height). It is possible that some glyphosate stores in rapidly growing sink tissue are bound, thus immobilized, as a dead-end complex with EPSPS and, therefore, are not detected by solution-state ^{31}P-NMR methods. [51] It follows that high S3P resonance amplitude and low glyphosate resonance amplitude commonly go hand-in-hand (Table 1), a finding consistent with the expected inhibition of EPSPS.

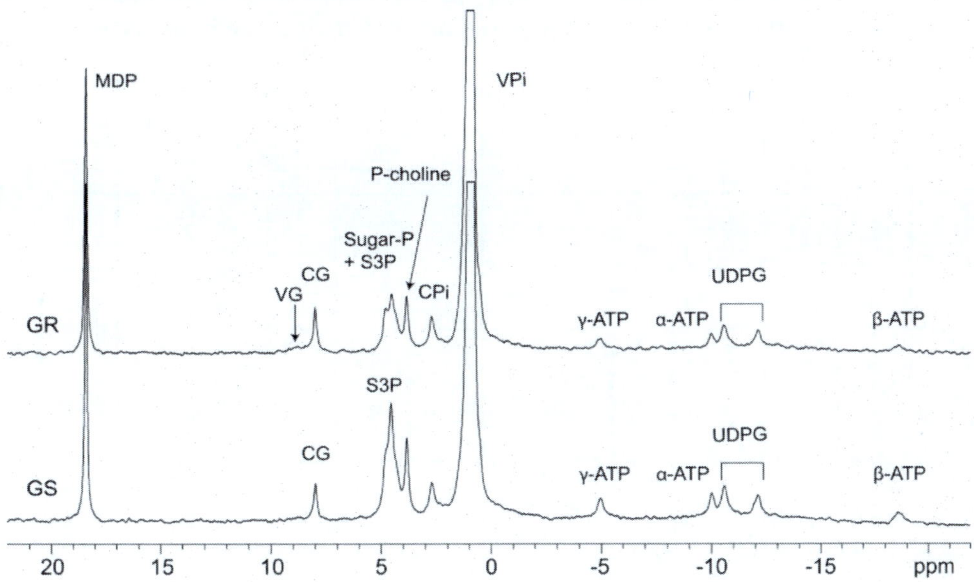

Figure 3. *In vivo* ^{31}P-NMR spectra of GR (top) and GS (bottom) horseweed upper shoot stem tissue (2-3 cm) collected from 18-cm tall plants. The upper shoot was protected during glyphosate spray treatment with aluminum foil and harvested 24 hours following treatment. Spectra result from nine hours of data averaging with oxygenated perfused tissue at 20 °C. Abbreviations are the same as in Figure 1 and 2 as well as phosphocholine (P-choline) that is clearly visible in this tissue.

Upper Shoot Stem: ^{31}P-NMR resonances of glyphosate (24 hours following spray treatment) and the resulting accumulation of S3P (more in GS than in GR) can easily be observed in horseweed upper stem tissue (harvested following spray treatment during bolting) from both biotypes (Figure 3). The GR upper shoot also clearly shows the RVS of glyphosate. This observation of vacuolar glyphosate in GR horseweed upper shoot indicates that vacuolar sequestration is not restricted to source leaf tissue but extends to other plant tissues as well. The ^{31}P-NMR linewidths observed in upper stem are notably narrower (~20 Hz) than in leaf tissue. This likely reflects a reduction in magnetic-susceptibility-induced magnetic field inhomogeneities that results from orienting the upper shoot stem tissue such that its compact cellular structure is aligned along the magnetic field axis. [68]

Table 1. ^{31}P-NMR-determined relative total glyphosate and S3P cellular content from source leaf and sink leaf tissues harvested from horseweed GS and GR biotypes 24 hours following treatment. The values for total, cytosolic plus vacuole, glyphosate (*N*-phosphomethylglycine: PMG) and S3P are normalized to an external MDP concentration reference. Upper shoot stem tissue is harvested during the early bolting phase (18 cm tall) of growth. The apical meristem tissue originated from plants in the rosette phase. Young sink leaves, upper shoot, and apical meristem tissues are protected during spray treatment. The percentage of measured vacuole sequestration relative to total detected glyphosate for GR horseweed source, sink leaf, and upper stem tissue is 85%, 62% and 7.7% respectively. Glyphosate in the vacuole cannot be accurately assessed in apical meristem primarily because the signal-to-noise is too low. Values are expressed as the ratio of signal amplitudes divided by signal amplitudes of MDP. One standard deviation of this estimated ratio is given in parenthesis

	Upper Stem		Apical meristem		Source mature leaf		Sink young leaf		Taproot		Root tips	
	PMG	S3P	PMG	S3P	PMG	S3P	PMG	S3P	PMG	S3P	PMG	S3P
GS horseweed	0.18 (0.01)	0.98 (0.02)	0.02 (0.01)	1.60 (0.42)	0.71 (0.03)	–	0.09 (0.01)	1.37 (0.04)	–	1.43 (0.03)	0.07 (0.02)	1.16 (0.05)
GR horseweed	0.40 (0.01)	0.39 (0.03)	0.04 (0.01)	1.46 (0.04)	1.14 (0.05)	–	0.26 (0.03)	0.81 (0.04)	–	0.45 (0.03)	–	0.50 (0.02)

Interestingly, unlike leaf tissue, we have been unsuccessful in delivering glyphosate through infusion of circulating perfusion buffer to upper stem tissue, probably due to the lack of apoplastic air space leaving cell-to-cell diffusion as the only possible path. The apoplast air space represents about 18% of the leaf volume in mature leaf tissue as measured by air evacuation with water. The only successful way of delivering the herbicide to meristem is *via* the natural phloem translocation process following spray treatment of the source leaf tissue of whole plants. This observation suggests a potential pitfall of historical methods of determining herbicide absorption into plant tissue by using any excised plant component immersed in a buffered herbicide solution. [69]

Apical Meristem/Taproot and Root Tips: Apical meristem and the associated taproot represent the major hub for translocation from the source leaves to the root tips. Yet it is also the tissue region that has extensive metabolic activity important for plant development. [70] We observe very high S3P resonance amplitude (Table 1) in these tissues after glyphosate challenge, always greater in GS than the GR biotype, reflecting the importance of, and high flux through, the shikimate pathway. [71]

In vivo ^{31}P NMR of root tips reveals substantial S3P accumulation demonstrating that glyphosate has inhibited EPSPS after phloem translocation 24 hours following spraying (Figure 4, Table 1). Like apical meristem and taproot tissues, glyphosate in root tips is not consistently observed. Substantial line-broadening caused by paramagnetic-metal-driven transverse relaxation may markedly reduce glyphosate detection sensitivity in root tips, where metal ion concentrations can be significant.

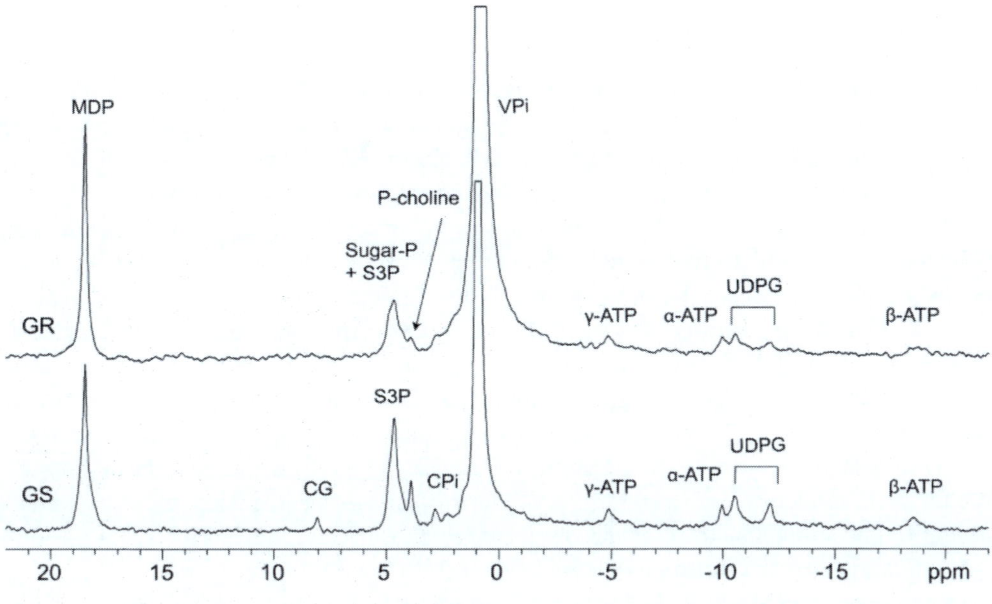

Figure 4. *In vivo* ^{31}P-NMR spectra of GR (top) and GS (bottom) horseweed root tips collected from plants 24 hours following treatment. Spectra result from nine hours of data averaging of oxygenated perfused tissue at 20°C.

Phosphocholine and Phosphoenolpyruvate: The ^{31}P-NMR resonance from phosphocholine is clearly observed in all sink tissues (note the presence in Figure 3 for meristem) and often at levels notable higher than in source leaf tissue. We have not investigated this difference in any detail. PEP (0.58 ppm chemical shift) is a key intermediate for the shikimate biosynthetic route and for other metabolic pathways. [72, 73] It is often obscured by the broad and intense vacuole Pi signal.

Summary: In vivo ^{31}P-NMR data from all plant tissue types following glyphosate spray treatment with both GS and GR horseweed biotypes show several clear trends. Some of these are highlighted in Table 1. First, the pH difference between cell vacuole and cytosol is roughly 1.4 units and this is the same for both biotypes. Second, significant differences in S3P content are observed, dependent upon tissue type. However, under identical treatment conditions, the amount of S3P observed in GS is always greater than in GR horseweed. Tissues showing the greatest S3P levels are also those where observed glyphosate levels are lowest, substantiating the higher sensitivity of these tissues to glyphosate inhibition. Plants succumb to death through glyphosate treatment, ultimately by shutting down the shikimate pathway. Our ^{31}P-NMR data agree completely with prior ^{14}C-glyphosate scintillation studies on horseweed biotypes by Feng *et al.* [57] RVS by source leaves in the GR biotype results in a limited pool of herbicide available for translocation. For glyphosate to be effective it must be delivered to the sink tissues. The greater accumulation of S3P in the GS horseweed biotype compared to GR, as quantified by ^{31}P NMR, is consistent with more effective inhibition of EPSPS on a per-dose basis. Vacuole sequestration appears to be a constitutive characteristic of GR horseweed as evidenced by the observation of a vacuole glyphosate component in source and sink leaves as well as in upper meristem tissue.

^{31}P-NMR Investigation of GR and GS Johnsongrass

Lack of Target Site Mechanism: Johnsongrass is considered one of the 10 worst, most-invasive weedy species in the world. [34] The subjects of our Johnsongrass studies are the Arkansas resistant and sensitive biotypes. The GR biotype shows moderate resistance, being controlled completely at 3.36 kg a.e. ha^{-1} or 4X normal field use rate of glyphosate. EPSPS extracted from the GS and GR biotype show similar activity suggesting enzyme point mutations that could affect glyphosate binding are not responsible for the observed resistance (unpublished data). Further, the amount of EPSPS detected in the sensitive and resistant biotypes is identical, supporting the absence of enzyme overproduction as an explanation for the observed resistance. Therefore, the observed glyphosate resistance is not explained by a target site mechanism in Johnsongrass. Studies on the South American version of GR Johnsongrass did show restricted translocation suggesting that again RVS might be involved.

Pulse Chase Protocol: ^{31}P-NMR studies were carried out on intact Johnsongrass leaf tissue to seek evidence for a non-target site resistance mechanism, such as vacuole sequestration, as an explanation for the observed glyphosate resistance. The method used in this investigation was the pulse-chase protocol whereby healthy source tissue is exposed to glyphosate-containing oxygenated perfusate during a10-hour pulse phase, followed by a glyphosate-free 2-hour wash phase, and then a 12-hour glyphosate-free chase phase. The advantage of the pulse-chase protocol is that glyphosate delivery can be carefully controlled. Further, ^{31}P-NMR observation throughout the protocol time course will reveal any dynamic processes or differences that may exist between the GR and GS biotypes. For vacuole sequestration, one anticipates a sequence where cell loading precedes vacuole loading. Hence, systematic loading of the cell vacuole should occur during the chase period at the expense of glyphosate that has loaded the cell cytosol during the initial pulse period. Studies on horseweed and ryegrass, where the vacuole is important for sequestration, clearly show this dynamic sequence of events using the pulse-chase protocol. [29-31]

^{31}P-NMR Observations: Results from studies on Johnsongrass (AR) are shown in Figure 5 where spectra obtained during the chase period are presented for both biotypes. These data reveal important differences between the GR and GS biotypes distinct from the horseweed studies. (i) Glyphosate net accumulation is notably more rapid (~2.5 times greater) for the GS compared to the GR biotype (Figure 6, Panel A). This ^{31}P-NMR observed difference indicates there is restricted net cellular entry for glyphosate in GR compared to the GS source leaf tissue. This net restricted entry could easily explain the observed difference in the herbicide control response of the two biotypes and the restricted translocation observed. [74] (ii) The notable accumulation of S3P observed for the GS compared to the GR biotype is consistent with what would be anticipated with the observed differences in glyphosate delivery to the

two biotypes, i.e., greater glyphosate delivered to the GS biotype leads to inhibition of EPSPS and thus greater buildup of S3P (Scheme 1). (iii) The vacuole pH for both GS and GR biotypes is notably more acidic (GR pH is ~4.5, GS pH is ~5.0) than the vacuole pH of ~5.5 observed for horseweed. (Recall that the compartmental pH is determined *via* the ^{31}P chemical shift of inorganic phosphate (Pi): 0.69-0.73 ppm in Johnsongrass biotypes compared to 0.85 ppm observed for horseweed. [29]) It is not known why the GR biotype has a more acidic vacuole but pH is known to vary across species and this is borne out through our studies (*vide infra*). (iv) While the vacuole chemical shift region, due to its lower pH, is somewhat obscured by residual glyphosate present in the perfusion buffer (maintained at pH 5) there is no evidence for vacuole sequestration in either Johnsongrass biotype, indicating RVS is not additionally contributing to the observed response to the herbicide. (v) The presence of a new, unassigned ^{31}P-NMR signal (labeled "?"), is more prominent in the GR biotype but is present in both GR and GS biotypes.

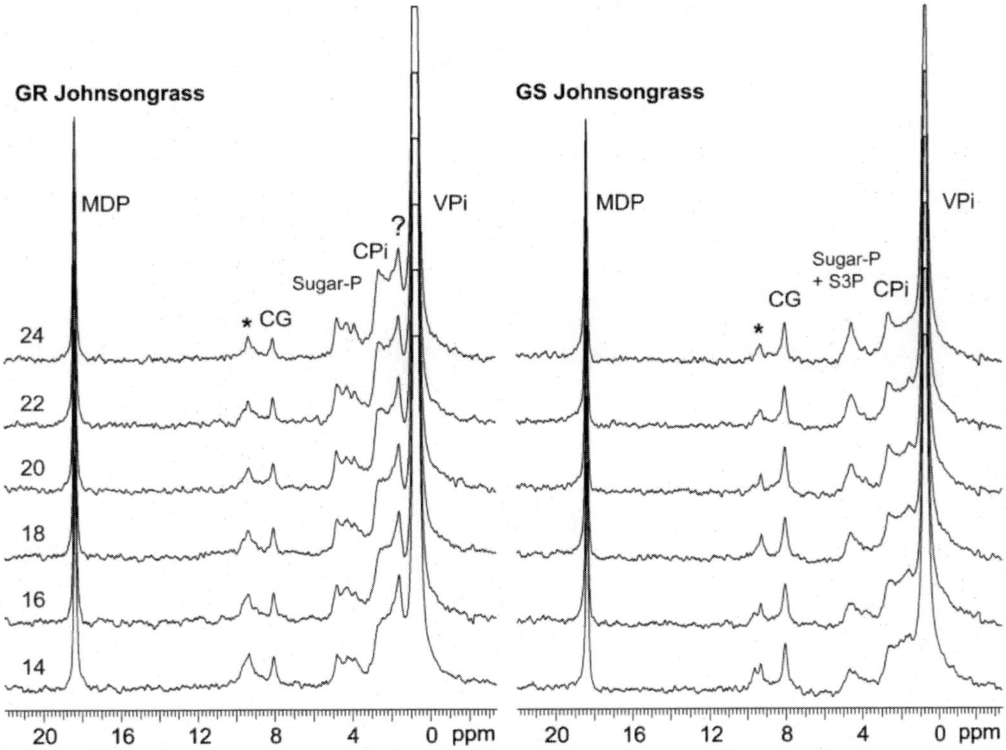

Figure 5. *In vivo* ^{31}P-NMR spectra of GR and GS Johnsongrass biotypes acquired during the chase phase following glyphosate delivery (during pulse phase). The chase phase data is collected in two-hour signal-averaging time blocks during hours 14-24 for mature leaf tissue following a 10-hour, 10-mM glyphosate-containing oxygenated buffer infusion. Abbreviations regarding resonance assignments are identical to earlier figures with the exception of (*), which represents residual buffer glyphosate, and (?), which indicates an unidentified phosphorous-containing compound.

^{14}C-Glyphosate Uptake: GR and GS biotypes of Johnsongrass were incubated with ^{14}C-radiolabeled glyphosate. Results from these trials (Figure 6, Panel B) show that_the rate of glyphosate uptake for the GS biotype is greater and more glyphosate accumulates over time

than for the GR biotype. These results are consistent with differences observed for Johnsongrass biotypes by ^{31}P NMR.

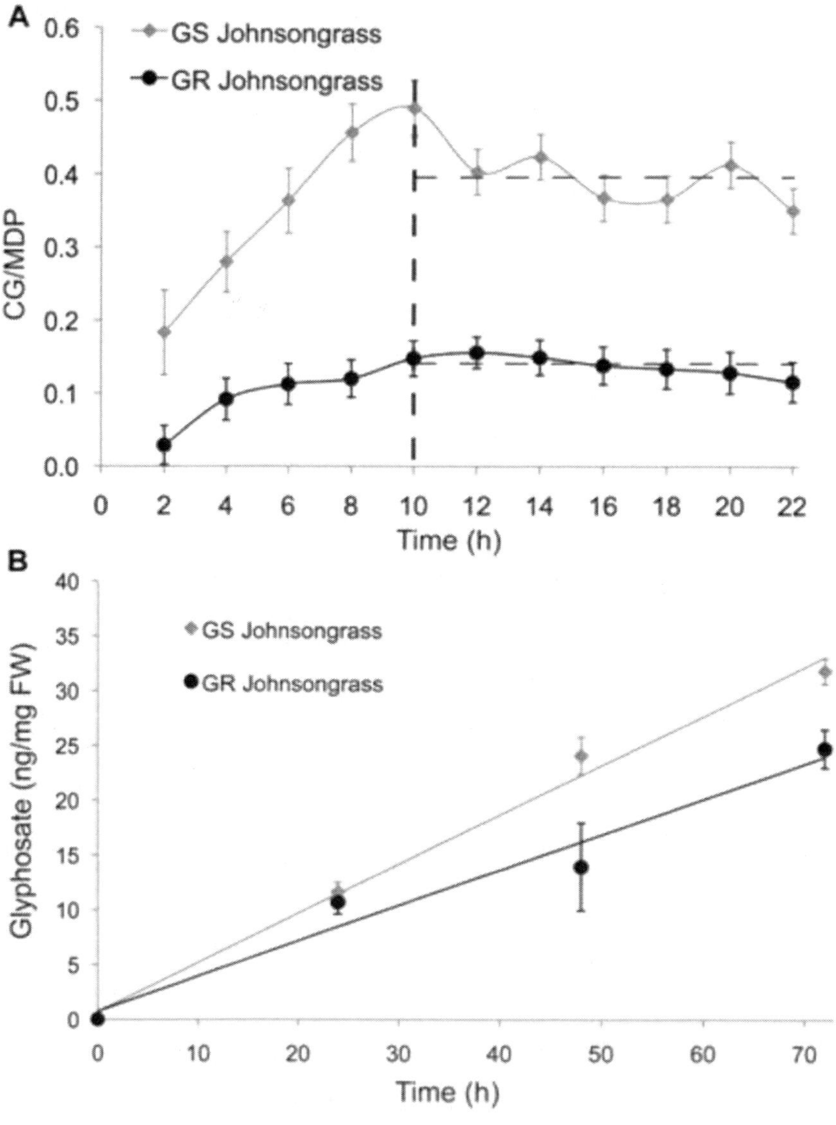

Figure 6. Glyphosate uptake in GR and GS Johnsongrass mature leaf tissue. Panel A) *In vivo* ^{31}P-NMR detected cytosolic glyphosate measured during the pulse phase of the pulse-chase experiment. Error bars are given as ± one standard deviation of uncertainty in the measurement. The GS biotype takes up glyphosate faster and accumulates 2.8 fold more glyphosate than the GR biotype. Panel B) nanograms of glyphosate as measured by ^{14}C label extracted from within intact leaf tissue as a function of incubation time. Data are given as mean ± one standard deviation. The rate of glyphosate uptake for the GS biotype is greater and more glyphosate accumulates over time than for the GR biotype.

Restricted Uptake Plays a Role, Vacuole Sequestration Does Not: The intervening two-hour wash phase is insufficient to remove all of the adventitious glyphosate from Johnsongrass, resulting in a small amplitude residual buffer glyphosate signal (Figure 5). This feature (at buffer pH 5) coupled with the more acidic pH of the vacuole in Johnsongrass (pH 4.5-5) makes it difficult to directly observe the presence of a small amplitude vacuole resonance were it to be present. However, it is clear that the cytosolic glyphosate signal remains relatively constant in amplitude (content) throughout the chase phase for both GR and GS biotypes (see Figure 6, Panel A).

The typical observation of vacuole sequestration involves both an increase in the ^{31}P resonance amplitude of glyphosate within the vacuole and concomitant decrease in the ^{31}P resonance amplitude of glyphosate within the cytosol, as glyphosate flows from the cytosol to the vacuole during the chase phase. [29] We do not observe this decrease in the GR (or GS) source leaf tissue suggesting that no vacuole sequestration takes place in Johnsongrass or is imperceptible on this time frame. The lack of a decrease in the cytosolic glyphosate signal shows that glyphosate is not pumped out or able to substantially diffuse out of the intracellular compartment(s) of Johnsongrass during this incubation period. Glyphosate uptake is unidirectional across the plasma membrane, as observed in horseweed and ryegrass. Further, there is no ^{31}P-NMR evidence that glyphosate is metabolized to an alternate chemical form (most commonly to AMPA), which would be observed directly should it be present. Hence, together with the characterization done by Riar *et al.*, [74] these observations suggest that the glyphosate resistant mechanism of the GR biotype is restricted entry to source leaf cells. This decreased uptake then explains the decreased phloem translocation of glyphosate observed in GR Johnsongrass from Creighton County, Arkansas, USA.

Corroborative evidence for restricted cellular uptake of glyphosate has been obtained with ^{14}C-radiolabeled-glyphosate incubations with Johnsongrass leaf disk samples. Reduced uptake of the radiolabeled substrate is observed (Figure 6, Panel B) in GR source leaf tissue in agreement with ^{31}P-NMR observations for the identical Arkansas Johnsongrass biotypes. The combined ^{14}C-radiolabeled incubation studies and ^{31}P-NMR investigations present data consistent with a restricted uptake mechanism contributing to, and perhaps dominating, Johnsongrass glyphosate resistance. [75, 76]

In Vivo ^{31}P-NMR Survey of Hard-to-Control Weeds

Survey Design: A number of weed species have historically been more difficult to control completely with glyphosate at normal field usage rates. These species have not evolved resistance like GR horseweed or GR Johnsongrass but have been designated hard-to-control from the onset of widespread glyphosate usage. With knowledge of the evolved resistance mechanisms for horseweed (vacuole sequestration) and Johnsongrass (restricted uptake), a series of the hard-to-control weed species was screened using the *in vivo* ^{31}P-NMR platform to see if these plants showed similar behavior that could explain their variable sensitivity to glyphosate. The list of weeds screened is included in Table 2. For comparison purposes, we also included data from GR horseweed (vacuole sequestration) as well as barley (reduced uptake yet controlled at well below the normal use rate of glyphosate).

Experimental Procedure: All plants were cultured under normal greenhouse conditions (see Methods), track sprayed at 4X the normal field usage rate of glyphosate, returned to the greenhouse, and harvested for ^{31}P-NMR studies 24 hours following treatment. All plants were handled identically (except GR horseweed, which was returned to the greenhouse for only 8 hours before tissue harvest). There may have been slight differences in leaf tissue mass or density monitored by ^{31}P NMR across the diversity of weeds included in the survey study. Harvested source leaf tissue was placed in glyphosate-free buffer, vacuum infiltrated, placed in the perfusion apparatus within the magnet, and the ^{31}P-NMR signal averaged over a period of three hours. The inorganic phosphate signals were quantified for resonance amplitude (content) and for chemical shift to establish compartmental pH (cytosol and vacuole). Glyphosate vacuole occupancy when present was noted and measured.

Table 2. Survey of *in vivo* ^{31}P-NMR screened weeds (source leaf) 24 hours following 3.4 kg a.e. ha^{-1} glyphosate spray treatment. Data for horseweed and for barley are also included. Values in parentheses are one standard deviation of uncertainty in the measurement

Common name	Scientific name	CPi/VPi	CG/MDP	VG/MDP	VG/(CG+VG)
GR horseweed	*Conyza canadensis*	5%	0.15 (0.02)	0.99 (0.03)	87%
GS horseweed	*Conyza canadensis*	5%	0.61 (0.03)	0.10 (0.02)	14%
Wild buckwheat	*Polygonum convolvulis*	7%	0.20 (0.02)	1.14 (0.08)	85%
Sicklepod	*Senna obtusifolia*	12%	0.69 (0.03)	1.20 (0.06)	63%
Prickly sida	*Sida spinosa*	5%	0.70 (0.04)	0.73 (0.07)	51%
Jungle rice	*Echinochloa colona*	21%	0.09 (0.01)	0.44 (0.01)	83%
Broadleaf signalgrass	Urochloa platyphylla	19%	0.35 (0.04)	0.42 (0.04)	54%
Shattercane	*Sorghum bicolor*	4%	0.69 (0.03)	0.18 (0.04)	21%
Morningglory	*Ipomoea lacunosa*	9%	0.31 (0.03)	–	
Cocklebur	*Xanthium strumarium*	8%	0.21 (0.02)	–	
Field bindweed	*Convolvulus arvensis*	6%	0.03 (0.01)	–	
Smallseed falseflax	*Camelina microcarpa*	16%	–	–	
Green foxtail	*Setaria viridis*	36%	0.16 (0.04)	–	
Barley	*Hordeum vulgare*	5%	0.14 (0.02)		
Wild oats	*Avena fatua*	5%	0.07 (0.03)	–	
Barnyardgrass	*Echinochloa crus-galli*	10%	0.04 (0.02)	–	
IL bundleflower	*Desmanthus illinoensis*	7%	3.25 (0.14)	–	
Lambsquarters	*Chenopodium album*	4%	0.52 (0.03)	–	

^{31}P-NMR Findings: The data in Table 2 show interesting trends and address the intended question of why these weedy species demonstrate variable control over a multi-year time period. The major observations are: (i) significant variability in glyphosate uptake with many species showing poor cellular accumulation after 24 hours, (ii) vacuole sequestration at moderate to significant levels in several of the weed species, and (iii) considerable variability in vacuole phosphate content within the species studied.

Monocots showing significantly restricted glyphosate uptake included barley, green foxtail, barnyard grass and wild oats. Dicotyledons showing reduced uptake are false flax, field bindweed and cocklebur. Some monocots have physical properties that are antagonistic to glyphosate spray application. For example, the orientation of the leaf tissue naturally grows vertically in some species, thus presenting a non-ideal leaf surface orientation to receive the herbicide during overhead spraying; [77] other species have source leaf tissue with a hydrophobic surface that repels the aerosol-like droplets of applied herbicide. [78, 79]

Illinois bundleflower [80] shows one of the highest glyphosate uptakes we have measured. Curiously though, 24 hours following treatment and in spite of the extreme uptake, this species does not accumulate any appreciable (^{31}P-NMR observable) S3P, the presence of which would indicate EPSPS interruption. This suggests glyphosate is somehow excluded from the chloroplast (the location of the shikimate biosynthetic pathway) and that glyphosate chloroplast exclusion serves as a resistance mechanism in the case of Illinois bundleflower. Figure 7 presents ^{31}P-NMR spectra that lead to the chloroplast exclusion hypothesis, contrasting Illinois bundleflower with barley, which is a well-controlled weed (crop) that shows low cytosolic glyphosate content but substantial S3P production.

Some weeds show a significant vacuolar glyphosate content including buckwheat, sicklepod, signalgrass and jungle rice. [81, 82] We speculate that the vacuole sequestration process removes a significant portion of the glyphosate from the translocatable pool for these weed species, contributing to their decreased sensitivity to glyphosate and thus the hard to control weed label.

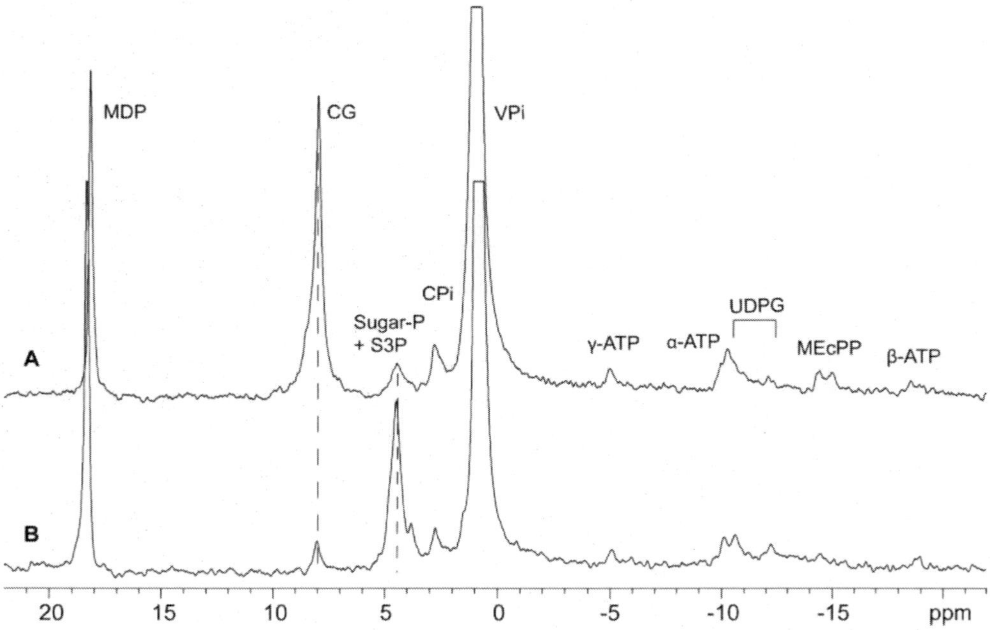

Figure 7. Representative *in vivo* ^{31}P-NMR spectra from a survey of hard-to-control weeds. A. Illinois bundleflower 24 hours following 3.4 kg a.e. ha^{-1} glyphosate spray treatment in the greenhouse. B. Barley 24 hours following 3.4 kg a.e. ha^{-1} glyphosate spray treatment in the greenhouse.

Discussion

^{31}P NMR has been applied to investigate many living systems over the past 35 years. NMR has the clear advantage over many destructive methods in that metabolic flux in various pathways can be investigated in serial, longitudinal, real-time fashion. Additionally, as NMR is quantitative, metabolite or substrate concentrations (content) and changes therein can be determined. The weakness of NMR is its limited signal-to-noise sensitivity compared to many other analytical techniques. Nevertheless, many phosphorous containing compounds are present at moderate concentrations *in vivo* and can be monitored and quantified *via* ^{31}P NMR.

High-energy phosphates and many sugar phosphates are commonly present at the 1-3 mM level and Pi present up to 100 mM. These substances can be readily observed. The pH dependence of the chemical shift is an additional advantage of *in vivo* ^{31}P NMR in that maintenance of cellular inter-compartmental pH gradients can be monitored and cell viability can be assessed continuously during experiments. Another advantage of *in vivo* ^{31}P NMR is that many phosphorous containing metabolites are inherently unstable absent their cellular environs. For example ATP and MEcPP are easily detected by *in vivo* ^{31}P NMR but have limited lifetimes in solution, especially in basic media. These metabolites can easily be missed, consumed, or partially degraded in extraction procedures or in sample manipulation where the tissue was not maintained intact and viable.

Applications of *in vivo* ^{31}P NMR to study plant tissue are well represented in the literature. However, reports employing NMR to monitor glyphosate in plant tissue and applications to understand glyphosate resistance mechanisms have not, until recently, been a focus. The work presented here, coupled with our prior published investigations, provides insights that, when coupled with information from supplementary methods, is highly informative toward understanding evolved or inherent glyphosate resistance in weedy species. The opportunity to monitor and quantify vacuole sequestration of glyphosate – compartmentally distinguishable through its pH-dependent ^{31}P chemical shift – demonstrates the value of NMR in elucidating the manner in which a plant can defend itself from the normal toxicity of a herbicide. Further, while not a subject of this contribution, we have shown that resistant plants can be made sensitive if this sequestration process is slowed or stopped [30] and this has been confirmed for ryegrass and Johnsongrass where even the plasma membrane restriction can be defeated. [83] In favorable cases, such knowledge can be applied to thwart the sequestration defense mechanism. Additionally, the pH-dependent chemical shift of glyphosate distinguishes signal originating from cytosolic environs (~pH 6.8) *vs.* perfusion media (~pH 5.0), making it straightforward to measure rate of uptake. That plant species show a range of herbicide uptake rates is not unexpected. However, identifying those species exhibiting restricted delivery with an outcome of poor herbicide control improves our understanding of resistance mechanisms. It follows that those species exhibiting both efficient vacuolar sequestration and restricted uptake of glyphosate will be especially tough to control.

In considering ^{31}P-NMR findings with living plant tissue that has received glyphosate by phloem translocation (i.e., sink tissue), trends emerge in comparing GR and GS horseweed biotypes that support prior work and also provide insight into metabolic flux in various cell types. First, significant S3P content is found in both GR and GS biotypes in all of the sink tissues investigated but the greatest amount is consistently found in GS plant tissue. This observation is consistent with vacuole sequestration competing with translocation of glyphosate in the GR biotype, thereby limiting translocation, and is also consistent with earlier ^{14}C-radiography data. That both GR and GS sink tissues show significant levels of S3P, thus, inhibition of EPSPS, implies considerable fidelity in the shikimate biosynthetic pathway. The GR biotype is, indeed, stunted in growth from glyphosate treatment but it survives, even at a 4X dosage. Presumably this is because only a fraction of the available

EPSPS is titrated by the herbicide and the shikimate pathway is able to function at a fraction of its normal flux. The GS biotype consistently shows greater S3P content in all sink tissues and presumably is unable to survive because any remaining shikimate pathway flux is too low to sustain life. It is known that GR horseweed can be controlled at 8X normal field use rate (6.9 kg a.e. ha^{-1}), where presumably even vacuole sequestration is insufficient to effectively shield the chloroplast.

A second observation from the horseweed tissue inventory study is evidence of vacuole sequestration being constitutive. In addition to source leaf tissue, upper stem (Figure 3) and unexpanded leaf tissue [29] show vacuole sequestration in GR horseweed. We also have examples of root tips where vacuole occupancy is present in the GR biotype (absent in GS) but this observation is not always reproducible. That glyphosate is progressively sequestered plant-wide agrees with the prior observation that the translocated herbicide is less toxic (on an equal content basis) in the GR *vs.* GS biotype. [57] A portion of the translocated herbicide is additionally sequestered within the cell vacuole in sink tissue and is unavailable for chloroplast delivery. The typically much smaller vacuole in young developing tissue could partially explain the increased sensitivity of these tissues to glyphosate.

A final observation from studies of tissue types in horseweed is that shikimate pathway flux, as measured by S3P buildup upon exposure to glyphosate, is more active in sink tissue. This is consistent with the metabolically more active, more rapidly growing sink tissue and its increased need for assimilates and the end product of the shikimate pathway, phenylalanine. It is estimated that up to roughly 30% of all of the plant carbon is sourced through the shikimate biosynthetic pathway. Clearly the success of glyphosate is, in part, due to the life-critical importance of its targeted pathway.

The observation that glyphosate is concentrated and retained in the vacuole suggests the presence of transporters in GR horseweed that effectively pump glyphosate across the tonoplast membrane and into the vacuole. Support for unidirectional transporters and the lack of passive diffusion comes from the observation that, as long as the plant tissue is healthy, once glyphosate is in the vacuole it remains there for at least a period of days. (As noted earlier, glyphosate uptake is virtually identical at pH 5 and 8, providing no support for the acid-trapping passive transport mechanism.)

Likewise, once glyphosate is delivered to the plant cell, in either horseweed biotype and across multiple species, as long as the plant tissue is viable it does not efflux out into the perfusate. Conversely, multiple perfusion trials have shown that when compartmental pH values (as determined by cytosolic and vacuole Pi chemical shifts) become compromised, commonly in concert with declining ATP levels, normal function is lost and glyphosate and other substrates can then efflux from the cell into the perfusate media. These observations are all consistent with unidirectional pumps being present on the cell plasma membrane as well the vacuole membrane and probably the chloroplast membrane.

Presumably these transporters are active to variable degrees depending upon species and biotype, thus explaining the limited glyphosate uptake in GR Johnsongrass compared to the GS biotype as well as the limited uptake observed in many of the difficult-to-control weed species. We argue that glyphosate delivery into the plant cell is an active process and the presence or absence of glyphosate-accepting transmembrane pumps controls the rate of uptake. Extending the argument for active transporters yet further to the chloroplast membrane provides an explanation of the unusual results observed for Illinois bundleflower and also for barley. We speculate that the difficult-to-control Illinois bundleflower represents an example where chloroplast exclusion exists due to the low efficiency of a glyphosate-accepting transporter(s). Barley, on the other hand, shows modest glyphosate delivery into the cell (same argument for the plasma membrane) but has no apparent restriction to chloroplast delivery, apparently explaining the extreme sensitivity of barley compared to Illinois bundleflower. Genetic control of classes of plasmalemma transporters is anticipated to be plant species, biotype, organelle, and possibly tissue specific, based on observations to date. [62]

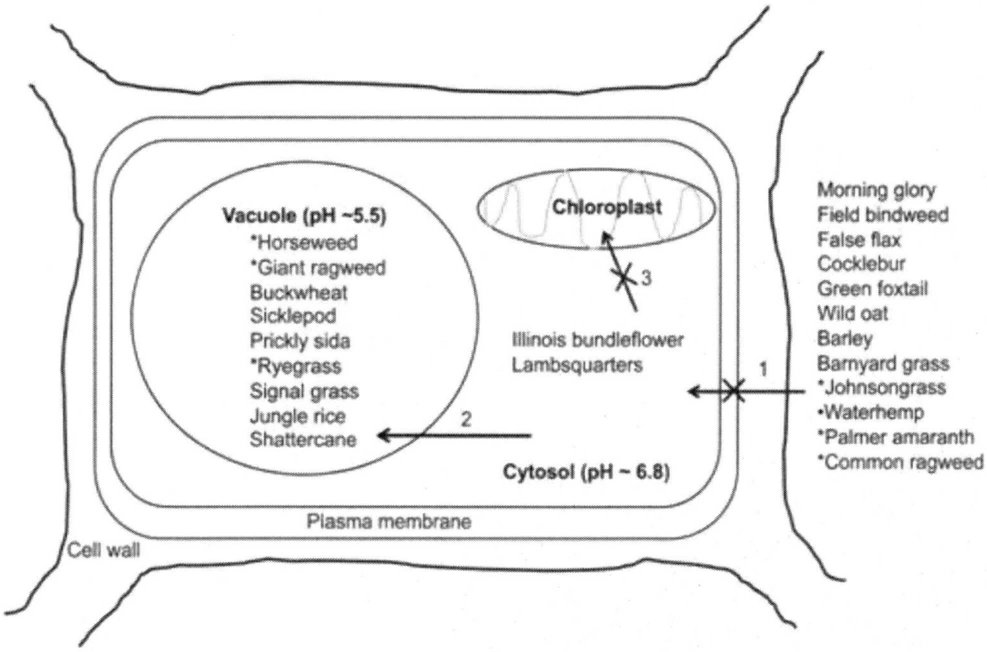

Figure 8. Cartoon describing plant cell transport of glyphosate (*via* putative active transporters) for those weed species surveyed by *in vivo* ^{31}P NMR. (1) Glyphosate entry to the plant cell is restricted for the species listed outside of the cell. It is hypothesized that this exclusion is controlled by a plasma membrane transporter labeled "1". (2) Glyphosate is vacuole sequestered for the species listed in the vacuole. It is hypothesized that this sequestration is facilitated by a tonoplast transporter labeled "2". (3) Glyphosate restriction from the chloroplast is presumed to occur for the species listed in the cell cytoplasm. The hypothesized transporter associated with chloroplast entry is labeled "3". The weeds species indicated with the star (*) are evolved GR biotypes. The glyphosate compartmental distribution at the cellular level can be represented as $G_{total} = G_{apoplast} + G_{cytosol} + G_{chloroplast} + G_{vacuole}$. $G_{apoplast}$ is immobilized and not visible by solution-state ^{31}P NMR. Further, it is diluted in perfusion media. $G_{cytosol}$ and $G_{chloroplast}$ have similar chemical shifts (similar pH values) and cannot be distinguished from each other. $G_{vacuole}$ can be distinguished from $G_{cytosol}$ and $G_{chloroplast}$ because the more acidic vacuole pH yields a significantly different ^{31}P chemical shift. Inhibition of glyphosate entry into cell cytosol and/or chloroplast, or glyphosate sequestration with the vacuole will reduce the amount of glyphosate available for translocation and binding with EPSPS in the chloroplast.

The data so gathered, together with arguments supporting the presence of unidirectional membrane transporters, allows a picture to emerge consistent with cellular traffic control of glyphosate. This is schematically illustrated in Figure 8, which shows a cartoon of the plant cell together with unidirectional membrane transporters that govern glyphosate movement into and within the plant cell. There is considerable precedent for this model from studies where evolved exclusion has led to drug resistance in bacteria, parasites, and cancer cells, all presumably through the presence of active membrane transporters, albeit these are exporters. [84-86] ^{31}P-NMR findings with difficult-to-control weeds reveal a number of species that show restricted uptake into the cell. These include those indicated exterior to the cell in Figure 8.

Included in this list are additional GR biotypes of waterhemp, some biotypes of Palmer amaranth and common ragweed (unpublished data not shown).

The degree of vacuole sequestration is variable for the difficult-to-control weedy species but approaches that observed in GR horseweed (Table 2) for junglerice, wild buckwheat and sicklepod. We have previously published vacuole sequestration for ryegrass and include this species in Figure 8 as a plant that derives resistance from, in large part, vacuole sequestration. Evidence for chloroplast exclusion comes from a lack of ^{31}P-NMR detected S3P coupled with the presence of a moderate to large cytoplasmic glyphosate content; we include lambsquarters along with Illinois bundleflower in this category as they both show no accumulation of S3P.

While there are reports of glyphosate metabolism in the literature, including the proposal that AMPA production is a detoxification mechanism, [33, 52] we have no evidence from ^{31}P-NMR studies that glyphosate metabolism is taking place to any appreciable extent under the time scale and conditions of our studies. Were AMPA to be produced through metabolism to any appreciable extent, its chemical shift would appear in a spectral region clear of other resonances. If AMPA is produced, the levels are below our detection threshold. Reddy *et al.* has reported that Illinois bundleflower metabolized 50% of delivered glyphosate 7 days following treatment. [33] However, our studies at 24 hours following treatment demonstrate that any metabolite present in Illinois bundleflower is below our detection limit. [29-31]

The widespread use of crops engineered with the glyphosate tolerant trait, makes glyphosate weed resistance a challenge, both to farming and food security. Understanding weed resistance is critical to efforts toward developing effective control strategies. *In vivo* ^{31}P NMR has provided important insights into the mechanisms of glyphosate resistance. Vacuole sequestration, shown by *in vivo* ^{31}P NMR initially in GR horseweed, is an effective mechanism. It shields the chloroplast from lethal levels of glyphosate and has been adapted to variable extent by other weed species (Figure 8). Restricted uptake is also shown by *in vivo* ^{31}P NMR to be a contributor to resistance and the source of problematic control for a number of weedy species. *In vivo* ^{31}P-NMR findings suggest chloroplast exclusion appears to be an additional potential resistance mechanism and source of variable sensitivity to glyphosate. All three of these mechanisms serve to prevent glyphosate from accessing chloroplast-targeted EPSPS and they can work together. Weed biotypes with multiple glyphosate resistance mechanisms undoubtedly present the greatest challenge to the field.

Genetic diversity in the face of intense selective pressure brings to the fore traits that ensure survival. Multiple drug resistant pumps are the bane of cancer therapy and, in like manner, weed biotypes have responded to glyphosate treatment. Whether multi-herbicide-resistance engineered traits can overcome the counter-options conferred by weed genetic diversity remains to be seen. Nevertheless, there is a pressing need for new weed suppression strategies as glyphosate resistance continues to increase. *In vivo* ^{31}P NMR will play an important role in developing and evaluating new strategies to overcome glyphosate resistance.

ACKNOWLEDGMENTS

Support from The Monsanto Company is gratefully acknowledged.

REFERENCES

[1] James, C. (2011) Global status of commercialized biotech/GM crops: 2011, Global Status of Commercialized Biotech/GM Crops: 2011. *ISAAA Brief* No. 43. ISAAA: Ithaca, NY.

[2] Gianessi, L. P. (2008) Economic impacts of glyphosate-resistant crops, *Pest Management Science* 64, 346-352.

[3] Gianessi, L. P. (2013) The increasing importance of herbicides in worldwide crop production, *Pest Management Science*, DOI: 10.1002/ps.3598.

[4] Duke, S. O., and Powles, S. B. (2008) Glyphosate: a once-in-a-century herbicide, *Pest Management Science* 64, 319-325.

[5] Dill, G. M., Sammons, R. D., Feng, P. C. C., Kohn, F., Kretzmer, K., Mehrsheikh, A., Bleeke, M., Honegger, J. L., Farmer, D., Wright, D., and Haupfear, E. A. (2010) *Glyphosate: discovery, development, applications, and properties, in glyphosate resistance in crops and weeds*, pp 1-33, John Wiley & Sons, Inc.

[6] Hollander, H., and Amrhein, N. (1980) The site of the inhibition of the shikimate pathway by glyphosate .I. Inhibition by glyphosate of phenylpropanoid synthesis in buckwheat (*Fagopyrum-Esculentum Moench*), *Plant Physiology* 66, 823-829.

[7] Amrhein, N., Deus, B., Gehrke, P., and Steinrucken, H. C. (1980) The site of the inhibition of the shikimate pathway by glyphosate .II. Interference of glyphosate with chorismate formation in vivo and in vitro, *Plant Physiology* 66, 830-834.

[8] Weaver, L. M., and Herrmann, K. M. (1997) Dynamics of the shikimate pathway in plants, *Trends in Plant Science* 2, 346-351.

[9] Bromilow, R. H., Chamberlain, K., and Evans, A. A. (1990) Physicochemical aspects of phloem yranslocation of herbicides, *Weed Science* 38, 305-314.

[10] Bromilow, R. H., and Chamberlain, K. (2000) The herbicide glyphosate and related molecules: Physicochemical and structral factors determining their mobility in phloem, *Pest Management Science* 56, 368-373.

[11] Perez, A., Alister, C., and Kogan, M. (2004) Absorption, translocation and allocation of glyphosate in resistant and susceptible Chilean biotypes of Lolium multiflorum, *Weed Biology and Management* 4, 56-58.

[12] Shaner, D. L. (2009) Role of translocation as a mechanism of resistance to glyphosate, *Weed Science* 57, 118-123.

[13] Baerson, S. R., Rodriguez, D. J., Tran, M., Feng, Y., Biest, N. A., and Dill, G. M. (2002) Glyphosate-resistant goosegrass. Identification of a mutation in the target enzyme 5-enolpyruvylshikimate-3-phosphate synthase, *Plant Physiology* 129, 1265-1275.

[14] Wakelin, A. M., and Preston, C. (2006) A target-site mutation is present in a glyphosate-resistant Lolium rigidum population, *Weed Research* 46, 432-440.

[15] Gaines, T. A., Zhang, W., Wang, D., Bukun, B., Chisholm, S. T., Shaner, D. L., Nissen, S. J., Patzoldt, W. L., Tranel, P. J., Culpepper, A. S., Grey, T. L., Webster, T. M., Vencill, W. K., Sammons, R. D., Jiang, J., Preston, C., Leach, J. E., and Westra, P. (2010) Gene amplification confers glyphosate resistance in Amaranthus palmeri, *Proceedings of the National Academy of Sciences of the United States of America* 107, 1029-1034.

[16] GM crops: A story in numbers, *Nature* 497, 22-23 (2013).

[17] Pratley, J., Urwin, N., Stanton, R., Baines, P., Broster, J., Cullis, K., Schafer, D., Bohn, J., and Krueger, R. (1999) Resistance to glyphosate in Lolium rigidum. I. Bioevaluation, *Weed Science* 47, 405-411.

[18] Powles, S. B., Lorraine-Colwill, D. F., Dellow, J. J., and Preston, C. (1998) Evolved resistance to glyphosate in rigid ryegrass (*Lolium rigidum*) in Australia, *Weed Science* 46, 604-607.

[19] Service, R. F. (2007) A growing threat down on the farm, *Science* (New York, N.Y.) 316, 1114-1117.

[20] Green, J. M., and Owen, M. D. (2011) Herbicide-resistant crops: utilities and limitations for herbicide-resistant weed management, *Journal of Agricultural and Food Chemistry* 59, 5819-5829.

[21] Wilson, R. G., Young, B. G., Matthews, J. L., Weller, S. C., Johnson, W. G., Jordan, D. L., Owen, M. D., Dixon, P. M., and Shaw, D. R. (2011) Benchmark study on glyphosate-resistant cropping systems in the United States. Part 4: Weed management practices and effects on weed populations and soil seedbanks, *Pest Management Science* 67, 771-780.

[22] Powles, S. B., and Yu, Q. (2010) Evolution in action: plants resistant to herbicides, *Annual Review of Plant Biology* 61, 317-347.

[23] Lee, L. J., and Ngim, J. (2000) A first report of glyphosate-resistant goosegrass (*Eleusine indica* (L) Gaertn) in Malaysia, *Pest Management Science* 56, 336-339.

[24] Vangessel, M. J. (2001) Glyphosate-resistant horseweed from Delaware, *Weed Science* 49, 703-705.

[25] Salas, R. A., Dayan, F. E., Pan, Z., Watson, S. B., Dickson, J. W., Scott, R. C., and Burgos, N. R. (2012) EPSPS gene amplification in glyphosate-resistant Italian ryegrass (*Lolium perenne* ssp. *multiflorum*) from Arkansas, *Pest Management Science* 68, 1223-1230.

[26] Perez-Jones, A., Park, K. W., Colquhoun, J., Mallory-Smith, C., and Shaner, D. (2005) Identification of glyphosate-resistant Italian ryegrass (*Lolium multiflorum*) in Oregon, *Weed Science* 53, 775-779.

[27] Jasieniuk, M., Ahmad, R., Sherwood, A. M., Firestone, J. L., Perez-Jones, A., Lanini, W. T., Mallory-Smith, C., and Stednick, Z. (2008) Glyphosate-resistant Italian ryegrass (*Lolium multiflorum*) in California: distribution, response to glyphosate, and molecular evidence for an altered target enzyme, *Weed Science* 56, 496-502.

[28] Kaundun, S. S., Dale, R. P., Zelaya, I. A., Dinelli, G., Marotti, I., McIndoe, E., and Cairns, A. (2011) A novel P106L mutation in EPSPS and an unknown mechanism(s) act additively to confer resistance to glyphosate in a South African Lolium rigidum population, *Journal of Agricultural and Food Chemistry* 59, 3227-3233.

[29] Ge, X., d'Avignon, D. A., Ackerman, J. J., and Sammons, R. D. (2010) Rapid vacuolar sequestration: the horseweed glyphosate resistance mechanism, *Pest Management Science* 66, 345-348.

[30] Ge, X., d'Avignon, D. A., Ackerman, J. J. H., Duncan, B., Spaur, M. B., and Sammons, R. D. (2011) Glyphosate-resistant horseweed made sensitive to glyphosate: low-temperature suppression of glyphosate vacuolar sequestration revealed by ^{31}P NMR, *Pest Management Science* 67, 1215-1221.

[31] Ge, X., d'Avignon, D. A., Ackerman, J. J., Collavo, A., Sattin, M., Ostrander, E. L., Hall, E. L., Sammons, R. D., and Preston, C. (2012) Vacuolar glyphosate-sequestration correlates with glyphosate resistance in ryegrass (*Lolium* spp.) from Australia, South America, and Europe: a ^{31}P NMR investigation, *Journal of Agricultural and Food Chemistry* 60, 1243-1250.

[32] Nandula, V. K., Reddy, K. N., Rimando, A. M., Duke, S. O., and Poston, D. H. (2007) Glyphosate-resistant and -susceptible soybean (*Glycine max*) and canola (*Brassica napus*) dose response and metabolism relationships with glyphosate, *Journal of Agricultural and Food Chemistry* 55, 3540-3545.

[33] Reddy, K. N., Rimando, A. M., Duke, S. O., and Nandula, V. K. (2008) Aminomethylphosphonic acid accumulation in plant species treated with glyphosate, *J Agric Food Chem* 56, 2125-2130.

[34] Heap, I. (2012) International Survey of Herbicide Resistant Weeds, Online. Internet. December 31, 2012: http://www.weedscience.org.

[35] Roberts, J. K., Ray, P. M., Wade-Jardetzky, N., and Jardetzky, O. (1980) Estimation of cytoplasmic and vacuolar pH in higher plant cells by ^{31}P NMR, *Nature* 283, 870-872.

[36] Ratcliffe, R. G., and Shachar-Hill, Y. (2001) Probing Plant Metabolism with NMR, *Annual Review of Plant Physiology and Plant Molecular Biology* 52, 499-526.

[37] Gout, E., Bligny, R., Genix, P., Tissut, M., and Douce, R. (1992) Effect of glyphosate on plant cell metabolism. ^{31}P and ^{13}C NMR studies, *Biochimie* 74, 875-882.

[38] Danova-Alt, R., Dijkema, C., P, D. E. W., and Kock, M. (2008) Transport and compartmentation of phosphite in higher plant cells--kinetic and P nuclear magnetic resonance studies, *Plant, Cell & Environment* 31, 1510-1521.

[39] Roberts, J. K. (1984) Study of Plant Metabolism in vivo Using NMR Spectroscopy, *Annual Review of Plant Physiology* 35, 375-386.

[40] Bligny, R., and Douce, R. (2001) NMR and plant metabolism, Current Opinion in Plant Biology 4, 191-196.

[41] Lundberg, P., and Lundquist, P. O. (2004) Primary metabolism in N_2-fixing Alnus incana-Frankia symbiotic root nodules studied with ^{15}N and ^{31}P nuclear magnetic resonance spectroscopy, *Planta* 219, 661-672.

[42] Cegelski, L., and Schaefer, J. (2005) Glycine metabolism in intact leaves by *in vivo* ^{13}C and ^{15}N labeling, *The Journal of Biological Chemistry* 280, 39238-39245.

[43] Masclaux-Daubresse, C., Reisdorf-Cren, M., Pageau, K., Lelandais, M., Grandjean, O., Kronenberger, J., Valadier, M. H., Feraud, M., Jouglet, T., and Suzuki, A. (2006) Glutamine synthetase-glutamate synthase pathway and glutamate dehydrogenase play distinct roles in the sink-source nitrogen cycle in tobacco, *Plant Physiology* 140, 444-456.

[44] Jacob, G. S., Schaefer, J., Stejskal, E. O., and McKay, R. A. (1985) Solid-state NMR determination of glyphosate metabolism in a Pseudomonas sp, *The Journal of Biological Chemistry* 260, 5899-5905.

[45] Gard, J. K., Feng, P. C., and Hutton, W. C. (1997) Nuclear magnetic resonance timecourse studies of glyphosate metabolism by microbial soil isolates, *Xenobiotica; the fate of foreign compounds in biological systems* 27, 633-644.

[46] Lipok, J., Wieczorek, D., Jewgiński, M., and Kafarski, P. (2009) Prospects of *in vivo* ^{31}P NMR method in glyphosate degradation studies in whole cell system, Enz*yme and Microbial Technology* 44, 11-16.

[47] Komossa, D., Gennity, I., and Sandermann, H., Jr. (1992) Plant metabolism of herbicides with C-P bonds: glyphosate, *Pesticide Biochemistry and Physiology* 43, 85-94.

[48] Ge, X., d'Avignon, D. A., Ackerman, J. J. H., and Sammons, R. D. (2012) Observation and identification of 2-C-methyl-d-erythritol-2,4-cyclopyrophosphate in horseweed and ryegrass treated with glyphosate, *Pesticide Biochemistry and Physiology* 104, 187-191.

[49] Mongelard, G., Seemann, M., Boisson, A. M., Rohmer, M., Bligny, R., and Rivasseau, C. (2011) Measurement of carbon flux through the MEP pathway for isoprenoid synthesis by ^{31}P-NMR spectroscopy after specific inhibition of 2-C-methyl-d-erythritol 2,4-cyclodiphosphate reductase. *Effect of light and temperature, Plant, Cell & Environment.*

[50] Rivasseau, C., Seemann, M., Boisson, A. M., Streb, P., Gout, E., Douce, R., Rohmer, M., and Bligny, R. (2009) Accumulation of 2-C-methyl-D-erythritol 2,4-cyclodiphosphate in illuminated plant leaves at supraoptimal temperatures reveals a bottleneck of the prokaryotic methylerythritol 4-phosphate pathway of isoprenoid biosynthesis, *Plant, Cell & Environment* 32, 82-92.

[51] Castellino, S., Leo, G. C., Sammons, R. D., and Sikorski, J. A. (1989) Phosphorus-31, nitrogen-15, and carbon-13 and NMR of glyphosate: comparison of pH titrations to the herbicidal dead-end complex with 5-enolpyruvoylshikimate-3-phosphate synthase, *Biochemistry* 28, 3856-3868.

[52] Duke, S. O. (2011) Glyphosate Degradation in Glyphosate-Resistant and -Susceptible Crops and Weeds, *Journal of Agricultural and Food Chemistry* 59, 5835-5841.

[53] Ostrovsky, D., Kharatian, E., Malarova, I., Shipanova, I., Sibeldina, L., Shashkov, A., and Tantsirev, G. (1992) Synthesis of a new organic pyrophosphate in large quantities is induced in some bacteria by oxidative stress, *BioFactors* 3, 261-264.

[54] Ostrovsky, D., Shashkov, A., and Sviridov, A. (1993) Bacterial oxidative-stress substance is 2-C-methyl-D-erythritol 2,4-cyclopyrophosphate, *The Biochemical journal* 295 (Pt 3), 901-902.

[55] Weaver, S. E. (2001) The biology of Canadian weeds. 115. Conyza canadensis, *Canadian Journal of Plant Science* 81, 867-875.

[56] Gougler, J. A., and Geiger, D. R. (1984) Carbon Partitioning and Herbicide Transport in Glyphosate-Treated Sugarbeet (*Beta vulgaris*), *Weed Science* 32, 546-551.

[57] Feng, P. C. C., Tran, M., Chiu, T., Sammons, R. D., Heck, G. R., and Cajacob, C. A. (2004) Investigations into glyphosate-resistant horseweed (*Conyza canadensis*): retention, uptake, translocation, and metabolism, *Weed Science* 52, 498-505.

[58] Koger, C. H., and Reddy, K. N. (2005) Role of absorption and translocation in the mechanism of glyphosate resistance in horseweed (*Conyza canadensis*), *Weed Science* 53, 84-89.

[59] Dinelli, G., Marotti, I., Bonetti, A., Minelli, M., Catizone, P., and Barnes, J. (2006) Physiological and molecular insight on the mechanisms of resistance to glyphosate in *Conyza canadensis* (L.) Cronq. biotypes, *Pesticide Biochemistry and Physiology* 86, 30-41.

[60] Anil Shrestha, K. J. H., and Neil, V. (2007) Growth stage influences level of resistance in glyphosate-resistant horseweed, *California Agriculture* 61, 67-70.

[61] Vangessel, M. J., Scott, B. A., Johnson, Q. R., and White-Hansen, S. E. (2009) Influence of glyphosate-resistant horseweed (*Conyza canadensis*) growth stage on response to glyphosate applications, *Weed Technology* 23, 49-53.

[62] Conte, S. S., and Lloyd, A. M. (2011) Exploring multiple drug and herbicide resistance in plants--spotlight on transporter proteins, *Plant Science* 180, 196-203.

[63] Bretthorst, G. L. (1990) Bayesian analysis. III. Applications to NMR signal detection, model selection, and parameter estimation, *Journal of Magnetic Resonance* 88, 571-595.

[64] Ratcliffe, R. G. (1997) In vivo NMR studies of the metabolic response of plant tissues to anoxia, *Annals of Botany* 79, 39-48.

[65] Sterling, T. M., Balke, N. E., and Silverman, D. S. (1990) Uptake and accumulation of the herbicide bentazon by cultured plant cells, *Plant Physiology* 92, 1121-1127.

[66] Feng, P. C. C., Chiu, T., and Douglas Sammons, R. (2003) Glyphosate efficacy is contributed by its tissue concentration and sensitivity in velvetleaf (*Abutilon theophrasti*), *Pesticide Biochemistry and Physiology* 77, 83-91.

[67] Shaner, D. L., Nadler-Hassar, T., Henry, W. B., and Koger, C. H. (2005) A rapid in vivo shikimate accumulation assay with excised leaf discs, *Weed Science* 53, 769-774.

[68] Kime, M. J., and Ratcliffe, R. G. (1982) The Application of ^{31}P Nuclear Magnetic Resonance to Higher Plant Tissue, *Journal of Experimental Botany* 33, 670-681.

[69] Devine, M. D., O., D. S., and Fedtke, C. (1993) Physiology of Herbicide Action, 441.

[70] Murray, J. A., Jones, A., Godin, C., and Traas, J. (2012) Systems analysis of shoot apical meristem growth and development: integrating hormonal and mechanical signaling, *The Plant Cell* 24, 3907-3919.

[71] Haslam, E. (1994) Shikimic Acid-Metabolism and Metabolites, *Journal of Natural Products* 57, 1470.

[72] Sweetlove, L. J., Beard, K. F., Nunes-Nesi, A., Fernie, A. R., and Ratcliffe, R. G. (2010) Not just a circle: flux modes in the plant TCA cycle, *Trends in Plant Science* 15, 462-470.

[73] Kruger, N. J., and Ratcliffe, R. G. (2012) Pathways and fluxes: exploring the plant metabolic network, *Journal of experimental botany* 63, 2243-2246.

[74] Riar, D. S., Norsworthy, J. K., Johnson, D. B., Scott, R. C., and Bagavathiannan, M. (2011) Glyphosate resistance in a Johnsongrass (*Sorghum halepense*) biotype from Arkansas, *Weed Science* 59, 299-304.

[75] Vila-Aiub, M. M., Balbi, M. C., Gundel, P. E., Ghersa, C. M., and Powles, S. B. (2007) Evolution of glyphosate-resistant johnsongrass (*Sorghum halepense*) in glyphosate-resistant soybean, *Weed Science* 55, 566-571.

[76] Vila-Aiub, M. M., Balbi, M. C., Distefano, A. J., Fernandez, L., Hopp, E., Yu, Q., and Powles, S. B. (2012) Glyphosate resistance in perennial *Sorghum halepense* (Johnsongrass), endowed by reduced glyphosate translocation and leaf uptake, *Pest Management Science* 68, 430-436.

[77] Michitte, P., De Prado, R., Espinoza, N., Ruiz-Santaella, J. P., and Gauvrit, C. (2007) Mechanisms of resistance to glyphosate in a ryegrass (*Lolium multiflorum*) biotype from Chile, *Weed Science* 55, 435-440.

[78] Schonherr, J. (2002) A mechanistic analysis of penetration of glyphosate salts across astomatous cuticular membranes, *Pest Management Science* 58, 343-351.

[79] Franz, J. E., Mao, M. K., and Sikorski, J. A. (1997) *Glyphosate: A Unique, Global Herbicide*, 688.

[80] Grichar, W. J., and Ocumpaugh, W. R. (2007) Bundleflower (*Desmanthus bicornutus*) response to postemergence herbicides, *Weed Technology* 21, 1089-1092.

[81] Thomas, W. E., Pline-Srnic, W. A., Viator, R. P., and Wilcut, J. W. (2005) Effects of glyphosate application timing and rate on sicklepod (*Senna obtusifolia*) fecundity, *Weed Technology* 19, 55-61.

[82] Alarcon-Reverte, R., Garcia, A., Urzua, J., and Fischer, A. J. (2013) Resistance to glyphosate in Junglerice (*Echinochloa colona*) from California, *Weed Science* 61, 48-54.

[83] Vila-Aiub, M. M., Gundel, P. E., Yu, Q., and Powles, S. B. (2013) Glyphosate resistance in Sorghum halepense and Lolium rigidum is reduced at suboptimal growing temperatures, *Pest Management Science* 69, 228-232.

[84] Staub, J. M., Brand, L., Tran, M., Kong, Y., and Rogers, S. G. (2012) Bacterial glyphosate resistance conferred by overexpression of an E. coli membrane efflux transporter, *Journal of Industrial Microbiology & Biotechnology* 39, 641-647.

[85] Martin, R. E., Marchetti, R. V., Cowan, A. I., Howitt, S. M., Broer, S., and Kirk, K. (2009) Chloroquine transport via the malaria parasite's chloroquine resistance transporter, *Science* 325, 1680-1682.

[86] van de Ven, R., Oerlemans, R., van der Heijden, J. W., Scheffer, G. L., de Gruijl, T. D., Jansen, G., and Scheper, R. J. (2009) ABC drug transporters and immunity: novel therapeutic targets in autoimmunity and cancer, *Journal of Leukocyte Biology* 86, 1075-1087.

In: Handbook on Herbicides
Editors: Daiki Kobayashi and Eito Watanabe

ISBN: 978-1-62948-053-4

Chapter 4

LETHAL AND SUBLETHAL GLYPHOSATE EFFECTS ON NON-TARGET FISH SPECIES: BIOMARKERS RESPONSES IN *CNESTERODON DECEMMACULATUS*

Renata J. Menéndez-Helman[1], Alfredo Salibián[2] and Maria dos Santos Afonso[1,*]
[1]CONICET-INQUIMAE, Facultad de Ciencias Exactas y Naturales, Universidad de Buenos Aires, Ciudad Universitaria-Pabellón II, (C1428EHA) - Buenos Aires, Argentina
[2]PRODEA-INEDES, Universidad Nacional de Luján, (B6700ZBA)-Luján, Argentina

ABSTRACT

Environmental degradation in Latin America has increased in the last three decades. The agricultural frontier of Argentina is spreading out, and this phenomenon has involved a significant increase in the use of biocides. Glyphosate (N-phosphonomethylglycine, PMG) is the most widely used herbicide in this country. Particularly the application of formulated glyphosate increased from 70 million kg in 1999 to more than 210 million in 2011. Although this herbicide is considered as "environmentally friendly" with moderate persistence, the mobility and fate will be conditioned by the chemical and biological profile of the environment, and regarding that, several authors have determined environmentally relevant levels of glyphosate and its metabolites in surface water. The risk associated with the use of PMG, could be assessed by the evaluation of the impact on a non-target species: *Cnesterodon decemmaculatus,* a Neotropical endemic teleost, widely distributed in the region. Experimental results would indicate that glyphosate (active ingredient, analytical grade, acid form) does not present a high level of toxicity (LC_{50}-96h > 100 $mg.L^{-1}$) on *C. decemmaculatus*. However, the toxicity of one of the glyphosate-based herbicide formulation was found to be significantly higher (LC_{50}-96h = 29 $mg.L^{-1}$). These results are consistent with other evidences that indicate a remarkable

[*] Correspondence to: Maria dos Santos Afonso; Phone/Fax: (+54) 11-45763378 ext.125; E-mail address: dosantos@qi.fcen.uba.ar.

increase in the toxicity of formulations regarding to the active ingredient in freshwater teleosts.

Several scientists have focused their research on the identification of biomarkers of environmental stress in non-target species. These could be specific and early signals of the toxic-organism interaction. In this regard, the utility of morphological and biochemical parameters as quantitative bioindicators of xenobiotic exposure has been shown. Within this framework, we measured the impact of sublethal exposure of *Cnesterodon decemmaculatus* to PMG on antioxidant defenses (catalase activity) and neurotoxicity (acetylcholinesterase activity). After acute exposure, PMG caused changes in the enzyme activities: an increase in the activity of catalase and an inhibition of the acetylcholinesterase activity. These results are the first report indicating that sublethal PMG affect these biomarkers in a native teleost species.

Keywords: Glyphosate, PMG, N-phosphonomethylglycine, Toxicity, *Cnesterodon decemmaculatus*, Freshwater fish, Biomarkers, Acetylcholinesterase, AChE, Catalase

1. General Introduction

Environmental degradation in Latin America has increased in recent decades. The main pressures on the environment and natural resources are the population growth, increasing urbanization processes, inequality of incomes of workers, limited planning and high dependence on natural resource exploitation in many economies (PNUMA, 2002).

The agricultural frontier in the region is spreading out. This phenomenon has been led by a number of temporary changes in international prices of agricultural products and, in turn, was based on the massive use of technologies such as no-tillage and chemical and biological nutrition, intensive application of biocides (herbicides, insecticides and fungicides) and the use of genetically modified (GM) seeds, all designed to reduce costs and increase production yields. In this process, determined by the increase of agricultural production and planted area, soybean cultivation has played a fundamental role as it represents today about 50% of total production in the sector.

Certainly, the introduction of a GM glyphosate-tolerant variety of soybean to the Argentine market in 1996 was an inflexion point. The Argentine Government authorized "the production and marketing of seed and the products and byproducts of soybean glyphosate herbicide tolerant 40/3/2 line containing the CP4 EPSPS gene". Thus, direct seeding (no-tillage soil) and GM soya constituted the technological package that supported the advancement of the crop. Currently, for commercial purposes the only cultivated soybean variety is the glyphosate resistant, known as RR (Roundup Ready resistant formulated) one of the most used in the region (CONICET, 2009). However, this process of agricultural industrialization has been developed without any real planning, regardless of the market control, without considering the socioeconomic and environmental impacts, and confirming the role assigned to Latin America in the international division of labor as exporters of raw materials.

In this sense there is an increase regarding the consequences of the phenomenon of soybean expansion (*"sojizacion"*) and the subsequent expansion of the agricultural frontier (Pengue, 2008; Carrasco et al., 2012; López et al., 2012).

There are several important issues to analyse concerning *social impacts*:

a) The collapse of rural employment caused by intensive herbicide use, replacing manual or mechanical removal of weeds and the substitution of conventional tillage by the no-tillage system (avoiding the plow).
b) Human migration processes that have led to greater concentration of population in urban and suburban areas.
c) The high concentration of land ownership, with the disappearance of a large number of smallholder farmers (the Argentina National Agricultural Census of 2002 recorded a reduction of 29.4% in the number of farms compared to the previous census, 1988).
d) The displacement of traditional crops and others productive activities (e.g. wheat, corn, sunflower, sweet potato, sugar cane, cotton, dairy farms, fruits, rice fields, livestock, etc.) because of an economic model based on soya monoculture mainly for export.
e) The consequent conflicts over land tenure with communities of small farmers and aboriginal peoples.

In relation to the *environmental impact assessment* of the herbicide, attention was attracted on adverse effects on biodiversity of local agroecosystems (Weyland et al., 2008) as well as on the increasing pressure on forests with the consequence of a widespread deforestation processes, with the concomitant loss and fragmentation of habitats and reduction of the biodiversity, desertification and soil depletion due to intensive usage and lack of crop rotation. Currently, soybean expansion process in our country is expressed on several extensive areas of the Chaco Austral Region, the Misiones jungle, piedmont areas of the yungas, the semi-arid grasslands, and even some sectors of the transition between Andean forests and the steppe, as well as the NE areas of the Rolling Pampa Region (Adámoli et al., 2011).

Associated with the above, another aspect is the marked increase in the use of agrochemicals (Bedmar, 2011). Between 2009 and 2010, 18 million Ha of soybeans were planted in Argentina, which is three times higher than the 1996-1997 period for the same crop, according to data from the Argentine Agriculture, Livestock and Fishing Office (Ministerio de Agricultura, Ganadería y Pesca de la Nación, MAGyP, 2012). Along with this process, the use of biocides in general has increased (see Figure 1) and in particular, glyphosate-based formulations has augmented from 12 million liters in 1996 to 197 million liters in 2012 (CASAFE, 2012).

It was established that approximately 0.1% of the amount of pesticide applied actually reaches the target species, while the remaining is spread through the environment, comprising various environmental compartments (soil, water and biota) in a complex dynamic pollution process (Carvalho et al., 1998; Martinez-Ghersa, 2011).

A significant amount of pesticides applied reach the ground due to the different application methods and environmental conditions (rain, wind, etc.). When they interact with soil constituents can be adsorbed, transported to streams, rivers and lakes, leached into groundwater, or chemically and / or biologically degraded. Some of the degradation products could be volatile and transported into the atmosphere.

The Food and Agriculture Organization of the United Nations (FAO) has defined the pesticides as "any substance or mixture of substances intended for preventing, destroying or controlling any pest" (FAO, 2002). Among these compounds, organophosphorus pesticides (OPP) are the most worldwide used group in agricultural production.

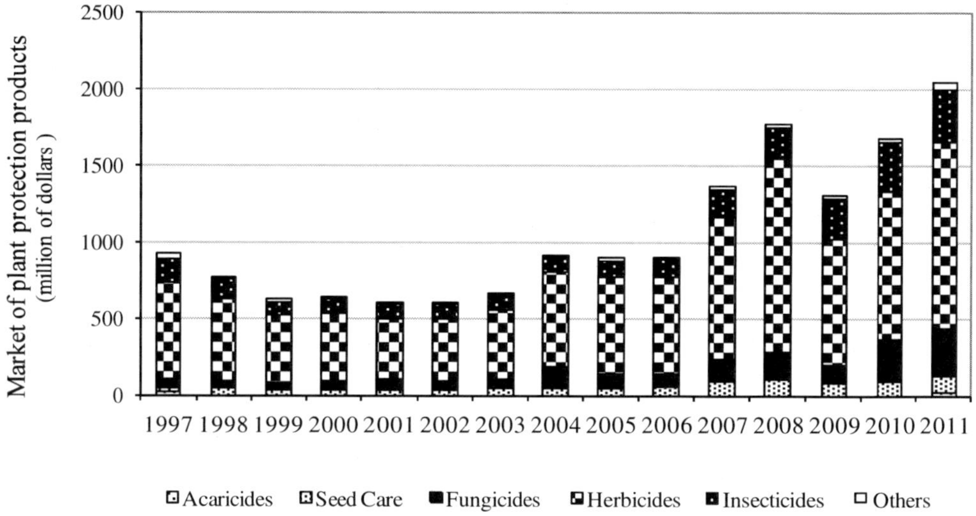

Figure 1. Market of plant protection products of Argentina evolution (CASAFE, 2012).

There is a large number of OPs. Many of them are used in agriculture as pesticides and have largely diminished the application rate of organochlorine pesticides (OCP). This was due to the high toxicity, persistence and bioaccumulation of the OCP in the food chains (Ecobichon, 2005). In general, OP compounds are not bioaccumulated and are readily dissipated in the environment (Chambers, 1992; Racke, 1992). However, it is now accepted that they are more environmentally hazardous than initially predicted.

Glyphosate

Among the pesticides, selective herbicides can eliminate weeds, while those crops of agricultural interest are preserved. In some cases these species have been genetically modified to resist these biocides.

Main Features

Glyphosate (N-phosphonomethylglycine; PMG, $C_3H_8NO_5P$, CAS 1071-83-6) (Figure 2) is at present the most widely used herbicide in Argentina.

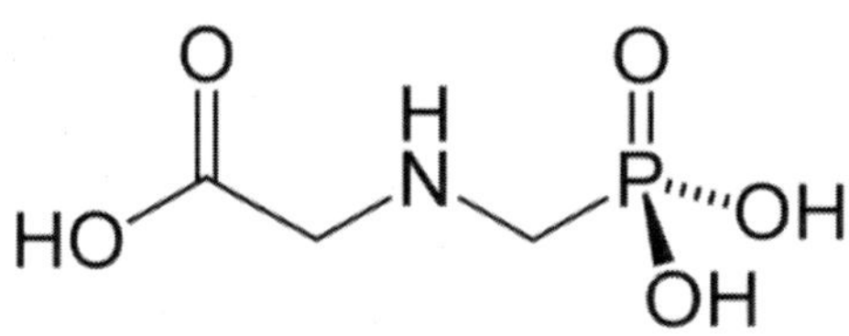

Figure 2. Molecular structure of N-phosphonomethylglycine, PMG.

PMG exhibits the following properties:

- Postemergence: it is applied directly to the foliage.
- Systemic: it is rapidly absorbed by foliage and translocated through the vascular system of the plant to the developing apical and root meristems.
- Broad spectrum: effective for the control of annual and perennial grasses, broadleaf weeds and woody species.

Plant growth is quickly inhibited followed by general foliar chlorosis and necrosis, through the inhibition of 5-enolpyruvyl shikimate-3-phosphate synthase (EPSPS), an enzyme involved in the biosynthesis pathway of an intermediate in plant aminoacids with a resulting reduction in protein synthesis. The mechanism of action is based on blocking the synthesis of essential aromatic amino acids (phenylalanine, tyrosine and tryptophan) by enzyme inhibition of metabolic pathways in higher plants and microorganisms.

It has been shown that certain isozymes of EPSPS do not bind glyphosate; this behaviour has enabled the development, by genetic engineering, of transgenic plants tolerant to glyphosate which may be treated continuously with this herbicide for weed control.

Chemical Equilibria, Mobility and Environmental Fate

Glyphosate is an organophosphorus herbicide belonging to the phosphonates chemical group. This compound is an organic acid condensing glycine group and phosphonomethyl groups, with zwitterionic properties because is a derivative of an amino acid. This ionic characteristics give it a relative solubility in aqueous media (T=25°C, 15.7 g/L at pH=7 and 11.6 g/L at pH=2.5) and is practically insoluble in organic solvents. The octanol/water partition varies between $3.22 > pK_{ow} > 2.77$ (Duke, 1988; Ahrens, 1994).

PMG is a tetraprotic acid, exists mostly as monoanion or dianion in the pH range of natural environments. pKa values are 0.78 (first phosphonate group proton), 2.29 (carboxyl group proton), 5.96 (second phosphonate group proton) and 10.98 (amino group) (Barja and dos Santos Afonso, 1998) (see Figure 3).

Monoanionic salts are used in the commercial glyphosate based formulations, in which the PMG is mixed with other chemical products (co-adjuvants). The recommended application rate is in the range of 0.84 to 4.20 kg / ha (Duke, 1988).

Phosphonate compounds have the ability to coordinate and form strong complexes with free metals, such us copper, iron, aluminum ions among others (Barja et al., 2001; Subramaniam and Hoggard, 1988). Thus, the ability of PMG to coordinate with metal ions places it in a special class of strongly chelating herbicides (Morillo et al., 1997), in this sense PMG forms complexes with dissolved metals (Barja et al., 2001; Trinelli et al., 2013).

Moreover, several studies have shown that glyphosate is strongly adsorbed by soil components such as clays, iron oxides and humic acids. Adsorption seems to be the main process that regulates the herbicide mobility in soils and particulate matter (Barja and dos Santos Afonso, 2005; Gimsing and dos Santos Afonso, 2005; Pessagno and dos Santos Afonso, 2006; Damonte et al., 2007; Pessagno et al., 2008; Khoury et al., 2010).

The adsorption process occurs via surface complexes formation by two predominating monodentate and bidentate complexes that are formed by phosphonate coordination to the surface, while the carboxylate and amino group are not coordinated (Barja and dos Santos Afonso, 2005; Pessagno et al., 2008; Khoury et al., 2010). Thus, the adsorption process should be characterized by the following reactions:

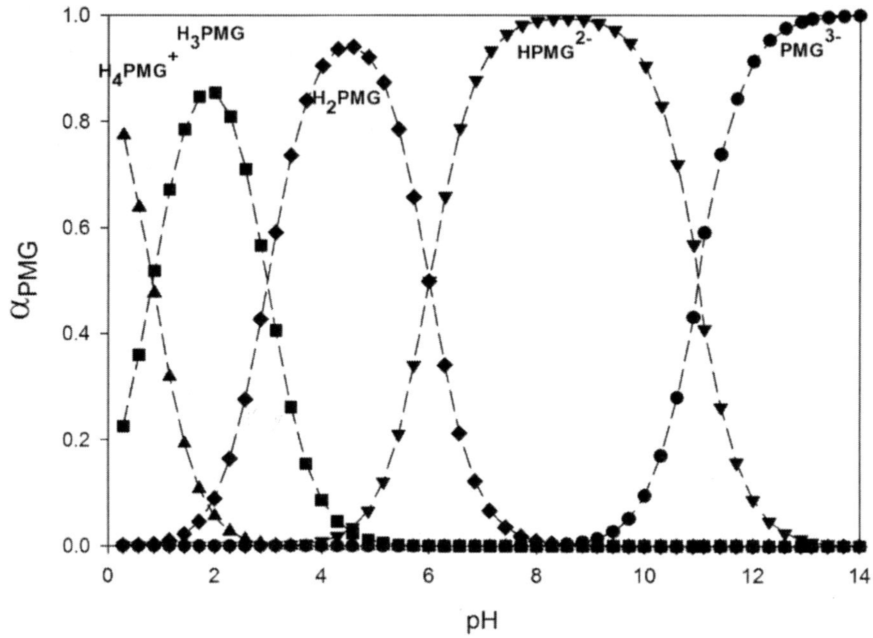

Figure 3. Speciation Diagramme of N-phosphonomethylglycine.

$\equiv$MeOH + R-PO(OH)-O$^-$ + H$^+$ $\equiv$MeO-PO(OH)-R + H_2O K^s_1

$\equiv$MeOH + R-PO(OH)-O$^-$ $\equiv$MeO-P(O)$_2$-R$^-$ + H_2O K^s_2

2 $\equiv$MeOH + R-PO(OH)-O$^-$ + H$^+$ ($\equiv$MeO)$_2$-PO-R + 2 H_2O K^s_3

where $\equiv$MeOH denotes a surface site, R is the $^-OCCH_2(NH_2)^+CH_2$ group and K^s_1, K^s_2 and K^s_3 are the respective surface complexes formation constants.

Thereby, glyphosate has low mobility in soils which is consistent with strong adsorption. Nevertheless, it has been asserted that PMG have not great persistence nor the ability to bioaccumulate or biomagnify in the food chain (WHO, 1994; Giesy et al., 2000; Solomon and Thompson, 2003). Also, PMG can suffer chemical and photochemical degradation (Trinelli et al., 2012).

Microbiological degradation occurs throughout different pathways being aminomethylphosphonic acid (AMPA) the major metabolite (Rueppel et al., 1977). According to these data, its presence should not be expected in high concentrations in the surface and / or in groundwater. However, it has been described that the adsorption varies greatly depending on the soil type, and can release from 15% to 80% of the adsorbed herbicide (Piccolo et al., 1994; Pessagno et al., 2005; 2008). There is much controversy about half life and persistence of PMG. Half life values in soil between 45 to 60 days and persistence of 170 days to more than 300 days have been reported (WHO, 1994; USEPA, 1999; Peruzzo et al., 2008; CONICET, 2009).

Once applied to the crop, by terrestrial or atmospheric way, the herbicide may remain adsorbed to soil particles until it is degraded by microorganisms, or may be mobilized in the agroecosystem by influence of the wind, rain or irrigation, factors that increase infiltration and surface run-off (Peruzzo et al., 2008). In this manner it can reach adjacent environments and adversely affect its biota. Therefore, several authors have reported levels of glyphosate and AMPA in sediment, soil and water, particularly in the surroundings of cultivation areas (Peruzzo et al., 2008; Demetrio, 2012), as well as in plants and soybeans (Arregui et al., 2004; Lorenzatti et al., 2004).

Toxicological Aspects

Glyphosate was introduced, after its discovery in 1974, in agricultural practices for direct sowing of transgenic herbicide-resistant seeds. From the outset it was assured that this was a highly selective inhibitor of the EPSPS enzyme (which is not expressed in animals) (Mousdale and Coggins, 1984; Rubin et al., 1984; Malik et al., 1989), so it was assumed that the use was completely safe for animals and humans. At the same time, there was no evidence that the herbicide (both active ingredient and commercial formulations) could significantly bioaccumulate in tissues or cultures.

The U.S. Environmental Protection Agency (USEPA) classifies herbicides in four categories: I, II, III and IV (in decreasing order of toxicity). Studies conducted for USEPA had not shown evidence of mutagenic, carcinogenic, teratogenic or allergenic in a wide range of trials and tests, being the reason why it was ranked as slightly toxic in Class III (mild irritant). Similarly, glyphosate has been classified in the lowest toxicological risk category, Class IV "unlikely to present hazard in normal use" according to the criteria of the World Health Organization (WHO) and the Food and Agriculture Organization (FAO), adopted by Argentina Regulation of the MAGyP.

However, recent toxicological studies suggested that glyphosate has been incorrectly described as "toxicologically benign" for both human health and environment (CONICET, 2009; López et al., 2012).

It is important to note that formulations contain in addition to the active ingredient, other substances whose purpose is to facilitate the herbicide handling or enhance its effectiveness. Most of the commercial formulations include a surfactant that helps the emulsion of the active ingredient to increase the penetration of the herbicide through the plant cuticle. In general, these substances are identified as "inert" and can be present as constituents of the commercial products, in concentration close to 50 % of the formulation (Dinehart et al., 2009), being not specified on the labels. However, in many cases those ingredients tend to have higher toxicity than the active ingredient of the herbicide. Thus, the toxicity is associated primarily with the type and concentration of the formulation adjuvants, as well as physical and chemical profile of the environment in which the herbicide is poured. Roundup®, for example, contained the non ionic surfactant polyoxyethylene-amine (POEA). Tsui and Chu (2003), among others authors, reported the comparative toxicity of the formulation, the active ingredient and the adjuvant POEA for a number of organisms (algae, bacteria, protozoans, molluscans and crustaceans) determining the following order of toxicity: POEA> Roundup> glyphosate (acid) > Glyphosate (salt). On the other hand, the toxicity data will vary because it is dependent on the ratio of glyphosate to surfactant, as well as on their different environmental persistence (Mann et al., 2009). In addition, POEA was reported to be more toxic to fish than glyphosate (Folmar et al., 1979).

Several reports had described adverse glyphosate impacts on biota, but mainly due to the formulations (CONICET, 2009). In amphibians, the herbicide produces a numerous adverse effects (Mann et al., 2009), such as delays in the development and inhibition of larval growth, malformations, increased mortality, decreased biodiversity (Lajmanovich et al., 2003; Howe et al., 2004; Relyea, 2005; Paganelli et al., 2010; Brodeur, 2011) and even disturbances in the selection of oviposition sites (Takahashi, 2007; Agostini, 2013; López et al., 2012).

Also, some reports have shown effects of Roundup on periphyton and phytoplankton communities, that there are the basis of the food chain (Pérez et al., 2007; Vera et al., 2010), on *Eisenia fetida* (Casabé et al., 2007; Verrell and van Buskirk, 2004) as well as teratogenic effects and endocrine disruption in laboratory mammals (Dallegrave et al., 2003; Guilherme et al., 2012; Romano et al., 2010; 2012). The in vitro genotoxicity of AMPA, one of the major environmental breakdown product of glyphosate, was reported in cells of mice and humans (Mañas et al., 2009).

In relation to the effects of glyphosate on fish, it can be noted that in most reports formulated glyphosate effects were assessed, while those related to the active ingredient are more scarce. Evasion of harmful concentrations (Hildebrand et al., 1982), histopathological changes (Domitrovic, 1997; Szarek *el al.*, 2000; Jiraungkoorskul et al., 2003), metabolic changes associated with oxidative stress (Cattaneo et al., 2011; Glusczak et al., 2007; Langiano and Martínez, 2008), genotoxicity (Cavas and Könen, 2007; Kier and Kiekland, 2013), malformations (Kelly et al., 2010), hematologic changes (Alvarez et al., 2012) and even alterations of hormonal profiles (Soso et al., 2007), decrease in cell mediated immune response (El-Gendy et al., 1998), among others, have been reported as sublethal effects of the herbicide in fish.

It is noteworthy that bioaccumulation of PMG has also been reported in different organisms (Wang et al., 1994; Casabé et al., 2007; Contardo-Jara et al., 2009). Considering this fact, the long persistence and the wide use in the region, it becomes necessary to characterize the fate and unforeseen toxicity of these pesticides on non-target species to assess with certainty the risk associated with their use (see López et al., 2012; Carrasco et al., 2012).

2. Acute Toxicity of Glyphosate (Active Ingredient and Formulated) for Freshwater Fish

The aquatic environment is the final receiver of natural and xenobiotic contaminants, the latter being synthetic substances whose chemical structure are not found naturally and cannot be biodegraded.

The concept of *environmental stress* (ES) is inherent to ecotoxicological studies in aquatic environments. ES can be defined as any quantifiable alteration of the normal state of the biological systems induced by an external or environmental change (natural or anthropogenic) (Boudou and Ribeyre, 1997; Walker et al., 2006; Newman and Clements, 2008; Timbrell, 2009). As a result, adverse effects can be observed in different levels of organization, from subcellular, to individual and population and, by extension, to the structure and functionality of the ecosystem.

Toxicity bioassays are adequate tools for environmental diagnostic that allow to determine under specific and controlled conditions, the physical and / or chemical effect on an *test* organism. A classic study to characterize the potential toxicity of a substance is the acute toxicity test, in which the LC_{50}-96h is determined as the concentration of a substance that provokes the death of 50% population in 96 h (Castillo Morales, 2004).

Cnesterodon decemmaculatus is one of the species used for biomonitoring of aquatic environments in the Southern region of South America (Ferrari et al., 1998; Gómez et al., 1998; García et al., 1999; Villar et al., 2000). This Cypridontiform teleost species has a wide distribution between 18 and 38 degrees South latitude involving five countries in the basin of the Rio de la Plata (Ringuelet et al., 1967). It is endemic of Neotropical America and ubiquitous in freshwater bodies of the Buenos Aires Province, Argentina.

This species has a successful reproductive strategy characterized by internal fertilization and ovoviviparity. Because of their small size, rapid growth, short life and easy reproduction and captive breeding (Mermoz and Salibián, 1992; Somma et al., 2011) is suitable for use as a test species for toxicity tests (Molero and Pisano, 1986; de la Torre et al., 2002; de la Torre et al., 2007) and biomonitoring of freshwater environments (Ferrari et al., 1998; Salibián, 2006).

In order to determine the range of acute and sublethal PMG concentrations for *Cnesterodon decemmaculatus,* toxicity tests were performed. The percentage of mortality was not significant for PMG, active ingredient, even in the highest concentration tested, only exceeding 10% of the control group (no significant differences) (Figure 4).

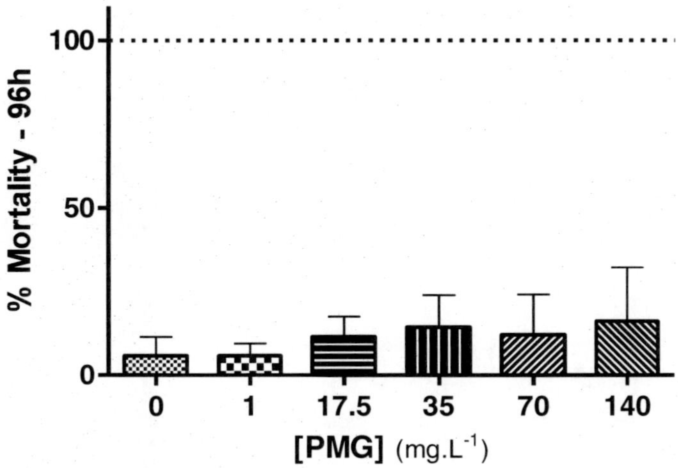

Figure 4. Mortality of *C. decemmaculatus* exposed for 96 h to glyphosate (active ingredient) and control group. Data as average mortality in percentage ± standard error of the mean (SEM), n = 25-35.

There are numerous studies about the toxic effects of glyphosate-based commercial formulations on aquatic animals (Giesy et al. 2000), but few related to the acute toxicity of solutions of glyphosate (active ingredient). The results indicate that *C. decemmaculatus* exposed to 140 $mg.L^{-1}$ of PMG (active ingredient) for 96 h has a survival rate of above 80%, so it follows that the LC_{50}-96h must be above the maximum concentration tested. The same parameter has been reported for other teleosts: *Oncorhynchus mykiss,* a species of high sensitivity, 140 $mg.L^{-1}$ (Folmar et al., 1979); *Odontesthes bonariensis*, 163 $mg.L^{-1}$ (Carriquiriborde, 2011); *Poecilia reticulata*, > 400 $mg.L^{-1}$ (Alvarez et al., 2012) and *Cyprinus carpio,* a high tolerance species, 620 $mg.L^{-1}$ (Neskovic et al., 1996). These results suggest that the active ingredient (PMG) does not present a significant level of acute toxicity to freshwater fish (LC_{50}-96h > 100 $mg.L^{-1}$).

The acute toxicity of formulated Glyphosate II (Atanor) in *C. decemmaculatus* was also analyzed, using the same experimental design. The mortality rate at 24 h and 96 h of exposure are shown in Figure 5.

C. decemmaculatus showed 100% mortality for formulated concentration of 35 mg.L^{-1} (equivalent PMG) resulting significantly different from the control group ($p < 0.05$). Instead, as was previously shown (Figure 4), the equivalent concentration of the active ingredient caused only 9% mortality over control after 96 h of exposure.

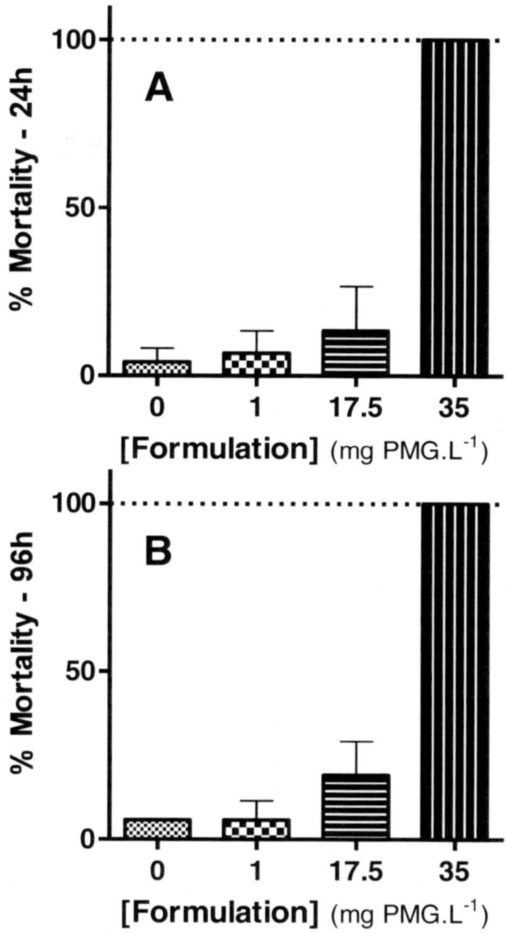

Figure 5. Mortality of *C. decemmaculatus* exposed for 96 h to glyphosate based formulation (Glyphosate II - Atanor) and control group (0). A: % cumulative mortality at 24 h. B: % cumulative mortality at 96 h. Data as average mortality in percentage ± standard error of the mean (SEM), n = 10-25.

It is noteworthy that 100% mortality occurred during the first 24 h of exposure (Figure 5 - A) for the formulated concentration corresponding to 35 mg.L^{-1}. This behaviour was reproduced in all trials. These results indicate a marked increased in toxicity of the formulation, when compared to the active ingredient.

Several authors have noted that PMG formulations containing the active principle together with coadjuvants as POEA present a significant increase in toxicity to aquatic organisms (algae, zooplankton, amphibians and fish) (Pérez et al. 2011).

The LC_{50}-96h of formulated glyphosate (Roundup) for various fish species are available: *Rhamdia quelen,* 7.3 $mg.L^{-1}$ (Kreutz et al., 2008); *Oncorhynchus mykiss,* 8.2-27 $mg.L^{-1}$ (Giesy et al., 2000); *Cyprinus carpio,* 10 $mg.L^{-1}$ (Giesy et al., 2000); *Prochilodus lineatus,* 13.7 $mg.L^{-1}$ (Langiano and Martinez, 2008); *Gambusia yucatana,* 17.8 $mg.L^{-1}$ (Osten et al., 2005); *Leporinus obtusidens,* >100 $mg.L^{-1}$ (Glusczak et al., 2006). Although, the chemical composition of the formulation of Glyphosate II (Atanor) was unknown, the results presented are consistent with the literature, confirming the higher acute toxicity of the formulation when compared to the active ingredient. It is interesting that in this case the LC_{50}-96h for *C. decemmaculatus* exposed to the formulation was 29 $mg.L^{-1}$, while the same parameter for the active ingredient was >140 $mg.L^{-1}$ (Menéndez-Helman, 2013).

C. decemmaculatus was placed among species with higher LC_{50}-96h, which means, the more tolerant for both the active ingredient and the formulation. This fact is consistent with the wide distribution faced by the fish waterbodies associated to agroecosystems of the Rolling Pampa Region where Peruzzo et al. (2008) reported herbicide concentrations between 0.1 and 0.7 $mg.L^{-1}$ and Ronco (2011) informed 10.9 $mg.L^{-1}$ on tributary of Pescado Stream, Buenos Aires Province.

It also worth mentioning that analytical concentration of PMG was established by ion cromatography in different trials. These concentrations were very close to those expected, and no significant differences were found in the concentrations in the course of the assays, from the initial time. Consequently, the herbicide (active ingredient or formulated) is stable in the media testing and the experimental conditions. Thus, hydrolysis or bioaccumulation processes at a magnitude capable of diminishing the concentration of the compound in these conditions was discarded.

3. Sublethal Effects of Glyphosate on Biomarkers of Environmental Stress in Fish

The pollutants effects on aquatic biota, particularly on fish, are varied: carcinogenic, hepatotoxic, neurotoxic, cytotoxic, oxidative stress promoters, endocrine disruptors, among others. These are revealed via a set of symptoms that are together known as stress expression, and which are able to keep the organisms away from the normal homeostatic levels. These alterations are reflected through changes in biomarkers. The term *biomarker* (WHO, 1993) includes, in a broad sense, parameters whose alteration reflects the interaction between a particular biological system and an environmental stressor of different character (chemical, physical, biological, including synergy of several of them) (Timbrell et al., 1996; Lagadic et al., 1997; Conti, 2008; Schlenk et al., 2008). At the same time, it is possible to quantify the degree of stress from the magnitude of changes in selected variables (biomarkers). Thus disturbances in these parameters become reliable warning signals of an impairment degree of a particular environment, the result being integrated responses of multiple changes that occur in the *test* organisms.

Biomarkers of Oxidative Stress and Antioxidant Defenses

The organisms are exposed to a large number of xenobiotics, that once absorbed by the body, can be bioaccumulated and affect its functional balance. As the absorption and distribution, biotransformation phenomena also occur through cellular mechanisms, using the same biochemical machinery that metabolize the endogenous compounds (sometimes with chemical structures similar to those of environmental toxics). The aim of the biotransformation is the xenobiotic metabolizing, in almost all cases enzymatically catalyzed, for conversion into a less toxic or more easily extractable substance or molecular complex.

Reactive oxygen species (ROS) are often generated during the biotransformation processes, ROS are also called reactive oxygen intermediaries, oxide free radicals or radicals that are the product of the reduction of molecular oxygen (O_2) to superoxide radicals ($O_2^{\bullet}$), hydrogen peroxide (H_2O_2) and hydroxyl radical ($OH^{\bullet}$). The latter intermediate is a strong oxidant capable of reacting with cellular macromolecules causing various effects such as enzymatic inactivation, membrane lipid peroxidation (mitochondria, lysosomes, endoplasmic reticulum), DNA damage and ultimately cellular death (Helmut, 1991; Mitchelmore et al., 1998; van der Oost et al., 2003). All these effects are generally known as *oxidative stress* (Lushchak, 2011).

Oxidative stress is a cellular response to stressors, caused by an imbalance between the production of those high oxidative potential molecules, oxygen-derived, and the ability of a particular biological system to quickly detoxify the intermediates generated or repair the resulting damage (Repetto Jiménez and Repetto Kuhn, 2009; Di Giulio and Meyer, 2008).

The antioxidant defense system which tends to inhibit the formation of oxyradicals, include antioxidant enzymes such as superoxide dismutase (SOD), catalase (CAT), glutathione peroxidase (GPX) and glutathione reductase (GRED) (van der Oost et al., 2003). Other nonenzymatic defenses include liposoluble vitamins (N-tocopherol and β-carotene), and low molecular weight molecules (ascorbic acid, catecholamines, glutathione) (Helmut, 1991; Winston and Di Giulio, 1991; van der Oost et al., 2003).

The oxidative stressors effect can be assessed by monitoring changes in the magnitude of some critical biomarkers. These may be those that show the effects of oxidative stress on the macromolecules or antioxidant response, such as enzymes of the antioxidant system whose synthesis or activity may be modified as a consequence of the imbalanced intracellular redox state.

There are some reports relating to the effects of oxidative stress and antioxidant responses in fish exposed to PMG formulated (Table 1). Some of them describe the effects on catalase (CAT) activity, an enzyme of the antioxidant defense system. Lushchak et al. (2009) found an increase in CAT in liver and kidney of *Carassius auratus* exposed to Roundup (2.5-20 $mg.L^{-1}$), while Langiano and Martínez (2008) found that the exposure to 10 $mg.L^{-1}$ of Roundup also caused increased liver CAT in *Prochilodus lineatus*.

Table 1. Effects of oxidative stress and antioxidant defenses for acute exposure of fish to PMG formulated

Bibliographic citations	***Test* species**	**Formulation** (tested concentrations)	**Oxidative stress**		**Antioxidant Defenses**						
			Lipid Peroxidation	DNA Damage	Tissue	SOD	CAT	GPx	GST	GRED	GSH
Glusczak *et al.* (2007)	*Rhamdia quelen*	Roundup (0.2 and 0.4 mg.L^{-1})	No changes - Liver ↑ Muscle ↓ Brain								
Langiano and Martinez (2008)	*Prochilodus lineatus*	Roundup (7.5 and 10 mg.L^{-1})			Liver		↑ (24 h)		No change		
Lushchak *et al.* (2009)	*Carassius auratus*	Roundup (2.5 – 20 mg.L^{-1})	No changes - Liver ↑ Kidney No changes - Brain		Liver Kidney Brain	↓ ↓ ↓	↑ ↑ No change		↓	↓ ↓ ↓	↑ No change ↓
Modesto and Martinez (2010a)	*Prochilodus lineatus*	Roundup Transorb (1 and 5 mg.L^{-1})	↑ (6 h) Liver		Liver	↓ (6 h)	↓ (6 and 24 h)	↑ (24 and 96 h)	↓ (6 and 24 h)		↓24h ↑96h
Modesto and Martinez (2010b)	*Prochilodus lineatus*	Roundup (10 mg.L^{-1})			Liver	↓ (24 h)	No change	↓ (6 and 24 h)	↑ (24 and 96 h)		↑ (6 and 24 h)
Guilherme *et al.* (2010)	*Anguilla anguilla*	Roundup (0.058 and 0.116 mg.L^{-1})	↑ (24 h) erythrocytes	+	Blood		No change	No change	No change	No change	No change
Ortiz-Ordoñez *et al.* (2011)	*Goodea atripinnis*	Yerbimat (4 and 8 mg.L^{-1})	↑ Liver ↑ Gills		Liver Gills		↑ ↓				
Rossi *et al.* (2011)	*Astyanax sp*	Roundup (1 and 2 mg.L^{-1})	↑ Liver	+	Liver		No change		No change		
Cattaneo *et al.* (2011)	*Cyprinus carpio*	Roundup (0.5 – 10 mg.L^{-1})	↑ Brain								

Roundup Transorb (a PMG formulation) showed a reduction of CAT in the liver after a short exposure time (6 h), in parallel with the appearance of lipid peroxidation in juvenile *P. lineatus* (Modesto and Martínez, 2010a). Also, a PMG formulation widely used in Mexico caused CAT reduction in gill and increased liver enzyme in *Goodea atripinnis* (Ortiz-Ordoñez et al., 2011). In addition, *Astyanax sp* exposure to Roundup (1 and 2 mg.L^{-1}) did not produce significant changes in liver CAT (Rossi et al., 2011). At the same time, the overall analysis of the literature cited suggests that the response of antioxidant defenses is highly variable and strongly dependent on the species, the exposure time and analysed tissue.

Menendez-Helman (2013) measured the CAT activity in the midsection (M) of the body of *Cnesterodon decemmaculatus*; it was 36 ± 9 *μmoles.min^{-1}.(mg protein)$^{-1}$* (mean ± SEM) for the control specimens. The results obtained by exposure of *C. decemmaculatus* to PMG (active ingredient) are consistent with the response of the liver CAT observed by other authors for acute exposures to formulate. PMG exposure caused an increase in the enzyme activity within the range of 47-70% in M CAT, resulting in a significant increase at 35 mg PMG.L^{-1} (Figure 6). It is noteworthy that, despite an observed tendency of increased CAT activity at the lowest concentration (1 mg.L^{-1}), the differences were significant only for the highest tested concentration (35 mg.L^{-1}). Probably some formulations effects are due to the active substance, and that adjuvants have also an important role as promoters of oxidative stress and antioxidant responses. However it should be noted that there are no previous reports in relation to an antioxidant response in fish by acute exposure to PMG.

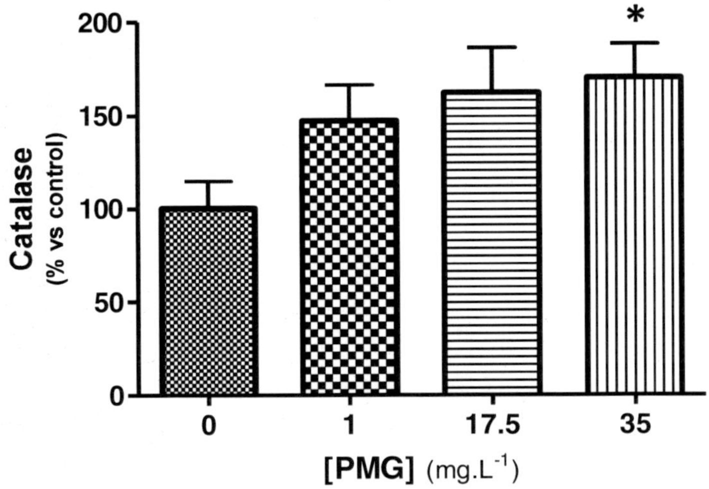

Figure 6. Specific activity of catalase in the midsection of *C. decemmaculatus* exposed to PMG (active ingredient) in acute test (96 h). Values are expressed as average ± standard error of the mean (SEM) (n=9-10) as a percentage of the control group. * $p < 0.05$ versus control group.

Biomarkers of Neurotoxicity: Acetylcholinesterase

Acetylcholinesterase (AChE), belongs to the family of esterases, and is responsible for the hydrolysis of acetylcholine, a neurotransmitter present in the synapses of the nervous system in both vertebrates and invertebrates. It has been adopted as a biomarker of changes occurred in synaptic junctions in the CNS, neuromuscular cholinergic and sympathetic, and may affect locomotion and balance of animals. AChE activity depends on the amount and the specific tissue modulation by phosphorylation of serine in the active site, the phosphorylated enzyme is very stable. It is noteworthy that there are no known endogenous inhibitors of this enzyme.

Furthermore, Behra et al. (2002) have demonstrated a nonclassical function of the enzyme, it is required for neuronal and muscular development in zebrafish embryos (*Danio rerio*), suggesting that exposure of fish to toxicants that alters functionality of AChE during early development, could over time generate alterations in the structure of the population.

Whereas AChE plays an extremely important role for many physiological functions of fish, its determination is useful in ecotoxicological studies (Bradbury et al., 2008). The changes in the activity of AChE, which provides information about the effects of a stressor on the nervous system, may help to explain some abnormal behavior of fish such as loss of balance and variations in the pattern of locomotion (Tierney et al., 2007).

This enzyme is used as a sensitive biosensor of OPP exposure (particularly those containing the phosphate group in the structure) and carbamates. These herbicides are inhibitors of the enzyme activity by covalent bonding to the active sites thereof (Thompson, 1999; Soreq and Seidman, 2001; van Dyk and Pletschke, 2011). The inhibition leads to an accumulation of acetylcholine in the synaptic cleft, thereby causing over-stimulation of the postsynaptic membrane, a process which can lead to death.

Recently, it has been described that acute exposure to glyphosate formulations can alter the activity of AChE. In *Leporinus obtusidens* exposed for 96 h to Roundup concentrations, in the range of 3-20 $mg.L^{-1}$, AChE activity was significantly decreased in the brain, whereas no changes were found in the muscle (Glusczak et al., 2006). In a chronic exposure study (90 days) of *Leporinus obtusidens* to Roundup, the concentration of 5 $mg.L^{-1}$ formulated produced brain AChE inhibition, while the muscle enzyme was not affected (Salbego et al., 2010). This differential response could be interpreted as a consequence of the different affinity of the enzyme in each tissue. A similar result was reported for *Rhamdia quelen* after exposure to 0.2 and 0.4 $mg.L^{-1}$ Roundup (Glusczak et al., 2007). No variations were determined by Rossi et al. (2011) in muscle of *Astyanax sp* exposed to 1 and 2 $mg.L^{-1}$ Roundup. In *Prochilodus lineatus*, AChE activity in the brain was significantly reduced by acute exposure to 1 and 5 $mg.L^{-1}$ RoundupTransorb formulated while in the muscle a significant decrease was only established for the highest concentration of the formulation (Modesto and Martínez, 2010a). In this species exposed to 10 $mg.L^{-1}$ of the formulation Roundup, the inhibition in both tissues was determined (Modesto and Martínez, 2010b), whereas the same pattern was found in *Cyprinus carpio* after acute exposures to Roundup between 0.5-20 $mg.L^{-1}$ (Cattaneo et al., 2011).

Although the PMG as active ingredient does not present a significant level of lethal toxicity to *Cnesterodon decemmaculatus* its important to analyze the existence of sublethal effects, specially considering that different studies reported sublethal effects of their formulations, as previously described. AChE activity in anterior (A) (mainly corresponding to the brain), posterior (P) (mainly containing muscle tissue) and midsection (M) (viscera) of the control group was 257 ± 19, 452 ± 85 and 343 ± 27 $nmoles.min^{-1}.(mg\ protein)^{-1}$ (mean ± SEM), respectively.

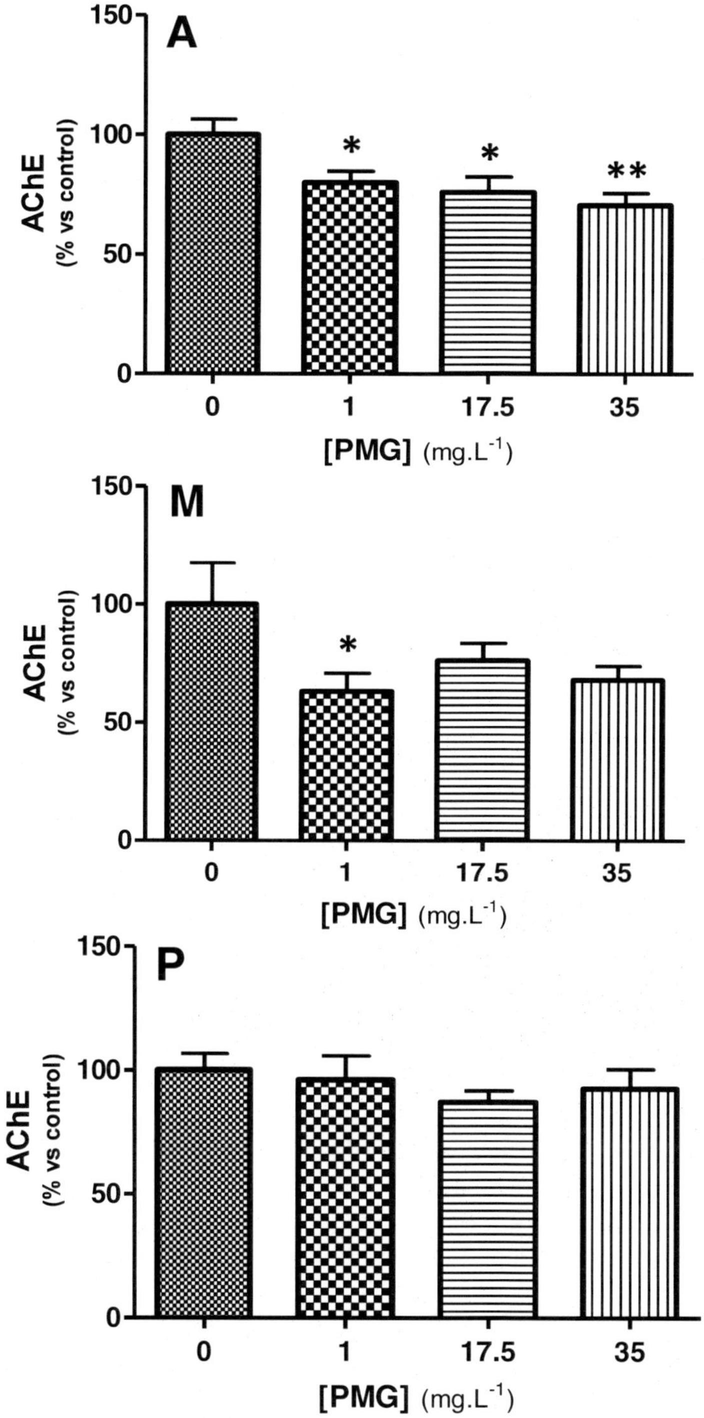

Figure 7. Specific activity of AChE in anterior (A), middle (M) and posterior (P) sections of *C. decemmaculatus* exposed to PMG (active ingredient) for 96 h. Values are expressed as average ± standard error of the mean (SEM) (n =14-15) as a percentage of the control group. *$p<0.05$, **$p<0.01$ versus control group.

Figure 7 shows the specific activities of AChE for each third of the animals exposed to different concentrations of PMG, expressed as percentages relative to controls.

Sections A and M showed a trend in the decrease of the AChE activity by exposure to PMG. In M, there was a decrease of AChE around 24 - 37%. However, the specific activity values showed greater variability in this section, which could explain why the observed differences were not statistically significant in all cases. In A, the activity decrease was in the range of 20-30% compared to controls by exposure to PMG, and differences were significant for all concentrations tested, confirming results previously published (Menendez-Helman et al., 2012). In contrast, the herbicide did not affect the AChE activity in P, which could suggest a different sensitivity respect to glyphosate, depending on the studied tissue.

It should be noted that in all of these acute PMG exposure bioassays its analytical concentrations were also determined in test media, evidencing that herbicide concentrations did not change significantly and confirming the compound stability in test conditions.

The objective proposed in this work was to discuss the ecotoxicological aspects associated with the expansion of the agricultural frontier, which has developed with technology package involving direct seeding, the introduction of genetically modified seeds and the massive use of chemicals (fertilizers and pesticides). Particularly, glyphosate (PMG), the herbicide mostly used today in Argentina and its impacts on *Cnesterodon decemmaculatus,* a freshwater fish, were studied. The results denoted low toxicity of the active ingredient, suggesting that PMG would not present a significant level of acute toxicity to freshwater fish (LC_{50}-96h > 100 $mg.L^{-1}$). However, the formulation (Glyphosate II Atanor) provoked 100% fish mortality after a few hours of applications for the concentration of 35 $mg.L^{-1}$. The LC_{50}-96h determined was 29 $mg.L^{-1}$, in agreement with the results obtained by other authors that indicate a significant increase in acute toxicity in relation to the active ingredient. In addition, the PMG effects on biomarkers, are complementary to those mentioned by previous reports, and also meant the first antecedent of the inhibitory effect of the PMG (active ingredient) on the AChE activity and on the antioxidant defences in *Cnesterodon decemmaculatus*.

REFERENCES

Adámoli J., Ginzburg R. & Torrela S. -colaboracion de Narbaiza C.- (2011). Informe "Escenarios productivos y ambientales del Chaco Argentino: 1977-2010" FCEyN - UBA y Fundación Producir Conservando. http://www.producirconservando.org.ar/documentos/escenarios_productivos_ambientales_chaco.pdf

Agostini, M.G. (2013). *Ecotoxicologia de anfibios en agroecosistemas del noreste de la region Pampeana*. Doctoral Dissertation. Faculty of Exact Sciences, National University of La Plata, Argentina.

Ahrens W.H. (Editor) (1994). Herbicide Handbook 7th Edition. Weed Science Society of America. Illinois, USA.

Alvarez M., Gimenez I.T., Saitua H., Enriz R.D. & Giannini F.A.(2012). Toxicidad en peces de herbicidas formulados con glifosato. *Acta Toxicológica Argentina*, 20, 5-13.

Arregui M.C., Lenardón A., Sanchez D., Maitre M.I., Scotta R. & Enrique S. (2004). Monitoring glyphosate residues in transgenic glyphosate-resistant soybean. *Pest Management Science*, 60, 163-166.

Barja B.C. & dos Santos Afonso M. (1998). An ATR-FTIR study of glyphosate and its Fe(III) complex in aqueous solution. *Environmental Science and Technology*, 32, 3331-3335.

Barja B.C. & dos Santos Afonso M. (2005). Aminomethylphosphonic acid and glyphosate adsorption onto goethite: A comparative study. *Environmental Science and Technology*, 39, 585-592.

Barja B.C., Herszage J. & Dos Santos Afonso M. (2001). Iron(III)-phosphonate complexes. *Polyhedron*, 20, 1821-1830.

Bedmar F. (2011). ¿Qué son los plaguicidas?. *Ciencia Hoy* 21, 10-16.

Behra M., Cousin X., Bertrand C., Vonesch J.L., Biellmann D., Chatonnet A. & Strähle U. (2002). Acetylcholinesterase is required for neuronal and muscular development in the zebrafish embryo. *Nature Neuroscience*, 5, 111-118.

Boudou A. & Ribeyre F. (1997). Aquatic ecotoxicology: from the ecosystem to the cellular and molecular levels. *Environmental Health Perspectives,* 105 (Suppl 1), 21-35.

Bradbury S.P., Carlson, R.W., Henry, T.R., Padilla, S. & Cowden, J. (2008).Toxic responses of the fish nervous system. In: Di Giulio, R.T., Hinton, D.E. (Eds), *The Toxicology of Fishes*, CRC Press-Taylor & Francis Group, Boca Raton, FL, 417-455.

Brodeur J.C. (2011). La soja y sus agroquímicos: evaluando impactos en anfibios. In: Camino M, Aparicio V (Eds), *Aspectos ambientales del uso del glifosato.* Ediciones INTA (Argentina), 69-76.

Carrasco A.E., Sánchez N.E. & Tamagno L.E. (2012). Modelo agrícola e impacto socioambiental en la Argentina: monocultivo y agronegocios. Serie monográfica Sociedad y Ambiente, N° 1. Ed. Asociación de Universidades Grupo Montevideo-UNLP (Uruguay-Argentina).

Carriquiriborde P. (2011). Toxicidad de glifosato en peces autóctonos: Estudios de laboratorio y campo. In: Camino M, Aparicio V (Eds), *Aspectos ambientales del uso del glifosato.* Ediciones INTA (Argentina), 57-68.

Carvalho F., Zhong N., Tavarez & Klaine S. (1998). Rastreo de plaguicidas en los trópicos. Boletín del OEIA 40.

Casabé N., Piola L., Fuchs J., Oneto M.L., Pamparato L., Basack S., Giménez R., Massaro R., Papa J.C. & Kesten E. (2007). Ecotoxicological assessment of the effects of glyphosate and chlorpyrifos in an Argentine soya field. *Journal of Soils and Sediments*, 7, 232-239.

CASAFE. (2012). Informe de Mercado Argentino de Fitosanitarios, año 2011. Cámara de Sanidad Agropecuaria y Fertilizantes. http://www.casafe.org/pdf/Informemercadofito sanitarios2011.pdf

Castillo Morales G. (Editora) (2004). Ensayos toxicológicos y métodos de evaluación de calidad de aguas. Estandarización, intercalibración, resultados y aplicaciones. México-IMTA/Canadá-IDRC, 1-189.

Cattaneo R., Clasen B., Loro V.L., De Menezes C.C., Pretto A., Baldisserotto B., Santi A. & De Avila L.A. (2011). Toxicological responses of *Cyprinus carpio* exposed to a commercial formulation containing glyphosate. *Bulletin of Environmental Contamination and Toxicology*, 87, 597-602.

Çavaş T. & Könen S. (2007). Detection of cytogenetic and DNA damage in peripheral erythrocytes of goldfish (*Carassius auratus*) exposed to a glyphosate formulation using the micronucleus test and the comet assay. *Mutagenesis*, 22, 263-268.

Chambers W.H. (1992). Organophosphorus Compounds: An Overview. In: J. E. Chambers, Levi, P.E. (Eds), *Organophosphates, Chemistry, Fate, and Effects*. San Diego, Academic Press: 3–17.

CONICET. Comisión Nacional de Investigación sobre Agroquímicos. Consejo Científico Interdisciplinario creado en el ámbito del Consejo Nacional de Investigaciones Científicas y Técnicas. (2009). "Informe - evaluación de la información científica vinculada al glifosato en su incidencia sobre la salud humana y el ambiente". Ciudad Autónoma de Buenos Aires.

Contardo-Jara V., Klingelmann E. & Wiegand C. (2009). Bioaccumulation of glyphosate and its formulation Roundup Ultra in *Lumbriculus variegatus* and its effects on biotransformation and antioxidant enzymes. *Environmental Pollution*, 157, 57-63.

Conti M.E. (2008) – Biomarkers for environmental monitoring. In: Conti M.E. (Editor), *Biological monitoring: Theory and applications.* Wit Press, 25-46.

Dallegrave E., DiGiorgio Mantese F., Soares Coelho R., Drawans Pereira J. Dalsenter P.R. & Langeloh A. (2003). The teratogenic potential of the herbicide glyphosate-Roundup in Wistar rats. *Toxicology Letters*, 142, 45-52.

Damonte M., Torres Sánchez R.M. & dos Santos Afonso M. (2007). Some aspects of the glyphosate adsorption on montmorillonite and its calcined form. *Applied Clay Science*, 36, 86-94.

de la Torre F.R., Ferrari L. & Salibián A. (2002). Freshwater pollution biomarker: response of brain acetylcholinesterase activity in two fish species. *Comparative Biochemistry and Physiology* 131 *C*, 271-280.

de la Torre F.R., Salibián A. & Ferrari L. (2007). Assessment of the pollution impact on biomarkers of effect of a freshwater fish. *Chemosphere*, 68, 1582-1590.

Demetrio P.M. (2012). Estudio de efectos biológicos de plaguicidas utilizados en cultivos de soja RR y evaluación de impactos adversos en ambientes acuáticos de agroecosistemas de la región pampeana. Doctoral Dissertation. Faculty of Exact Sciences, National University of La Plata, Argentina.

Di Giulio, R.T. & Meyer, J.N. (2008). Reactive oxygen species and oxidative stress. In: Di Giulio, R.T., Hinton, D.E. (Editors), *The Toxicology of Fishes*, CRC Press-Taylor & Francis Group, Boca Raton, FL, 273-324.

Dinehart S.K., Smith L.M., McMurry S.T., Anderson T.A., Smith P.N. & Haukos D.A. (2009). Toxicity of a glufosinate- and several glyphosate based herbicides to juvenile amphibians from Southern High Plains, USA. *Science of the Total Environment* 407: 1065-1071.

Domitrovic H.A. (1997). Toxicidad y respuesta histopatológica en *Aequidens portalegrensis* (Pisces, Cichlidae) expuestos a Glifosato en ensayos de toxicidad aguda y en ensayos subletales. Univ Nac Nordeste (Argentina), *Actas Reunión Comunicaciones Científicas Tecnológicas*. Tomo II: 4-29/4-32.

Duke, S.O. (1988). Glyphosate. In: Kartney P.C. & Kaufman D.D. (Eds.), *Herbicides: Chemistry, degradation and mode of action*. Marcel Dekker, New York, 1-70.

Ecobichon D.J. (2005). Efectos tóxicos de los pesticidas. In: Klaassen CD, Watkins JB, (Editores). Madrid: McGraw-Hill. p. 339-53.

El-Gendy K.S., Aly N.M. & El-Sebae A.H. (1998). Effects of Edifenphos and glyphosate on the immune response and protein biosynthesis of bolti fish (*Tilapia nilotica*). *Journal of Environmental Science and Health* B 33: 135-149.

FAO (2002). Manual on development and use of FAO and WHO specifications for pesticides. Rome. Plant Production and Protection Paper N° 173. http://www.fao.org/ag/agp/agpp/pesticid/

Ferrari L., García M.E., de la Torre F.R. & Demichelis S.O. (1998). Evaluación ecotoxicológica del agua de un río urbano mediante bioensayos con especies nativas. *Revista Museo Argentino de Ciencias Naturales "Bernardino Rivadavia"*, Nueva Serie N° 148, 1-16.

Folmar L.C., Sanders H.O. & Julin A.M. (1979). Toxicity of the herbicide glyphosate and several of its formulations to fish and aquatic invertebrates. *Archives of Environmental Contamination and Toxicology*, 8, 269-278.

García M.E., Cappelletti C.A. & Salibián A. (1999). Sublethal maternal pre-exposure of fish to cadmium. Effect on the survival of the newly hatched alevins. *Archives of Physiology and Biochemistry*, 107, 152-158.

Giesy J.P., Dobson S. & Solomon K.R. (2000). Ecotoxicological risk assessment for Roundup® herbicide. *Reviews of Environmental Contamination and Toxicology*, 167, 35-120.

Gimsing A.L. & dos Santos Afonso M. (2005). Glyphosate. In: van Briesen J.M. & Nowack B. (Eds), *Biogeochemistry of Chelating Agents*, ACS Symposium Series Volume 910, 263-277.

Glusczak L., dos Santos Miron D., Crestani M., Braga da Fonseca M., Araújo Pedron F.d., Duarte M.F. & Vieira V.L.P. (2006). Effect of glyphosate herbicide on acetylcholinesterase activity and metabolic and hematological parameters in piava (*Leporinus obtusidens*). *Ecotoxicology and Environmental Safety*, 65, 237-241.

Glusczak L., Miron D.d.S., Moraes B.S., Simões R.R., Schetinger M.R.C., Morsch V.M. & Loro V.L. (2007). Acute effects of glyphosate herbicide on metabolic and enzymatic parameters of silver catfish (*Rhamdia quelen*). *Comparative Biochemistry and Physiology,* 146C, 519-524.

Gómez S., Villar C. & Bonetto C. (1998). Zinc toxicity in the fish *Cnesterodon decemmaculatus* in the Paraná River and Río de La Plata Estuary. *Environmental Pollution*, 99, 159-165.

Guilherme S., Gaivão I., Santos M.A. & Pacheco M. (2010). European eel (*Anguilla anguilla*) genotoxic and pro-oxidant responses following short-term exposure to Roundup® - A glyphosate-based herbicide. *Mutagenesis*, 25, 523-530.

Guilherme S., Santos M.A., Barroso C., Gaivvao I. & Pacheco M. (2012). Differential genotoxicity of Roundup formulation and its constituents in blood cells of fish (*Anguilla anguilla*): considerations on chemical interactions and DNA damaging mechanisms. *Ecotoxicology* 21, 1381-1390.

Helmut S. (1991). Oxidative Stress, Oxidants and Antioxidants. Academic Press, London.

Hildebrand L.D., Sullivan D.S. & Sullivan T.P. (1982). Experimental studies of rainbow trout populations exposed to field applications of Roundup® herbicide. *Archives of Environmental Contamination and Toxicology*, 11, 93-98.

Howe C.M., Berrill M., Pauli B.D., Helbing C.C., Werry K. & Veldhoen N. (2004). Toxicity of glyphosate-based pesticides to four North American frog species. *Environmental Toxicology and Chemistry*, 23, 1928-1938.

Jiraungkoorskul W., Upatham E.S., Kruatrachue M., Sahaphong S., Vichasri-Grams S. & Pokethitiyook P. (2003). Biochemical and histopathological effects of glyphosate herbicide on Nile tilapia (*Oreochromis niloticus*). *Environmental Toxicology*, 18, 260-267.

Kelly D.W., Poulin R., Tompkins D.M. & Townsend C.R. (2010). Synergistic effects of glyphosate formulation and parasite infection on fish malformations and survival. *Journal of Applied Ecology,* 47, 498-504.

Khoury G.A., Gehris T.C., Tribe L., Torres Sánchez R.M. & dos Santos Afonso M. (2010). Glyphosate adsorption on montmorillonite: An experimental and theoretical study of surface complexes. *Applied Clay Science*, 50, 167-175.

Kier L.D. & Kirkland D.J. (2013). Review of genotoxicity studies of glyphosate and glyphosate-based formulations. *Critical Reviews in Toxicology* 43, 283-315.

Kreutz L.C., Barcellos L.J.G., Silva T.O., Anziliero D., Martins D., Lorenson M., Marteninghe A. & Da Silva L.B. (2008). Acute toxicity test of agricultural pesticides on silver catfish (*Rhamdia quelen*) fingerlings. *Ciencia Rural*, 38, 1050-1055.

Lagadic L., Caquet Th. & Amiard J.C. (1997). Biomarqueurs en écotoxicologie: principles et definitions. En: Lagadic et al. *Biomarqueurs en Écotoxicologie. Aspects Fondamentaux.* Masson, Paris, 1-9.

Lajmanovich R.C., Sandoval M.T. & Peltzer P.M. (2003). Induction of mortality and malformation in *Scinax nasicus* tadpoles exposed to glyphosate formulations. *Bulletin of Environmental Contamination and Toxicology*, 70, 612-618.

Langiano V.d.C. & Martinez C.B.R. (2008). Toxicity and effects of a glyphosate-based herbicide on the Neotropical fish *Prochilodus lineatus*. *Comparative Biochemistry and Physiology,* 147C, 222-231.

López S.L., Alassa D., Benítez-Leite S., Lajmanovich R., Mañas F., Poletta G., Sánchez N., Simoniello M.F. & Carrasco A.E. (2012). Pesticide used in South American GMO-based agriculture: a review of their effects on humans and animal models. *Advances in Molecular Toxicology* 6: 41-75.

Lorenzatti E., Maitre M.I., Argelia L., Lajmanovich R., Peltzer P. & Anglada M. (2004). Pesticide residues in immature soybeans of Argentina croplands. *Fresenius Environmental Bulletin*, 13, 675-678.

Lushchak O.V., Kubrak O.I., Storey J.M., Storey K.B. & Lushchak V.I. (2009). Low toxic herbicide Roundup induces mild oxidative stress in goldfish tissues. *Chemosphere,* 76, 932-937.

Lushchak O.V. (2011). Environmentally induced oxidative stress in aquatic animals. *Aquatic Toxicology*, 101, 13-30.

MAGyP, Ministerio de Agricultura, Ganadería y Pesca de la Nación. Sistema integrado de información agropecuaria. (2012). Site: http://old.siia.gov.ar/

Malik J., Barry G. & Kishore G. (1989). The herbicide glyphosate. *BioFactors*, 2, 17-25.

Mann R.M., Hyne R.V., Choung C.B. & Wilson S.P. (2009). Amphibians and agricultural chemicals: Review of the risks in a complex environment. *Environmental Pollution*, 157, 2903-2927.

Mañas F., Peralta L., Raviolo J., García Ovando H., Weyers A., Ugnia L., González Cid M., Larripa I. & Gorla N. (2009). Genotoxicity of AMPA, the environmental metabolite of glyphosate, assessed by the Comet assay and cytogenetic tests. *Ecotoxicology and Environmental Safety* 72, 834-837.

Martínez-Ghersa M.A. (2011). Consecuencias ambientales del uso de pesticidas. *Ciencia Hoy* 21, 30-35.

Menéndez-Helman R.J., Ferreyroa G.V., dos Santos Afonso M. & Salibián A. (2012) Glyphosate as an Acetylcholinesterase Inhibitor in *Cnesterodon decemmaculatus*. *Bull Environ Contam Toxicol* 88:6–9.

Menéndez-Helman R.J. (2013). *Efectos del glifosato en peces dulceacuícolas. Biomarcadores de estrés ambiental.* Doctoral Dissertation. Faculty of Exact and Natural Sciences, University of Buenos Airres, Argentina.

Mermoz M.L.A. & Salibián A. (1992). Dinámica de una población de alevinos de *Cnesterodon sp.* criada en cautiverio: resultados preliminares. *V Jornada Científica Ecología, Depto Cs Básicas-UNLuján*. Serie Resúmenes 2 (2).

Mitchelmore C.L., Birmelin C., Chipman J.K. & Livingstone D.R. (1998). Evidence for cytochrome P-450 catalysis and free radical involvement in the production of DNA strand breaks by benzo[a]pyrene and nitroaromatics in mussels (*Mytilus edulis*) digestive gland cells. *Aquatic Toxicology*, 41, 193-212.

Modesto K.A. & Martinez C.B.R. (2010a). Effects of Roundup Transorb on fish: Hematology, antioxidant defenses and acetylcholinesterase activity. *Chemosphere*, 81, 781-787.

Modesto K.A. & Martinez C.B.R. (2010b). Roundup® causes oxidative stress in liver and inhibits acetylcholinesterase in muscle and brain of the fish *Prochilodus lineatus*. *Chemosphere*, 78, 294-299.

Molero A.M. & Pisanó A. (1986). Different susceptibility to lindane toxicity between male and female of *Cnesterodon decemmaculatus*. *Comunicaciones Biológicas*, 4, 377-386.

Morillo E., Undabeytia T. & Maqueda C. (1997). Adsorption of glyphosate on the clay mineral montmorillonite: effect of Cu (II) in solution and adsorbed on the mineral. *Environmental Science and Technology* 31 (12), 3588-3592

Mousdale D.M. & Coggins J.R. (1984). Purification and properties of 5-enolpyruvylshikimate 3-phosphate synthase from seedlings of *Pisum sativum* L. Planta, 160, 78-83.

Nešković N.K., Poleksić V., Elezović I., Karan V. & Budimir M. (1996). Biochemical and histopathological effects of glyphosate on carp, *Cyprinus carpio*. *Bulletin of Environmental Contamination and Toxicology*, 56, 295-302.

Newman M.C. & Clements W.H. (2008). Ecotoxicology. A comprehensive treatment. CRC Press-Taylor & Francis Group. Boca Raton FL.

Ortiz-Ordoñez E., Uría-Galicia E., Ruiz-Picos R.A., Sánchez Duran A.G., Hernández Trejo Y., Sedeño-Díaz J.E. & López-López E. (2011). Effect of Yerbimat herbicide on lipid peroxidation, catalase activity, and histological damage in gills and liver of the freshwater fish *Goodea atripinnis*. *Archives of Environmental Contamination and Toxicology*, 61, 443-452.

Osten J.R. v-, Ortíz-Arana A., Guilhermino L. & Soares A.M.V.M. (2005). In vivo evaluation of three biomarkers in the mosquitofish (*Gambusia yucatana*) exposed to pesticides. *Chemosphere*, 58, 627-636.

Paganelli A, Gnazzo V, Acosta H, López SL & Carrasco AE. (2010). Glyphosate-based herbicides produce teratogenic effects on vertebrates by impairing retinoic acid signaling. *Chemical Research in Toxicology* 23, 1586-1595.

Pengue W.A. (compilador) (2008). *La apropiación y el saqueo de la naturaleza. Conflictos ecológicos distributivos en la Argentina del Bicentenario*, Lugar Editorial, Buenos Aires.

Pérez G.L., Torremorell A., Mugni H., Rodríguez P., Solange Vera M., Do Nascimento M., Allende L., Bustingorry J., Escaray R., Ferraro M., Izaguirre I., Pizarro H., Bonetto C., Morris D.P. & Zagarese H. (2007). Effects of the herbicide roundup on freshwater microbial communities: A mesocosm study. *Ecological Applications*, 17, 2310-2322.

Pérez G.L., Vera M.S. & Miranda L. (2011). Effects of Herbicide Glyphosate and Glyphosate-Based Formulations on Aquatic Ecosystems. In: Kortekamp A (Ed), *Herbicides and Environment*. InTech, 343-368.

Peruzzo P.J., Porta A.A. & Ronco A.E. (2008). Levels of glyphosate in surface waters, sediments and soils associated with direct sowing soybean cultivation in north pampasic region of Argentina. *Environmental Pollution*, 156, 61-66.

Pessagno R.C., dos Santos Afonso M. & Torres Sanchez R.M. (2005). N-phosphonomethylglycine interactions with soils. *Journal of the Argentine Chemical Society*, 93, 97-108.

Pessagno R.C. & dos Santos Afonso M. (2006). Estudio comparativo del impacto ambiental de tres herbicidas de uso común en cultivos de soja y trigo. In: Gallardo Lancho J.F.(Ed.), *Medioambiente en Iberoamérica. Visión desde la Física y la Química en los albores del siglo XXI*, Tomo III, 345-352.

Pessagno R.C., Torres Sánchez R.M. & dos Santos Afonso M. (2008). Glyphosate behavior at soil and mineral-water interfaces. *Environmental Pollution*, 153, 53-59.

Piccolo A., Celano G., Arienzo M. & Mirabella A. (1994). Adsorption and desorption of glyphosate in some European soils. *Journal of Environmental Science and Health*, 29B, 1105-1115.

PNUMA (2002). Perspectivas del medio ambiente mundial. GEO-3. Ed. Mundiprensa. ISBN: 84-8476-062-6.

Racke K.D. (1992). Degradation of organophosphorus insecticides in environmental matrices. In: Chambers JE, Levi PE (Eds.), *Organophosphates, Chemistry, Fate, and Effects*. San Diego, Academic Press, 47-78.

Relyea R.A. (2005). The impact of insecticides and herbicides on the biodiversity and productivity of aquatic communities. *Ecological Applications*, 15, 618-627.

Repetto Jiménez M. & Repetto Kuhn G. (2009). *Toxicología Fundamental*. Ediciones Díaz de Santos, Madrid.

Ringuelet R.A., Aramburu R.H. & Alonso de Aramburu A. (1967). *Los peces argentinos de agua dulce*. Ediciones CIC Provincia de Buenos Aires.

Romano M.A., Wisniewski P., Viau P., Romano R.M., Campos D.A., Bernardi M.M., Santos L.D., De Souza P.B., Nunes M.T. & De Oliveira C.A. (2012). Glyphosate impairs male offspring reproductive development by disrupting gonadotropin expression. *Archives of Toxicology*, 86, 663-673.

Romano R.M., Romano M.A., Bernardi M.M., Furtado P.V. & Oliveira C.A. (2010). Prepubertal exposure to commercial formulation of the herbicide glyphosate alters testosterone levels and testicular morphology. *Archives of Toxicology*, 84, 309-317.

Ronco A.E. (2011). Impacto de plaguicidas en ambientes acuáticos pampeanos: integración de estudios químico ecotoxicológicos en experimentos de campo y laboratorio, con especial énfasis al caso del glifosato. . In: Camino M, Aparicio V (Eds), *Aspectos ambientales del uso del glifosato.* Ediciones INTA (Argentina), 85-94.

Rossi S.C., Dreyer Da Silva M., Piancini L.D.S., Oliveira Ribeiro C.A., Cestari M.M. & Silva De Assis H.C. (2011). Sublethal effects of waterborne herbicides in tropical freshwater fish. *Bulletin of Environmental Contamination and Toxicology*, 87, 603-607.

Rubin J.L., Gainess C.G. & Jensen R.A. (1984). Glyphosate inhibition of 5-enolpyruvylshikimate–3-phosphate synthase from suspension-cultured cells of *Nicotina silvestris*. *Plant Physiology*, 75, 839-845.

Rueppel M.L., Brightwell B.B., Schaefer J. & Marvel J.T. (1977). Metabolism and degradation of glyphosate in soil and water. *Journal of Agricultural and Food Chemistry*, 25, 517-528.

Salbego J., Pretto A., Gioda C.R., De Menezes C.C., Lazzari R., Radünz Neto J., Baldisserotto B. & Loro V.L. (2010). Herbicide formulation with glyphosate affects growth, acetylcholinesterase activity, and metabolic and hematological parameters in Piava (*Leporinus obtusidens*). *Archives of Environmental Contamination and Toxicology*, 58, 740-745.

Salibián A. (2006). Ecotoxicological assessment of the highly polluted Reconquista River of Argentina. En: Ware GW (Editor), *Reviews of Environmental Contamination and Toxicology* 185, 35-65.

Schlenk D., Handy R., Steinert S., Depledge M.H. & Benson W. (2008). Biomarkers. En: Di Giulio R.T., Hinton D.E. (Eds), *The toxicology of fishes*. CRC Press – Taylor & Francis Group. Boca Raton. FL, 683-731.

Solomon K.R. & Thompson D.G. (2003). Ecological risk assessment for aquatic organisms from over-water uses of glyphosate. *Journal of Toxicology and Environmental Health,* 6B, 289-324.

Somma L.A., Mastrangelo M. & Ferrari L. (2011). *Manual de producción de Cnesterodon decenmaculatus (Jenyns, 1842) en Laboratorio*. Editorial Utopías, Buenos Aires.

Soreq H. & Seidman S. (2001). Acetylcholinesterase - New roles for an old actor. *Nature Reviews Neuroscience*, 2, 294-302.

Soso A.B., Barcellos L.J.G., Ranzani-Paiva M.J., Kreutz L.C., Quevedo R.M., Anziliero D., Lima M., Silva L.B.d., Ritter F., Bedin A.C. & Finco J.A. (2007). Chronic exposure to sub-lethal concentration of a glyphosate-based herbicide alters hormone profiles and affects reproduction of female Jundiá (*Rhamdia quelen*). *Environmental Toxicology and Pharmacology* 23, 308-313.

Subramaniam V. & Hoggard P.E. (1988). Metal Complexes of Glyphosate. *Journal of Agriculture and Food Chemistry*. 36 1326-1329

Szarek J., Siwicki A., Andrzejewska A., Terech-Majewska E. & Banaszkiewicz T. (2000). Effects of the herbicide Roundup(TM) on the ultrastructural pattern of hepatocytes in carp (*Cyprinus carpio*). *Marine Environmental Research*, 50, 263-266.

Takahashi M. (2007). Oviposition site selection: Pesticide avoidance by gray treefrogs. *Environmental Toxicology and Chemistry*, 26, 1476-1480.

Thompson H.M. (1999). Esterases as markers of exposure to organophosphates and carbamates. *Ecotoxicology*, 8, 369-384.

Tierney K., Casselman M., Takeda S., Farrell T. & Kennedy C. (2007). The relationship between cholinesterase inhibition and two types of swimming performance in chlorpyrifos-exposed coho salmon (*Oncorhynchus kisutch*). *Environmental Toxicology and Chemistry*, 26, 998-1004.

Timbrell J.A., Draper R. & Waterfield C.J. (1996). Biomarkers in toxicology: New uses for some old molecules? *Biomarkers*, 1, 1-11.

Timbrell J.A. (2009). *Principles of Biochemical Toxicology*, Informa Healthcare, New York.

Trinelli M.A., Areco M.M. & Dos Santos Afonso M. (2013) Co-biosorption of copper and glyphosate by *Ulva lactuca*. *Colloids Surf B Biointerfaces* 105: 251–258.

Trinelli M.A., Cantera G. & dos Santos Afonso M. (2012). Estudios de fotodegradación de glifosato en medio acuoso In: dos Santos Afonso, M and Torres Sanchez, R M (Eds.), *Ciencia y Tecnología Ambiental Un enfoque integrador*. Editorial AAPC. Buenos Aires.

Tsui M.T.K. & Chu L.M. (2003). Aquatic toxicity of glyphosate-based formulations: comparison between different organisms and the effects of environmental factors. *Chemosphere* 52, 1189-1197.

USEPA - U.S.Environmental Protection Agency (1999) Technical Fact Sheets on: Glyphosate. Primary Drinking Water regulations.

van der Oost R., Beyer J. & Vermeulen N.P.E. (2003). Fish bioaccumulation and biomarkers in environmental risk assessment: a review. *Environmental Toxicology and Pharmacology*, 13, 57-149.

van Dyk J.S. & Pletschke B. (2011). Review on the use of enzymes for the detection of organochlorine, organophosphate and carbamate pesticides in the environment. *Chemosphere*, 82, 291-307.

Vera M.S., Lagomarsino L., Sylvester M., Pérez G.L., Rodríguez P., Mugni H., Sinistro R., Ferraro M., Bonetto C., Zagarese H. & Pizarro H. (2010). New evidences of Roundup® (glyphosate formulation) impact on the periphyton community and the water quality of freshwater ecosystems. *Ecotoxicology*, 19, 710-721.

Verrell P. & van Buskirk E. (2004). As the worm turns: *Eisenia fetida* avoids soil contaminated by a glyphosate-based herbicide. *Bullettin of Environmental Contamination and Toxicology* 72, 219-224.

Villar C.A., Gómez S.E. & Bentos C.A. (2000). Lethal concentration of Cu in the neotropical fish *Cnesterodon decemmaculatus* (Pisces, Cyprinodontiformes). *Bulletin of Environmental Contamination and Toxicology*, 65, 465-469.

Walker C.H., Hopkin S.P., Sibly R.M. & Peakall D.B. (2006). *Principles of Ecotoxicology*. Taylor & Francis. Boca Raton, FL.

Wang Y.S., Jaw C.G. & Chen Y.L. (1994). Accumulation of 2,4-D and glyphosate in fish and water hyacinth. *Water, Air, and Soil Pollution*, 74, 397-403.

Weyland F., Poggio S.L. & Ghersa C.M. (2008). Agricultura y biodiversidad. *Ciencia Hoy* 18, 27-35.

WHO. (1993). Biomarkers and risk assessment: concepts and principles. *Environmental Health Criteria* 155. Geneva.

WHO. (1994). Glyphosate. *Environmental Health Criteria* 159. Geneva.

Winston G.W. & Di Giulio R.T. (1991). Prooxidant and antioxidant mechanisms in aquatic organisms. *Aquatic Toxicology,* 19, 137-161.

In: Handbook on Herbicides
Editors: Daiki Kobayashi and Eito Watanabe

ISBN: 978-1-62948-053-4

Chapter 5

PHENYLUREA HERBICIDES: CHEMICAL PROPERTIES AND GENOTOXIC EFFECTS

Concetta Federico,[1] Matteo Pappalardo,[2,3] Claudia Giovanna Leotta,[1] Zelica Minniti,[2,3] Vito Librando,[2,3] and Salvatore Saccone[1,3,*]

[1]Dipartimento di Scienze Biologiche, Geologiche e Ambientali – Sezione di Biologia Animale "M. La Greca", University of Catania, Catania, Italy
[2]Dipartimento di Scienze Chimiche, University of Catania, Catania, Italy
[3]Centro Universitario di Analisi, Monitoraggio e Metodologie di Minimizzazione del Rischio Ambientale
c/o Dipartimento di Scienze Chimiche, University of Catania, Catania, Italy

ABSTRACT

A high number of herbicides is used to protect important crops, but an intensive use of these chemical compounds produces a general increasing of environmental pollution, and could be hazardous for human health. Thus, it is very important the identification of their mode of action, and the evaluation of their genotoxic properties, especially to assess the exposition level of people that come into contact via inhalation, skin contact, or ingestion of these compounds, and working in the areas where herbicides are used. Results concerning genotoxic activity of chemical compounds could be obtained using *in vitro* short term tests such as the evaluation of cytogenetic effects on exposed mammalian cell lines. Phenylurea herbicides are selective agents, widely used in agriculture for the control of weeds, belonging to C2 group of the Herbicide Resistance Action Committee (HRAC) classification, and are endowed by a mode of action that determines inhibition of photosynthesis at photosystem II complex. The genotoxic properties of some of these compounds were evaluated by a number of tests, including cytogenetic effects (chromosomal aberrations, and sister chromatid exchanges) on Chinese hamster cell

* Corresponding author: Salvatore Saccone. Dipartimento di Scienze Biologiche, Geologiche e Ambientali, Sezione di Biologia Animale "M. La Greca", Via Androne, 81, 95124 Catania, Italy. Tel: +39.095.7306037; Fax: +39.095.7306067; E-mail: saccosal@unict.it.

lines. Here we present a review of chemical properties of the phenylurea herbicides, and their genotoxic effects evaluated with mutagenic assays.

INTRODUCTION

The application of herbicides on agricultural soils is a well-known practice to control weed growth. Among herbicides, phenyl-urea derivatives represent a prominent group of herbicides. Slow biodegradation of these lead to their passing to surface and ground waters. Many factors affect herbicidal persistence and degradation. This leads to the environment being exposed not only to mixtures of parent compounds but also to mixtures of parent compounds and to their corresponding transformation products. For evaluation of toxic effects and fate into the environment of the phenylurea herbicides, it is primarily necessary to have reliable information on their behavior in environmental compartments. Besides, extraction and analysis of degradation products is the second pivotal aspects. Phenylurea herbicides are typically extracted from complex matrixes of soil, or water with a step of extraction and concentration of target compounds from a sample matrix. After the extraction, a detection step is required, and the phenylurea herbicide (PUH) residues are mainly determined by chromatographic methods, but many other techniques are available.

The use of phenylurea herbicides to protect crops determines an exposure to these chemicals especially in the rural areas where they are used. A chemical compound, in order to be used in the safest way possible, should be analysed in its different biological properties that can be harmful to human health. One of these properties is the presence of mutagenic activity that could interfere to the correct expression of the genetic material of people who come into contact with these chemicals. Evaluation of genotoxic activity of chemical pollutants could be done with *in vitro* short term assays using cell cultures endowed by particular features. There are a number of cell lines that consent to obtain data on mutagenic or promutagenic activity of chemicals, such as a Chinese hamster ovary (CHO)-and a Chinese hamster epithelial liver (CHEL) cell lines. The CHO cells consent to have information on the direct mutagenic activity of the chemicals tested, on the other hand the CHEL cells are able to demonstrate the presence of promutagenic properties, namely the possibility of non mutagenic chemicals to be converted in mutagenic compounds by intracellular enzymes. These cell lines were recently used to identify the genotoxic effects of some phenylurea herbicides by evaluation of their cytogenetic effects on exposed cells.

CHEMICAL PROPERTIES

Every chemical substance that comes into contact with a natural ecosystem will be distributed into the solid, liquid or gaseous phase. In most cases, the transport of these compounds is linked to water, and the quantity transported from one phase to another depends on the initial concentration of a given molecule, and its equilibrium coefficient. For the evaluation of the ecotoxicologic effects and fate into the environment of a chemical it is necessary to have reliable information on its behaviour in water, air and soil, the last one plays a special role as it includes all other environmental compartments.

Thus, a broad variety of different interactions between a chemical and the soil ecosystem are possible, and it is therefore obvious, that the fate of a chemical substance in soil is governed by its soil sorption data. Consequently, the knowledge of soil sorption data is necessary for the evaluation of potential dangers caused by certain chemicals to man and nature [1]. Hydrophobicity plays an important role in the fate of organic compounds in the environment as well as their interactions with biological systems. Octanol-water and soil-water partitioning coefficients, Kow, and Koc, respectively, have been the most widely used hydrophobic parameters in evaluating the movement and persistence of pesticides in the environment and structure-activity relationship studies. In the subsequent table 1 and table 2, we reported main data from a recent study about such parameters [1-3]. For the study has been compared soil of various composition, but main aspect is that soil 1 contains a greater amount of organic matter respect to the soil 2.

FATE OF PHENYLUREA HERBICIDES INTO THE ENVIRONMENT

The application of herbicides on agricultural soils is an established practice to control weed growth. Among herbicides, phenyl-urea derivatives represent a prominent group of herbicides. Since their discovery in the early 1950s, the variety and amount of this herbicide group have increased markedly. Slow biodegradation of these [5] together with washing and leaching processes lead to their passing to surface and ground waters.

Table 1. Values of Freundlich parameter for some phenylurea herbicides (AR groups is the placeholder for the aryl group)

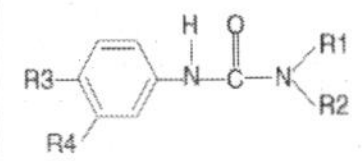	R1	R2	R3	R4	Values of Freundlich parameter					
					Soil 1			Soil 2		
					Kf	n	R^2	Kf	n	R^2
Chlorbromuron	O-CH3	CH3	Br	Cl	23.2	0.75	0.992	11	0.8	0.986
Chlorotoluron	CH3	CH3	CH3	Cl	7.3	0.82	0.996	3.7	0.77	0.988
Difenoxuron	CH3	CH3	O-AR-OCH3	H	2.7	1.08	0.97	1.88	0.62	0.99
Diuron	CH3	CH3	Cl	Cl	11.1	0.84	0.992	5.2	0.88	0.998
Fenuron	CH3	CH3	H	H	2	0.81	0.967	0.6	0.9	0.907
Fluometuron	CH3	CH3	CF3	H	2.7	0.89	0.991	1.8	1.06	0.981
Isoproturon	CH3	CH3	CH (CH3)2	H	6.4	0.86	0.99	4	1	0.997
Linuron	O-CH3	CH3	Cl	Cl	12	0.78	0.979	9	0.86	0.983

Table 1. (Continued)

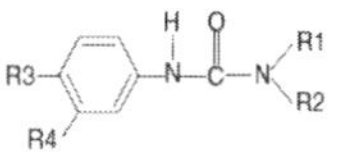	R1	R2	R3	R4	Values of Freundlich parameter					
					Soil 1			Soil 2		
					Kf	n	R^2	Kf	n	R^2
Metobromuron	O-CH3	CH3	Br	H	6	0.81	0.999	2.9	0.76	0.991
Metoxuron	CH3	CH3	O-CH3	Cl	6.1	0.76	0.996	2.4	0.78	0.854
Monolinuron	O-CH3	CH3	Cl	H	4.6	0.78	0.956	2.6	0.86	0.965
Monuron	CH3	CH3	Cl	H	4.2	0.71	0.982	1.7	0.75	0.832
Neburon	(CH3)3 CH3	CH3	Cl	Cl	41.8	0.78	0.996	17.8	0.69	0.789

Data from [1, 2, 4].

Table 2. Equilibrium time (teq), coefficients of adsorption (Kd), logartim of the coefficients of adsorption, Kow and Koc for some phenylureas on two soils (Soil 1 and Soil 2)

	Soil 1			Soil 2				
	t_{eq}	K_d	Log K_d	t_{eq}	K_d	Log K_d	Log K_{ow}	Log K_{oc}
Chlorbromuron	72	48	1.38	23.9	11.7	1.07	3.3	3.12
Chlorotoluron	48	48	0.86	7.3	3.7	0.57	2.38	2.81
Difenoxuron							2.43	2.68
Diuron	72	72	1.04	11	5.1	0.71	2.58	2.6
Fenuron	24	48	0.3	2	0.5	-0.3		1.43
Fluometuron	48	48	0.45	2.8	1.8	0.26		
Isoproturon	48	48	0.86	7.3	3.9	0.59	2.5	2.5
Linuron	72	72	1.12	13.3	8.5	0.93	3	2.83
Metobromuron	24	24	0.76	5.7	2.9	0.46	2.32	2.43
Metoxuron	48	72	0.76	5.7	3.2	0.51	1.54	2.01
Monolinuron	24	24	0.7	5	2.4	0.38	2.3	2.11
Monuron	48	72	0.62	4.2	2	0.3	1.89	2.26
Neburon	24	24	1.8	63.7	28.4	1.45	4.1	3.49

Data from [1-3].

The soil composition affects herbicidal persistence through herbicide adsorption, leaching, and volatilization. Isoproturon like any other phenylurea herbicide (PUH) has been shown to be stable to chemical degradation within the pH range of 4–10 [6]. When chemicals, such as pesticides, are released into the environment, they are subject to different transformation processes [6], namely abiotic and biotic. This leads to the environment being exposed not only to mixtures of parent compounds but also to mixtures of parent compounds and to their corresponding transformation products. If transformation products are more persistent and mobile than their parent compounds, they may be detected in even higher concentrations than their parent compound in the aquatic environment [6, 7]. Transformation products may still possess a similar mode of toxic action as the parent compound, may exhibit unexpected effects towards non-target organisms, or, even if they lose their specific toxicity, can still contribute as baseline toxicants to mixture toxicity [6, 8].

Diuron is mainly degraded via aerobic [9] and anaerobic [10] microbial degradation (figure 1). N-demethylation leads to the formation of 1-(3,4-dichlorophenyl)-3-methlyurea and 1-(3,4-dichlorophenyl)urea, whereas reductive dechlorination leads to 3-(3-chlorophenyl)-1,1-dimethylurea. The final known transformation product of the aerobic pathway is 3,4-dichloroaniline. Recent analysis led to the conclusion that all transformation products would contribute equally to the toxicity of the mixture with different impact respect to the receptor [11].

Triazines and phenylureas are neutral herbicides widely used in both agricultural and urban settings. Among these, atrazine, Diuron and their main degradation products esethylatrazine, desisopropylatrazine, and dichlorophenylmethylurea have often been found in waters and effluents of wastewater treatment.

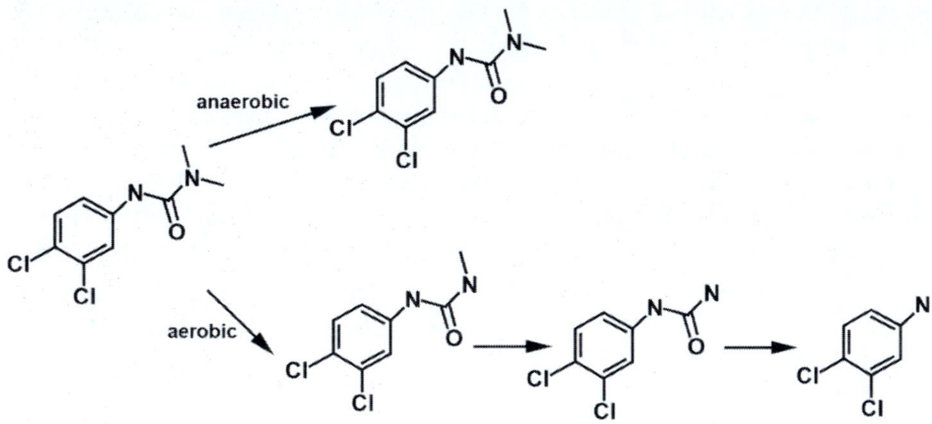

Figure 1. Aerobic and anaerobic microbial degradation pathways of diuron.

The high persistence of these molecules in the environment and their toxicity as potent endocrine disrupters have reinforced the need to determine their content in sludge [12]. Their identification and quantification could lead to both more comprehensive and global vision of risk assessment as well as more complete removal of these compounds from the environment [13, 14]. Therefore to have a complete overview of the fate and environmental toxicity, biological and chemical approaches have to be combined [15]. Therefore, the main degradation process is due to the action of microorganisms (biotic process). The microbial aspects of the soil environment include the types and abundance of soil microorganisms present in the soil, able to mineralize PUH. In the environment, the simultaneous or consecutive action of different microbial communities on the pesticide can lead to a wide range of metabolites. Therefore, the understanding of field sample analyses can be very difficult. Moreover, you have to know what you are looking for to determine your analysis procedure. A more comprehensive way to get information is to start working with pure strains. Several strategies can be developed to identify degradation products, to determine biodegradation pathways and evaluate the environmental impact of the active ingredient and its metabolites. From the soil they can migrate to crops and enter the food chain and, the herbicides can also reach ground waters [16].

Many other factors affect herbicide persistence into the soil, including composition, chemistry, and microbial activity. Soil composition is a physical factor determined by the relative amounts of sand, silt, and clay in the soil (the soil texture), as well as by the organic-matter content. Accumulating evidence points toward that pyrogenic carbon particles or black carbon (BC) may enhance sorption of organic pollutants to sediments and soils, leading to distribution coefficients that are several orders of magnitude higher than in sediments and soils without BC. Moreover, it may reduce bioaccumulation of organic pollutants by a factor of up to 100. BC is ubiquitously present in the environment and has been estimated to account for approximately 9% of the total organic carbon in marine and freshwater sediments and 4% in temperate soils. The molecular structure of many modern pesticides (monoaromatics, diphenylether, and biphenyl-analogues) suggests a potential high affinity to BC. Still, there are reports of a 600- to 1000-fold increase in sorption of Diuron to burned relative to unburned crop residues.

However, atrazine shows lo value of absorption, indeed they have similar octanol-water partitioning coefficients and polarizability but they have very different molecular structure.

Degradation of PUH in soil/water systems showed that the reaction rate, of this process, should be, correlated with the soil organic matter content. Though biodegradation seems to be the main cause of herbicide degradation, abiotic degradation can also be important for chemicals such as PUH, which are subject to catalyzed soil reactions.

Dissolved Organic Matter (DOM) is one of the central components of soil composition and plays central roles in many chemical processes in soils; it is composed by a mixture of small molecular organic acids and macromolecule compositions [17]. While DOM brings its benefits as compost, it has potential impact on environmental quality. Most of studies have focused on the influence of DOMs on toxicity of hydrophobic organic chemicals in aquatic organisms [18], but at now know little about the biological effect of DOMs on herbicide availability. Chlorotoluron (Chl) is a PUH and is widely used for controlling weeds. While it has brought great benefits to crop production, it has also resulted in contamination to ecosystem. As Chlorotoluron is normally soil-applied and relatively soluble in water, it readily accumulates in crops, accumulation of Chlorotoluron by plants resulted in stopping growth and metabolic disorder. Wheat treated with Chl in the presence of DOM accumulated less Chl than those treated with Chl alone. Moreover, the presences of DOM, decrease the bioconcentration factor while increase the translocation factors. Although it is unknown for the mechanism by which DOM regulates Chl damage in wheat, it also prevented cellular toxic effects by reducing Chl accumulation in wheat [19].

Among the various fractions of soil organic matter, humic substances are the most abundant, and are closely connected to soil fertility. Humic Acids are characterized by the presence of acidic functional groups such as carbonyls and phenols, and they could promote the degradation of PUH the effect of dissolved humic acids (HAs), normally present in natural waters, on the hydrolysis of PUH, and it kinetic model that takes into account the role of adsorption. The adsorption isotherms indicate that phenylurea–humic acid interaction can be considered in terms of a repartition-like equilibrium of phenylurea between water and HAs. Kinetic experiments show that the degradation rates of phenylureas increase with HA concentration [20].

Enhanced degradation is a phenomenon whereby, a soil-applied pesticide is rapidly degraded by microorganisms that have developed ability to use these compounds as energy, and/or nutrient source [21]. Recent studies on atrazine degradation in soil from one of the sugarcane field atrazine mineralization were enhanced by adding organic amendments [22].

Enhanced degradation has been observed for atrazine and Isoproturon in soils where they have been applied repeatedly and used for a long time with subsequent isolation of the bacterial strains, which metabolized the pesticides to get energy for growth [23]. But it has been noticed a slow mineralization of atrazine by native soil microbes that has been attributed to the halogen on the atrazine ring which prevent easy microbial metabolism [24, 25].

In these studies, the degradation pathway of Isoproturon has been found to proceed via initial co-metabolic steps followed by metabolic processes. The initial attack is demethylation resulting in the removal of one methyl group followed by removal of another methyl group from N of the urea side chain [26]. While many factors may affect degradation, as reported in table 3, degradation processes by bacteria are the most significant pathways responsible for the degradation of herbicides.

The populations of bacteria determine how quickly decomposition occurs, but they require certain chemico-physical conditions for optimal growth and utilization of any herbicide, like composition, temperature, pH, oxygen.

Very recent studies have reported the adaptation of soil micro flora to 3-(4-isopropylphenyl)-1,1-dimethylurea (IPU) degradation in response to repeated exposure to this herbicide over a long period [27, 28]. The phenylurea herbicide IPU was found rapidly mineralized in an agricultural soil that had been periodically exposed to IPU. Analysis of this soil evidences a bacterial strain able to mineralize IPU belonging to the phylogeny of the genus Sphingomonas.

The IPU degrading ability of strain SH was strongly influenced by pH with maximum degradation taking place at pH 7.5; moreover, it could not degrade other structurally related phenylurea herbicides (PUH) such as Diuron, Linuron, Monolinuron and Chlorotoluron or their aniline derivatives. The complete pathway for degradation of IPU is unknown but it has been suggested that the IPU metabolic pathway is initiated by two successive N-demethylation, followed by the cleavage of the urea side chain resulting in transient accumulation of 4-isopropylaniline and finally by the mineralization of the phenyl structure [29, 30].

Linuron is a globally used phenylurea herbicide, and a large number of studies have been made on the microbial degradation of the herbicide. Recent paper report a Linuron mineralizing bacterium that inhabits river sediment was successfully. Paper reports that isolated bacteria rapidly mineralize Linuron and a trace amount of 3,4-dichloroaniline were detected. Besides the bacteria also degraded Monolinuron, Metobromuron and Chlorobromuron, but not Diuron, Monuron or Isoproturon, probably due to recognition of the structural difference between *N*-methoxy-*N*-methyl and *N.N*-dimethyl substitution of various phenylurea herbicides [31]. The soil bacterial isolate *Variovorax* sp. strain SRS16 is able to mineralize the PUH Linuron.

Table 3. Factor-affecting degradation of some phenylurea herbicides

Factor affecting degradation	Primary effect	Secondary effect	Molecule degraded
Enhanced degradation	Rapid degradation		Atrazine. Isoproturon
Humic Acids	catalytic effect on the degradation		Isoproturon. Fenuron
Dissolved Organic Matter	reducing accumulation in plants	Reduce bioconcentration, increase translocation	Chlorotoluron
black carbon	increase in sorption	reduce bioaccumulation	Diuron

The proposed pathway of Linuron catabolism starts with amide hydrolysis to 3,4-dichloroaniline and *N.O*-dimethylhydroxylamine. 3,4-dichloroaniline is harmful and recalcitrant, while *N.O*-dimethyl hydroxylamine is not and degraded easily. Several Linuron-degrading *Variovorax* strains, in addition to mediating Linuron hydrolysis, are able to use *N.O*-dimethyl hydroxylamine as the sole carbon source and mineralize it [32].

DETERMINATION

Organic pollutants, like PUH, are typically extracted from samples by time-consuming methods and they need large amounts of solvents. In general this step consists of extraction and concentration of target compounds from a sample matrix. Liquid extraction (LLE) or solid-phase extraction (SPE), ultrasonic solvent extraction (USE) are efficient economic and miniaturized sample preparation. However LLE requires a large amount of organic solvent and is a time-consuming process. SPE is currently adopted for the analysis of liquid samples and the cost of equipment purchasing is very high. Matrix solid phase dispersion-capillary (MSPD) was introduced firstly by Barker [33] which has many advantages such as simple operation, high recovery rate and less sample loss and solvent consumption. In particular MSPD is more applicable to the pretreatment of solid samples rather than other sample pretreated techniques [34, 35]. Solid phase micro-extraction (SPME) supercritical fluid extraction (SFE) accelerated solvent extraction (ASE) and liquid-phase micro extraction (LPME) were developed as alternative techniques to the classical LLE and SPE. Online sample concentration using SPE or a trapping Liquid Chromatography column provides a convenient approach to automate an application as it has been applied to the analysis of micro-organic contaminants in environmental waters. LPME achieved extensive attention as a novel sample pretreatment technique due to its advantages of miniaturization simple operation, low cost, low consumption of sample solvent, swiftness, LPME may be performed by using a single drop of solvent. Recently, microwave-assisted extraction (MAE) has been developed. This technique is based on the absorption of the microwave energy by extraction solvents resulting in an increase of the temperature and pressure, thus, diffusion of the compounds from the matrix to the solvent can be achieved. Compared with traditional extraction methods such as ultrasonication MAE has many advantages: smaller volumes of solvents are needed the extraction time is shorter due to the direct heating of the solvents by microwaves and multiple samples (up to 14) can be extracted simultaneously. However the MAE has some drawbacks as well. The extraction solvent must be able to absorb the microwave energy. A clean-up step is needed due to co-extraction of matrix material in the sample, which can cause interferences in chromatographic separations. Moreover if water is used as extractant transfer of analytes into an organic solvent should be performed before GC/MS analysis.

Another extraction technique has been recently developed by Assadi et al. [36, 37]. They developed a novel liquid-phase micro extraction technique as a high-performance and powerful preconcentration method named as dispersive liquid–liquid micro extraction (DLLME). DLLME avoids instability and provides possibility for the enhancement of sensitivity and simplification of the extraction procedure compared to the conventional LPME. In table 4, we summarize all extraction techniques reported and a valuation of its own main feature.

After the extraction of complex solid samples, and cleanup step generally, detection step is required. Currently, PUH residues are mainly determined by chromatographic methods. Gas chromatography (GC) is commonly used because of its high sensitivity and selectivity, however, the thermal instability of most phenylurea pesticides requires preparation of stable derivatives prior detection, which complicates the analysis because of time consuming and tedious manipulations.

Table 4. Summary of the extraction methods here reported, with its own main features

Extraction tecniques	Solvent used	Efficiency	Cost	Analisys time	Laboratory work
Liquid liquid	high	good	low	high	high
Solid-phase	low	good	medium	high	high
Ultrasonic solvent	low	good	medium	medium	high
Solid phase micro-extraction	medium	high	high	medium	medium
Supercritical fluid	medium	high	high	low	low
Accelerated solvent	medium	good	very high	very low	medium
Liquid-phase micro extraction	very low	good	low	low	low
Microwave assisted	low	medium	very low	very high	medium
Dispersive liquid–liquid microextraction	high	good	medium	low	low
Matrix solid phase dispersion capillary	High	high		low	low

Therefore, high-performance liquid chromatography (HPLC) with UV diode array detection (DAD) is often preferred to GC. LC–MS/MS using triple quadrupole analyzers is an alternative of instrumentation for targeted environmental analysis because of the high-sensitivity achieved, the linearity over a several orders of magnitude and the determination of a wide range of analytes without the requirement of derivatization steps. The coupling of HPLC to mass spectrometry (MS) is now one of the most frequently used determination techniques in environmental chemistry. Phenylurea, *N*-methylcarbamates, and phenoxy acid pesticides can be determined relative easily by HPLC–MS or HPLC-DAD (diode array detection). The degradation products of the majority of the pesticides are, however, usually more polar than their parent compounds (i.e. demethyldiuron, deisopropylatrazine) and HPLC–MS is, therefore, appropriate [38]. Diuron degrades by *N*-demethylation under aerobic and anaerobic conditions to various metabolites including N-(3,4-dichlorophenyl)-N-methylurea, 1-(3,4-dichlorophenyl)urea, (3,4)-dichloroaniline, and *N*-(3-chlorophenyl)-*N*-methylurea. Their separation and identification is accomplished using GC–MS, high performance liquid chromatography–diode array detection (HPLC–DAD) or HPLC–atmospheric pressure chemical ionization (APSI)–MS. Although LC provides the advantage of direct analysis of PUHs without derivatization, it need to be coupled with detection methods, including UV detection, fluorescence detection (FD) and mass spectrometry (MS), which have some drawbacks, such as narrow linear ranges, high detection limits, expensive instrumentations, or high running costs [39, 40]. Although analytical methods for the analysis of contaminants are mainly based on chromatographic techniques, many methods have been reported for measuring pesticides and industrial chemicals in human matrices, most of them refer to a limited number of compounds or analytes belonging to the same chemical group. One of the main problems in multiresidue analysis of contaminants in real samples is the tediousness and complexity of the procedures required for the extraction, cleanup, and preconcentration of the matrix analytes. Most methods described in the literature for the determination of contaminants employ a combination of some extraction methods with organic solvents with one or several washing steps aimed at removing co-extractants before the samples are injected into the chromatographic system.

Efficient analytical methods for triazines and phenylureas in environmental matrices are scarce. While Triazines determination in sludge is simple, it is not true for the phenylureas, because derivatization is required and, furthermore, yields unstable products. Only in recent time, it has been developed a method that after a methanolic ultrasonic extraction, detect atrazine, Diuron, and their main degradation products with LC–ESI-MS–MS. Main advantage of the method is the high selectivity and sensitivity offered by LC–ESI-MS–MS, the internal standard calibration, and only one cleanup step is required [41]. The use of LC–ESI–MS/MS produced good analytical data with respect to specificity, linearity, detection limits, precision and accuracy. A study of the effect of modifying the resolution in low mass resolution triple quadrupole MS showed that an increase in signal/noise can be achieved by a modest increase in resolution. The recovery data of the pesticides studied using this method were satisfactory in both the rapid SPE clean up and the online column trapping. This method was successfully applied to environmental wastewater samples and should prove useful for environmental monitoring programs [42]. Although LC provides the advantage of direct analysis of PUHs without derivatization, it needs to be coupled with detection methods, like UV detection fluorescence detection and mass spectrometry. In recent times, capillary electrophoresis (CE) becomes an attractive alternative technique for their determination ability in herbicides analysis. Carabias-Martinez et al. [43] reported that CE method resulted in much more simple chromatogram profile and a flat baseline by comparing CE-UV with HPLC-UV methods for determination of five triazines herbicides. As the most common detection modes for CE, electro chemiluminescence (ECL) offers not only lower background noise and higher detection sensitivity, but also simple and inexpensive instrumentation. In particular, the ECL fit for determination of amine compounds, especially the secondary amine and the tertiary amine compounds. Meanwhile, all phenylurea herbicides have secondary or tertiary amine groups, and they can be sensitively detected by ECL system, which has advantages including good solubility, stability and reusability, broad linear range, sensitivity for the determination of amines. Recently, capillary electrophoresis (CE) becomes an attractive alternative technique for their determination ability in herbicides analysis. Dispersive liquid-phase micro extraction IL-DLPME coupled with HPLC-DAD was used to determine triazines and phenylurea polar herbicides in aqueous samples. The proposed procedure had some advantages such as combining the process of extraction and concentration into one step, shorter extraction time, better reproducibility, reducing the exposure danger to organic toxic solvent. The factors affecting the extraction efficiency such as the volume of IL, pH of samples, extraction time, and strength were optimized. IL-DLPME is proved too fast, simple and environment-friendly method [44].

Environmental Mutagenic Assays

Experimental studies to evaluate effects of environmental pollution on human health are important not only in large cities, generally endowed by a great amount of Polycyclic Aromatic Hydrocarbons (PAH) due to the high number of cars, but also in rural areas where pollution is associated with extensive use of chemical compounds. A number of genetic studies indicates that exposure to high levels of environmental pollutants is related with increased risk of cancer [45-47], thus *in vitro* assays to evaluate the genotoxic properties of many different chemical compounds are of great importance [48-53], especially before their commercialization. In fact, the identification of mutagenic activity of environmental pollutants was obtained in many large cities [54-57] as well as in rural areas where a number of herbicides and pesticides are largely used [58].

These chemical compounds are widely used, all over the world, to protect crops, but several of them showed mutagenic or clastogenic properties [59-61]. A number of studies revealed a higher risk level of particular cancers in farmers compared to the general population, although the average mortality due to cancers is generally lower in rural populations [62-64].

Many mutagenic tests were carried out to identify generic mutagenic properties on different matrixes (air, soil, water), but the knowledge of the genotoxic activity of individual herbicides is very useful and important especially to encourage the use of specific protective systems by the farm workers exposed to high amount of these molecules, being relevant the possibility to come in contact with these compounds by inhalation, skin contact or ingestion. Studies published recently indicate that the frequent indiscriminate use of many types of pesticides produced genotoxic damage in people occupationally exposed to those substances [58]. Many of these studies, assessing specific effects of exposure to genotoxic pollutants, in humans and other living species, used different biomarkers [65]. Among these, Chromosomal

Aberrations (CA), Sister Chromatid Exchanges (SCE) and Micronuclei (MN) were the most widely employed in *in vivo* and *in vitro* mutagenic tests.

In vivo tests are performed using several prokaryotic and eukaryotic organisms. The more famous of the former could be certainly mentioned the Ames test, which represents a milestone for the mutagenesis, that uses heterotrophic bacteria strains endowed by specific point mutations to identify chemicals that induce reversion to wild-type strains [66, 67]. A number of eukaryotic organisms (including mammals, such as rodents) were also employed for these purposes, but their use is very expensive for time and laboratory organization. Not secondary should be also considered ethical aspects in laboratory tests that use living organisms.

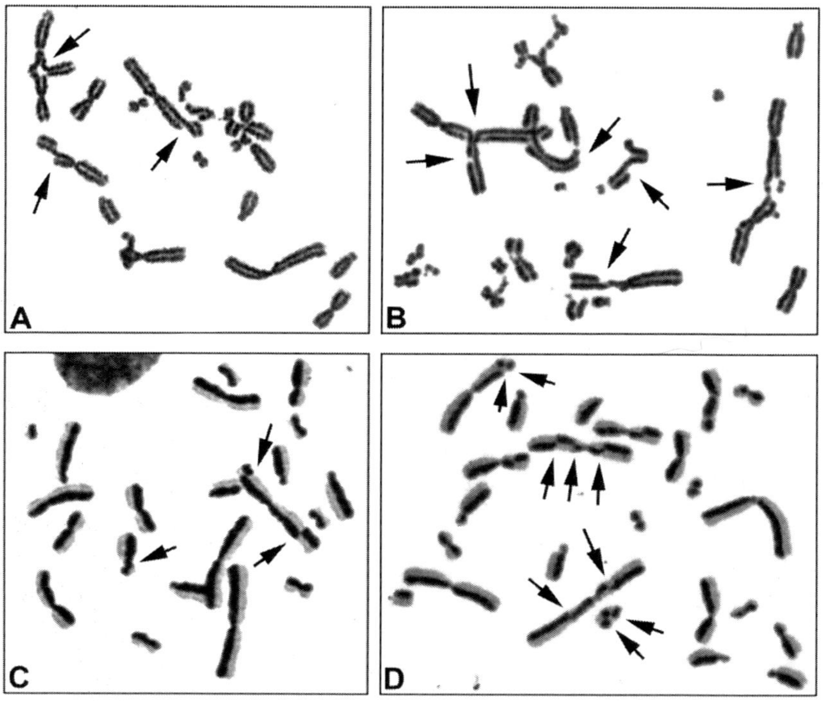

Figure 2. Metaphases from Chinese hamster ovary cells showing chromosomal aberrations (A, and B) and sister chromatid exchanges (C, and D). Arrows indicate some of these cytogenetic biomarkers used to identify the genotoxic effect of chemical compounds, such as herbicides or pesticides used to protect crops.

On the other hand, *in vitro* mutagenic tests with cell cultures exposed to chemicals, to investigate their genotoxic properties, are largely used. In this way, mutagenic effects of a chemical compound could be identified by analysis of specific cytogenetic biomarkers observed in exposed compared to non-exposed cells. The main cytogenetic biomarkers generally used are CA and SCE, whose analysis is performed on metaphase cells, or micronuclei, whose analysis is carried out on interphase cells. It should be stressed, that the same biomarkers can also be used in monitoring of human populations living in specific areas, or showing specific lifestyle, or having occupational exposure [51, 53, 54, 56, 58].

Chromosomal aberrations can be numerical or structural and are the result of breakage of chromosomes and their subsequent rearrangement in abnormal shapes such as dicentric or triradial chromosomes, and other chromosomal anomalies which are in particular insertions or deletions.

These biomarkers are visualized in cells at the metaphase stage, whose preparation follows conventional, and simple technical procedures, even if analyses must be performed by expert cytogeneticists. Also the SCE are due to DNA breakage visible in metaphase chromosomes with a specific experimental protocol that consents to see the sister chromatids stained with different intensity, being a chromatid very pale and the other very dark, and SCE are visible by a different stain along the same chromatid (figure 2).

Another kind of biomarker, that are successfully used to identify mutagenic properties of chemicals, is represented by micronuclei that could be observed, with higher frequency, in exposed cells respect to not exposed cells.

Both the preparation and the analysis of micronuclei are simpler respect to CA and SCE, being not necessary the metaphase chromosome preparation. The micronuclei are visualized in interphase cells, thus increasing the number of cells that could be analysed respect to the other two types of biomarkers.

The micronucleus assay consists in the analysis of binucleated cells, where cytogenetic damages are visible as small accessory nuclei formed from acentric chromosomal fragments or whole chromosomes that are not incorporated into the main nuclei.

Micronuclei can generate through a number of mechanisms, essentially related to action of clastogenic or aneuploidization agents, which cause direct chromosome breakages and/or dysfunction of the mitotic spindle [68, 69].

MUTAGENIC PROPERTIES OF PHENYLUREA HERBICIDES

Data on the mutagenic effects of PUH using *in vivo assays* are limited, and findings are not consistent. As an example, Diuron was tested using several methods, and it was indicated 'suspect genotoxic' (together to other nine PUH), with Mutatox test performed with the *Photobacterium phosphoreum*, directly and after S9 activation [70], but it was shown not mutagen using *Salmonella typhimurium* strain with or without metabolic activation [71]. On the other hand, data with the Pacific oysters *Crassostrea gigas* exposed to Diuron showed a significant increase of the aneuploidy level [72], and an early physiological effect involving gene expression and enzyme activities of proteins implicated in the immune responses [73]. Another PUH, the Linuron, was analysed using *in vivo* micronucleus tests in bone marrow of Wistar rats exposed to a mixture with paraquat under commercial formulations; in these conditions, the mixture of the two compounds was not proven to be genotoxic [74].

Recently, Diuron, Fenuron, Chlorotoluron, and Difenoxuron (figure 3) were tested for their mutagenic properties using *in vitro* short term tests with Chinese hamster ovary (CHO), and Chinese hamster epithelial liver (CHEL) cell lines [75]. This was done by analyses of their cytogenetic effects, namely by the increase of CA and SCE in exposed CHO, and CHEL cell lines (see next section for experimental procedure details).

Cytogenetic analysis of CHO, and CHEL cells exposed to PUH showed that they have mutagenic and pro-mutagenic properties, even if each herbicide showed different level and nature of genotoxic activity. When the not metabolising CHO cells were used, the increase of CA, respect to controls, was not statistically significant for all herbicides at the lowest dose levels (table 5) and that, among the four PUH analysed, Diuron showed the highest mutagenic activity [75]. In a more details, CA induction in CHO cells is addicted to doses, and each herbicide has its own lower dose-level (table 5). In fact, Diuron was the herbicide able to induce CA in CHO cells at low concentration (4.5 μg/ml), on the contrary to the others that induced an increase of CA at higher dose-levels: 48 μg/ml, 110 μg/ml, and 190 μg/ml for Chlorotoluron, Fenuron, and Difenoxuron, respectively. It should be noted that Difenoxuron shows statistically significant induction of CA only at a concentration of about 40 times higher than the Diuron [75]. Data obtained with CHEL cell line showed that all investigated herbicides, at all dose-levels, have brought about statistically significant increase of CA, also indicating a promutagenic activity, namely that these herbicides, into cells, are metabolically converted into other compounds having genotoxic properties.

Table 5. Statistical significance of the chromosomal aberrations induced in Chinese hamster ovary (CHO) cells exposed to phenylurea herbicides respect to controls

Herbicide	Dose-level (μg/ml) and statistical significance (p value)							
	1st		2nd		3rd		4th	
Fenuron	1.10	NS	11.00	NS	110.00	<0.05	220.00	<0.01
Chlorotoluron	0.48	NS	4.80	NS	48.00	<0.05	96.00	<0.05
Diuron	0.45	NS	4.50	<0.05	45.00	<0.05	90.00	<0.0001
Difenoxuron	0.95	NS	9.50	NS	95.00	NS	190.00	<0.05

NS: statistically not significant. Data from [75].

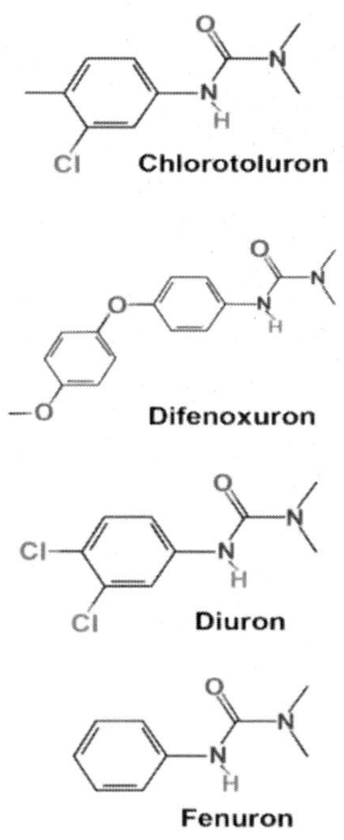

Figure 3. Phenylurea herbicides tested in *in vitro* mutagenic assays.

Analyses of SCE on CHEL cell line showed a statistical significant increases of frequencies of this cytogenetic biomarker, respect to the controls, for every dose level, and for all herbicides analysed [75]. Thus, information available up to now indicates that the above phenylurea herbicides could be assigned to the group of mutagenic chemicals, even if some of them produce cytogenetic effects only at high dose level.

Moreover, genotoxic effects following metabolic activation were observed also at low dose-levels, indicating a further problem using these compounds, due to the additional chemicals with genotoxic properties produced by cellular enzymes. In fact, data obtained using metabolising vs non metabolising cell lines indicated not only a direct mutagenic activity of phenylurea herbicides, but also a pro-mutagenic one, due to additional compounds with genotoxic properties possibly derived by intracellular metabolic activation of the tested herbicides. This is in agreement to a variety of intermediate compounds obtainable with some of the tested herbicides, such as 3,4-dichloroaniline (DCA).

Even if DCA was described only weakly genotoxic with Ames test [76], and being fast photodegraded in water after solar light exposition [77], in CHEL cells, could exert its putative genotoxic effect because it is produced inside cells.

EXPERIMENTAL PROCEDURE FOR CYTOGENETIC TESTS

An *in vitro* genotoxic assay could be performed using several mammalian cell cultures. This allows to avoid use of living animals to test the mutagenic properties of chemical compounds. Cell lines used in mutagenic tests are generally grown *in vitro* in a specific medium supplemented with Foetal Calf Serum (FCS), at 37°C in a 5% CO_2 atmosphere. Chinese hamster ovary (CHO), and Chinese hamster epithelial liver (CHEL) cell lines were recently employed to detect the mutagenic properties of phenylurea herbicides.

CHO and CHEL cell lines contain a small number of chromosomes (see figure 2), allowing easy cytogenetic analysis of cell cultures exposed to chemicals. These two cell lines differ one to each other in their internal metabolic activity: the first lack of a metabolic system, while the second express a number of enzymes that enable the cells to metabolise a number of chemical compounds [78]. Thus, the CHEL cell line is useful to identify those chemicals endowed by the absence of a direct mutagenic effect, but that can acquire a mutagenic activity after metabolic modification and activation by cellular enzymes. The use of both CHO and CHEL cell lines enable to understand whether a chemical compound has direct or indirect genotoxic activity, or it lack of these properties.

For the chromosomal aberration (CA) assay, approximately 24 hours before each experiment, cells are seeded in the flasks in order to ensure that they will be in exponential growth phase at the time of treatment with chemicals, that generally are performed for 16 hours. Chromosomes are prepared using standard cytogenetic procedure that involves use of the mitotic poison colcemid for the last three hours of incubation to increase the number of metaphase cells, and detachment of cells from the flask by using trypsin. Cells are treated with an hypotonic solution, and fixed with a freshly prepared methanol-acetic acid mixture. Finally, chromosome spreads, on air-dried slides, are stained by immersion in a Giemsa stain solution.

For sister chromatid exchange (SCE) assay, cell cultures at the exponential growth phase are incubated with 5-bromo-2'-deoxyuridine (BrdU), under safety light, until harvesting 26 hours later. Chromosome staining for SCE visualization is generally performed using Fluorophore Photolysis Giemsa (FPG) method, that involves an incubation of the metaphase plates in Hoechst-33258, exposition under UV light and staining by immersion in a Giemsa solution [79].

Micronuclei assay (MN) is a procedure to obtain information on genotoxic properties of chemical compounds without chromosomal preparation. In this case, the extra small nuclei present in binucleated cells are used as biomarker for chromosomal damages. Cells are artificially induced to block cell cycle just before the cytokinesis (obtained by incubation of cells in the presence of cytochalasin B), thus obtaining cells with two nuclei. The presence of additional micronuclei indicates that chromosomes were subjected to mutational events [68].

In mutagenic experiments to study genotoxic property, phenylurea herbicides are added into the cell culture medium at defined amount.

To do this is necessary to dissolve herbicides in the better solvent for these compounds. Acetone can be used to dissolve each of the phenylurea herbicides to obtain a high concentrated stock solution. The amount of the herbicide solution that could be added to a 5 ml of culture medium should not be greater than 5-10 μl, to obtain a final concentration of solvent not higher than 0.1-0.2%.

Negative and positive controls are generally included on each experiment. The negative controls are represented by untreated and solvent treated cell cultures. Positive controls are performed with Mitomycin-C (0.10 μg/ml), and 7,12-dimethylbenzanthracene (4 μg/ml) for CHO and CHEL cell lines, respectively, having direct mutagenic activity the first compound, and promutagenic the second.

Each slide is randomly coded by a person not subsequent involved in the slide evaluation. Thus, cytogenetic analysis of the slides is undertaken in a blind way with coded slides. First of all, evaluation of the mitotic index is performed by analysis of at least 1,000 cells, to evaluate that the amount of chemical used is not cytotoxic.

Analyses of the slides, with CA and SCE, are performed by scoring at least 200 and 50 metaphases per test point, respectively. CA are generally classified as chromatid and chromosome type, and the statistical significance in the increasing number of chromosomal aberrations and of cells bearing aberrations is evaluated with Fisher's exact test. For the SCE assays, after the identification of the number of SCE in 50 randomly selected metaphases, the statistical significance of the observed values is evaluated using two tails T-test.

CONCLUSION

Whereas prediction of the fate of phenylurea herbicides is still difficult for the complexity of environmental matrix involved, nowadays many data are available, and here reported them. Knowledge of adsorption data is fundamental into the assessment of these pollutants into the environment, so we collect and select most interesting data on a number of molecules.

The other side of research of phenylurea herbicides is the analytical question; many molecules are involved and many techniques should be adopted for its identification. While phenylurea herbicide extraction from complex environmental matrixes is fundamental, here we presented main extraction techniques highlighting strengths and weaknesses. Finally, traditional and new gas-chromatographic aspects have been overviewed for a simple consultation and assessment of the best technique on case by case.

Presently data on the genotoxic effects of phenylurea herbicides tested indicate that they are endowed with mutagenic property and this is addicted to doses. Moreover, phenylurea herbicides showed higher genotoxic effect following metabolic activation in the CHEL cell lines. This greater mutagenic effect is possibly due to the molecules themselves, and to the additional compounds generated by the cellular enzymes, that induce further cytogenetic effects.

The above information should be considered important for a number of aspects, including human health control. In fact, must be kept under control human exposure to these chemical compounds, especially in areas where a great amount of herbicides is present, such as in factories producing them, and in farms where workers use them to protect crops.

Concerning environmental aspects, they should be implemented control actions aimed to reduce the hazard derived to the intensive use of herbicides that could cause contamination of soil, of surface water, of groundwater and of agricultural products.

Human health and the quality of environment must be preserved by the action of chemicals potentially harmful. The knowledge of the toxic and/or genotoxic properties of chemical molecules used in several environment of work and life is essential, so that they can be used in aware and safe way.

REFERENCES

[1] Gawlik, B. M., Sotiriou, N., Feicht, E. A., SchulteHostede, S., Kettrup, A. (1997). Alternatives for the determination of the soil adsorption coefficient, K-oc, of non-ionicorganic compounds - A review. *Chemosphere* 34: 2525-2551.

[2] Blondel, A., Langeron, J., Sayen, S., Henon, E., Couderchet, M., Guillon, E. (2013). Molecular properties affecting the adsorption coefficient of phenylurea herbicides. *Environ. Sci. Pollut. Res.* 20: 6266-6281.

[3] Liu, J. B., Qian, C. F. (1995). Hydrophobic Coefficients of S-Triazine and Phenylurea Herbicides. *Chemosphere* 31: 3951-3959.

[4] Fouque-Brouard, C. M., Fournier, J. M. (1996). Adsorption-desorption and leaching of phenylurea herbicides on soils. *Talanta* 43: 1793-1802.

[5] Khadrani, A., Seigle-Murandi, F., Steiman, R., Vroumsia, T. 1999 Degradation of three phenylurea herbicides (chlortoluron, isoproturon and diuron) by micromycetes isolated from soil. *Chemosphere* 38: 3041-3050.

[6] Boxall, A. B., Sinclair, C. J., Fenner, K., Kolpin, D., Maund, S. J. (2004). When synthetic chemicals degrade in the environment. *Environ. Sci. Technol.* 38: 368A-375A.

[7] Ingoldsby, E. M., Shaw, D. S., Winslow, E., Schonberg, M., Gilliom, M., Criss, M. M. (2006). Neighborhood disadvantage, parent-child conflict, neighborhood peer relationships, and early antisocial behavior problem trajectories. *J. Abnormal Child Psychology* 34: 303-319.

[8] Sinclair, C. J., Boxall, A. B. (2003). Assessing the ecotoxicity of pesticide transformation products. *Environ. Sci. Technol.* 37: 4617-4625.

[9] Ellis, P. A., Camper, N. D. (1982). Aerobic degradation of diuron by aquatic microorganisms. *J. Environ. Sci. Health* B 17: 277-289.

[10] Attaway, H. H., 3rd, Paynter, M. J., Camper, N. D. (1982). Degradation of selected phenylurea herbicides by anaerobic pond sediment. *J. Environ. Sci. Health* B 17: 683-699.

[11] Neuwoehner, J., Zilberman, T., Fenner, K., Escher, B. I. (2010). QSAR-analysis and mixture toxicity as diagnostic tools: Influence of degradation on the toxicity and mode of action of diuron in algae and daphnids. *Aquat. Toxicol.* 97: 58-67.

[12] Ghanem, A., Bados, P., Perreau, F., Benabdallah, R., Plagellat, C., de Alencastro, L. F., et al. (2008). Multiresidue analysis of atrazine, diuron and their degradation products in sewage sludge by liquid chromatography tandem mass spectrometry. *Anal. Bioanal. Chem.* 391: 345-352.

[13] Librando, V., Sarpietro, M. G., Minniti, Z., Micieli, D., Castelli, F. (2006). Biomimetic approach to biomembrane models studies: Medium influence on the interaction kinetics of some phenylurea derivatives herbicides. *Environ. Sci. Technol.* 40: 2462-2468.

[14] Librando, V., Forte, S., Sarpietro, M. G. (2004). Structure effect on the interaction of phenylurea herbicides with model biomembrane as an environmental mobility parameter. *Environ. Sci. Technol.* 38: 503-507.

[15] Librando, V., Sarpietro, M. G., Cascone, C., Minniti, Z., Castelli, F. (2005). Time-dependent absorption evidence of phenylurea-derived herbicides on model biomembranes: a DSC study. *Environ. Chem.* 2: 63-70.

[16] Gil-Garcia, M. D., Martinez-Galera, M., Parrilla-Vazquez, P., Mughari, A. R., Ortiz-Rodriguez, I. M. (2008). Application of the Doehlert design to optimize the signal obtained in photochemically induced fluorescence for the determination of eight phenylureas. *J. fluorescence* 18: 365-373.

[17] Ravichandran, M. (2004). Interactions between mercury and dissolved organic matter--a review. *Chemosphere* 55: 319-331.

[18] Galle, T., Gregoire, C., Wagner, M., Bierl, R. (2005). Bioavailability of HOC depending on the colloidal state of humic substances: A case study with PCB-77 and Daphnia magna. *Chemosphere* 61: 282-292.

[19] Song, N. H., Zhang, S., Hong, M., Yang, H. (2010). Impact of dissolved organic matter on bioavailability of chlorotoluron to wheat. *Environ. Pollut.* 158: 906-912.

[20] Salvestrini, S., Capasso, S., Iovino, P. (2008). Catalytic effect of dissolved humic acids on the chemical degradation of phenylurea herbicides. *Pest. Management Science* 64: 768-774.

[21] Krutz, L. J., Shaner, D. L., Accinelli, C., Zablotowicz, R. M., Henry, W. B. (2008). Atrazine dissipation in s-triazine-adapted and nonadapted soil from Colorado and Mississippi: Implications of enhanced degradation on atrazine fate and transport parameters. *J. Environ. Quality* 37: 848-857.

[22] Getenga, Z. M. (2003). Enhanced mineralization of atrazine in compost-amended soil in laboratory studies. *Bull. Environ. Contam. Toxicol.* 71: 933-941.

[23] Getenga, Z. M., Dorfler, U., Schroll, R. (2009). Laboratory degradation studies of 14C-atrazine and -isoproturon in soil from sugarcane cultivated fields under Kenyan tropical conditions. *Bull. Environ. Contam. Toxicol.* 82: 678-682.

[24] Wackett, L. P., Sadowsky, M. J., Martinez, B., Shapir, N. (2002). Biodegradation of atrazine and related s-triazine compounds: from enzymes to field studies. *Appl. Microbiol. Biotechnol.* 58: 39-45.

[25] Sorensen, S. R., Bending, G. D., Jacobsen, C. S., Walker, A., Aamand, J. (2003). Microbial degradation of isoproturon and related phenylurea herbicides in and below agricultural fields. *FEMS Microbiology Ecology* 45: 1-11.

[26] Lehr, S., Glassgen, W. E., Sandermann, H., Beese, F., Scheunert, I. (1996). Metabolism of isoproturon in soils originating from different agricultural management systems and in cultures of isolated soil bacteria. *Int. J. Environ. An. Ch.* 65: 231-243.

[27] Bending, G. D., Lincoln, S. D., Sorensen, S. R., Morgan, J. A. W., Aamand, J., Walker, A. (2003). In-field spatial variability in the degradation of the phenyl-urea herbicide isoproturon is the result of interactions between degradative Sphingomonas spp. and soil pH. *Appl. Environ. Microb.* 69: 827-834.

[28] Sorensen, S. R., Aamand, J. (2001). Biodegradation of the phenylurea herbicide isoproturon and its metabolites in agricultural soils. *Biodegradation* 12: 69-77.

[29] Sorensen, S. R., Ronen, Z., Aamand, J. (2001). Isolation from agricultural soil and characterization of a Sphingomonas sp. able to mineralize the phenylurea herbicide isoproturon. *Appl. Environ. Microbiol.* 67: 5403-5409.

[30] Hussain, S., Devers-Lamrani, M., El Azhari, N., Martin-Laurent, F. (2011). Isolation and characterization of an isoproturon mineralizing Sphingomonas sp. strain SH from a French agricultural soil. *Biodegradation* 22: 637-650.

[31] Satsuma, K. (2010). Mineralisation of the herbicide linuron by Variovorax sp. strain RA8 isolated from Japanese river sediment using an ecosystem model (microcosm). *Pest. Management Science* 66: 847-852.

[32] Bers, K., Leroy, B., Breugelmans, P., Albers, P., Lavigne, R., Sorensen, S. R., et al. (2011). A novel hydrolase identified by genomic-proteomic analysis of phenylurea herbicide mineralization by Variovorax sp. strain SRS16. *Appl. Environ. Microbiol.* 77: 8754-8764.

[33] Long, A. R., Hsieh, L. C., Malbrough, M. S., Short, C. R., Barker, S. A. (1990). Matrix solid-phase dispersion (MSPD) isolation and liquid chromatographic determination of oxytetracycline, tetracycline, and chlortetracycline in milk. *Journal - Association of Official Analytical Chemists* 73: 379-384.

[34] Wang, Y. C., Xiao, L., Cheng, M. R. (2011). Determination of phenylureas herbicides in food stuffs based on matrix solid-phase dispersion extraction and capillary electrophoresis with electrochemiluminescence detection. *J. Chromatography* A 1218: 9115-9119.

[35] Barker, S. A. (2007). Matrix solid phase dispersion (MSPD). *J. Biochem. Bioph. Meth.* 70: 151-162.

[36] Berijani, S., Assadi, Y., Anbia, M., Milani Hosseini, M. R., Aghaee, E. (2006). Dispersive liquid-liquid microextraction combined with gas chromatography-flame photometric detection. Very simple, rapid and sensitive method for the determination of organophosphorus pesticides in water. *J. Chromatogr. A* 1123: 1-9.

[37] Assadi, Y., Ahmadi, F., Hossieni, M. R. M. (2010). Determination of BTEX Compounds by Dispersive Liquid-Liquid Microextraction with GC-FID. *Chromatographia* 71: 1137-1141.

[38] Martinez, K., Barcelo, D. (2001). Determination of antifouling pesticides and their degradation products in marine sediments by means of ultrasonic extraction and HPLC-APCI-MS. *Fresenius' J. Anal. Chemistry* 370: 940-945.

[39] Gatidou, G., Kotrikla, A., Thomaidis, N. S., Lekkas, T. D. (2004). Determination of two antifouling booster biocides and their degradation products in marine sediments by high performance liquid chromatography-diode array detection. *Anal. Chimica Acta* 505: 153-159.

[40] Gatidou, G., Zhou, J. L., Thomaidis, N. S. (2004). Microwave-assisted extraction of Irgarol 1051 and its main degradation product from marine sediments using water as the extractant followed by gas chromatography-mass spectrometry determination. *J. Chromatography* A 1046: 41-48.

[41] Ghanem, A., Bados, P., Perreau, F., Benabdallah, R., Plagellat, C., de Alencastro, L. F., et al. (2008). Multiresidue analysis of atrazine, diuron and their degradation products in sewage sludge by liquid chromatography tandem mass spectrometry. *Anal. Bioanal. Chemistry* 391: 345-352.

[42] Janniche, G. S., Mouvet, C., Albrechtsen, H. J. (2011). Vertical small scale variations of sorption and mineralization of three herbicides in subsurface limestone and sandy aquifer. *J. contaminant hydrology* 123: 167-177.

[43] Carabias-Martinez, R., Rodriguez-Gonzalo, E., Miranda-Cruz, E., Dominguez-Alvarez, J., Hernandez-Mendez, J. (2006). Comparison of a non-aqueous capillary electrophoresis method with high performance liquid chromatography for the determination of herbicides and metabolites in water samples. *J. Chromatogr. A* 1122: 194-201.

[44] Wang, S., Ren, L., Liu, C., Ge, J., Liu, F. (2010). Determination of five polar herbicides in water samples by ionic liquid dispersive liquid-phase microextraction. *Anal. Bioanal. Chem.* 397: 3089-3095.

[45] Toulomi, G., Katsouyanni, K., Zmirou, D., Schwartz, J., Spix, C., De Leon, A. P., Tobias, A., Quennel, P., Rabczenko, D., Bacharova, L., Bisanti, L., Vonk, J. M., Ponka, A. (1998). Short term effects of ambient oxidant exposure on mortality: a combined analysis within the APHEA project. Part 1. Air pollution and health: a European approach. *Am. J. Epidemiol.* 146: 177-185.

[46] Goldberg, M. S., Burnet, R. T., Bailar, III J. C., Brook, J., Bonvolot, Y., Tamblyn, R., Singh, R., Valois, M. F., Vincent, R. (2001). The association Between daily mortality and ambient air particle pollution in Montreal, Quebec. *Environ. Res.* 86: 26-36.

[47] Cirera, L., Cirarda, F., Palència, L., Estarlich, M., Montes-Martínez, A., Lorenzo, P., Daponte-Codina, López-Abente, G. (2013) Mortality due to haematological cancer in cities close to petroleum refineries in Spain. *Environ. Sci. Pollut. Res.* 20: 591-596.

[48] Courtois, Y. A., Min, S., Lachenal, C., Jacquot-Deschamps, J. M., Callais, F., Festy, B. (1988). Genotoxicity of organic extracts from atmospheric particles. *Ann. New York Acad. Sci.* 534: 724-740.

[49] Lewtas, J., Chuang, J., Nishioka, M., Peterson, B. (1990). Bioassays-directed fractionation of the organic extract of SRM1649 urban air particulate matter. *Int. J. Environ. Anal. Chem.* 39: 245-254.

[50] Pott, F., Heinrich, U. (1990). Relative significance of different hydrocarbons for the carcinogenic potency of emissions from various incomplete combustion processes. *IARC Sci. Public* 104: 245-254.

[51] Whyatt, R. M., Santella, R. M., Jedrychowski, W., Garte, S. J., Bell, D. A., Ottman, R. A., Gladek-Yarborough, A., Cosma, G., Young, T. L., Cooper, T. B., Randall, M. C., Manchester, D. K., Perera, F. P. (1998). Relationship between ambient air pollution and DNA damage in Polish mothers and newborns. *Environ. Health Perspect.* 106: 821-826.

[52] De Martinis, B. S., Kado, N. Y., De Carvalho, L. R., Gundel, R. A. (1999). Genotoxicity of fractionated organic material in airborne particles from Sao Paulo, Brazil. *Mutation Res.* 446: 83-94.

[53] Roma-Torres, J., Teixeira, J. P., Silva, S., Laffon, B., Cunhac, L. M., Mendez, J., Mayana, O. (2006) Evaluation of genotoxicity in a group of workers from a petroleum refinery aromatics plant. *Mutation Res.* 604: 19–27.

[54] Karahalil, B., Karakaya, A. E., Burgaz, S. (1999) The micronucleus assay in exfoliated buccal cells: application to occupational exposure to polycyclic aromatic hydrocarbons. *Mutation Res.* 442: 29–35.

[55] Motta, S., Federico, C., Saccone, S., Librando, V., Mosesso, P. (2004). Cytogenetic evaluation of extractable agents from airborne particulate matter generated in the city of Catania (Italy). *Mutation Res.* 561: 45-52.

[56] Hallare, A. V., Gervasio, M. K. R., Gervasio, P. L. G., Acacio-Claro, P. J. B. (2009). Monitoring genotoxicity among gasoline station attendants and traffic enforcers in the City of Manila using the micronucleus assay with exfoliated epithelial cells. *Environ. Monit. Assess.* 156: 331–341.

[57] Librando, V., Minniti, Z., Perrini, G., Pappalardo, M., Federico, C., Motta, S., and Saccone, S. (2010). Distribution of Nitro-PAHs in Size-Segregated Particulate Matter Fractions from an Elevated Car Transit Area. *Fresenius Environmental Bulletin* 19: 2282-2286.

[58] Martínez-Valenzuela, C., Gómez-Arroyo, S., Villalobos-Pietrini, R., Waliszewski, S., Calderón-Segura, M. E., Félix-Gastélum, R., Álvarez-Torres, A. (2009). Genotoxic biomonitoring of agricultural workers exposed to pesticides in the north of Sinaloa State, Mexico. *Environment International* 35: 1155–1159.

[59] Garrett, N. E., Stack, H. F., Waters, M. D. (1986). Evaluation of the genetic activity profiles of 65 pesticides. *Mutation Res.* 168: 1301–325.

[60] World Health Organization, Miscellaneous pesticides, IARC Monograph on Evaluation of Carcinogenic Risk to Humans, IARC Scientific Publication No. 53, World Health Organization, Lyon, 1991, 621 pp.

[61] Vigreux, C., Poul, J. M., Deslandes, E., Lebailly, P., Godard, T., Sichel, F., Henry-Amar, M., Gauduchon, P. (1998). DNA damaging effects of pesticides measured by the single cell gel electrophoresis assay (comet assay) and the chromosomal aberration test, in CHOK1 cells. *Mutat. Res.* 419: 79-90.

[62] Blair, A., Malker, H., Cantor, K. P., Burmeister, L. F., Wiklund, K. (1985). Cancer among farmers, *Scand. J. Work Environ. Health* 11: 397–407.

[63] Morrison, H. I., Wilkins, K., Semenciw, R., Mao, Y., Wigle, D. (1992) Herbicides and cancer: a review. *J. Natl. Cancer Inst.* 84: 1866–1874.

[64] Alavanja, M. C., Bonner, M. R. (2012). Occupational pesticide exposures and cancer risk: a review. *J. Toxicol. Environ. Health B Crit. Rev.* 15: 238-263.

[65] Albertini, R. J. (2001). Developing sustainable studies on environmental health. *Mutation Res.* 480–481: 317–331.

[66] Ames, B. N. (1973) Carcinogens are mutagens: their detection and classification. *Environ. Health Perspect.* 6: 115-118.

[67] Ames, B. N., McCann, J., Yamasaki, E. (1975) Methods for detecting carcinogens and mutagens with the Salmonella/mammalian microsome mutagenicity test. *Mutation Res.* 31: 347–364.

[68] Fenech, M. (1997). The advantages and disavantages of the cytokinesis-block micronulceus method. *Mutation Res.* 392: 11-18.

[69] Bonassi, S., Znaor, A., Ceppi, M., Lando, C., Chang, W. P., Holland, N., Kirsch-Volders, M., Zeiger, E., Ban, S., Barale, R., Bigatti, M. P., Bolognesi, C., Cebulska-Wasilewska, A., Fabianova, E., Fucic, A., Hagmar, L., Joksic, G., Martelli, A., Migliore, L., Mirkova, E., Scarfi, M. R., Zijno, A., Norppa, H., Fenech, M. (2007) An increased micronucleus frequency in peripheral blood lymphocytes predicts the risk of cancer in humans. *Carcinogenesis* 28: 625-631.

[70] Canna-Michaelidou, S., Nicolaou, A. S. (1996). Evaluation of the genotoxicity potential (by Mutatox test) of ten pesticides found as water pollutants in Cyprus. *Sci. Total Environ.* 193: 27–35.

[71] Seiler, J. P. (1978). Herbicidal phenylalkylureas as possible mutagens I. Mutagenicity tests with some urea herbicides. *Mut. Res. Genet. Toxicol.* 58: 353–359.

[72] Bouilly, K., Bonnard, M., Gagnaire, B., Renault, T., Lapègue, S. (2007). Impact of diuron on aneuploidy and hemocyte parameters in Pacific oyster, *Crassostrea gigas. Arch. Environ. Contam. Toxicol.* 52: 58–63.

[73] Luna-Acosta, A., Renault, T., Thomas-Guyon, H., Faury, N., Saulnier, D., Budzinski, H., Le Menach, K., Pardon, P., Fruitier-Arnaudin, I., Bustamante, P. (2012). Detection of early effects of a single herbicide (diuron) and a mix of herbicides and pharmaceuticals (diuron, isoproturon, ibuprofen) on immunological parameters of Pacific oyster (Crassostrea gigas) spat. *Chemosphere* 87: 1335-1340.

[74] Goumenou, M., Machera, K. (2004). Mixture of paraquat and linuron formulations examined for genotoxicity in Wistar rats by the in vivo micronucleus test evaluation by different statistical tests. *J. Environ. Sci. Health* B. 39: 75-87.

[75] Federico, C., Motta, S., Palmieri, C., Pappalardo, M., Librando, V., Saccone, S. (2011). Phenylurea herbicides induce cytogenetic effects on Chinese hamster cell lines. *Mutation Res.* 721: 89-94.

[76] Lyons, C. D., Katz, S. E., Bartha, R. (1985). Persistence and mutagenic potential of herbicide derived aniline residues in pond water. *Bull. Environ. Contam. Toxicol.* 35: 696–703.

[77] Duke, S. O. (1990) Overview of herbicide mechanisms of action. *Environ. Health Perspect.* 87: 263–271.

[78] Turchi, G., Carluccio, M. A., Oesch, F., Gemperlein, I., Glatt, H. R. (1987). Characterisation of an epithelial nearly diploid liver cell strain from Chinese hamster able to activate promutagens. *Mutagenesis* 2, 127-135.

[79] Perry, P., Wolff, J. (1974). New Giemsa method for the differential staining of sister chromatid. *Nature* 251: 156–158.

In: Handbook on Herbicides
Editors: Daiki Kobayashi and Eito Watanabe
ISBN: 978-1-62948-053-4

Chapter 6

Action Mode of Triazines and Toxic Effects on Vertebrates

G. D. C. Severi-Aguiar[1]*, E. C. M. Silva-Zacarin[2], F. D. Campos-Pereira[3] and A. A. Alves[1]

[1]Programa de Pós-Graduação em Ciências Biomédicas,
Centro Universitário Hermínio Ometto, UNIARARAS,
Araras, São Paulo, Brazil
[2]Laboratório de Biologia Estrutural e Funcional (LABEF),
Universidade Federal de São Carlos – UFSCAR,
Rodovia João Leme dos Santos, Sorocaba, São Paulo, Brazil
[3]Laboratório de Mutagênese Ambiental (LMA),
Universidade Estadual Paulista – UNESP, Rio Claro, São Paulo, Brazil

Abstract

Triazines are the family of herbicides that include atrazine, ametryn and simazine that are widely used in Brazil and can contaminate groundwater. Cattle can accumulate herbicides in their body through ingestion plants infested with these compounds and one of the ways, by which, human beings are exposed to atrazine is through cattle meat and milk consumption. The toxicity of these compounds can be explained mainly by their interaction with microsomal biotransformation processes. The herbicides per se or their metabolites or the secondary products of oxidative stress interact with biomolecules such proteins and DNA affecting a lot of cellular types. Cellular effects of chemicals might involve recruitment or de-repression of cell death mechanisms. Whether a cell survives or dies in the presence of a chemical insult is often determined by proliferative status, repair enzyme capacity, and the ability to induce proteins that either promote or inhibit the cell death process. In this chapter we will present an extensive bibliographical review about this herbicide class focusing its effects on vertebrates, looking for defense cellular mechanisms, at morphological and biochemical levels.

* Corresponding author. G.D.C. Severi-Aguiar at Centro Universitário Hermínio Ometto, UNIARARAS, Av. Maximiliano Barutto nº 500, Jardim Universitário, 13607-339, Araras, SP, Brazil. Tel.: +55 (19) 3543 1474; fax: +55 (19) 3543 1412. *E-mail address:* grasielaguiar@uniararas.br (G.D.C. Severi-Aguiar).

Introduction
Triazine Herbicides – Absorption and Biotransformation

Triazine herbicides are used to control broad weeds and grasses at crops with great economic importance. Their herbicide action mechanism is mainly based on inhibition of Photosystem II (PSII) of chloroplast electron chain [1].

Up to 380,000 libs of ametryn active ingredient are used per year. Sixty per cent of the annual use of ametryn is used with corn, 20% with pineapple and 20% with sugarcane. 76.4 million pounds of Atrazine was used, only in USA, per year, 86% on corn, 10% on sorghum and 10% on sugar cane crops. Simazine is used on artichokes, asparagus, beans and other deep-rooted crops [2].

Technical grade ametryn (95%) pure is slightly soluble in water (18.5 mg/L) and presents an octanol/water partition coefficient of 676 at pH 7.0 at 25°C (EPA, 2005). Atrazine presents water solubility 28 mg/L (20°C) and partition coefficient 2.3404 [3]. Simazine presents water solubility 3.5 mg/L and log of partition coefficient 1.94^{-3} [4].

Figure 1 shows the chemical structures of triazine herbicides. All compounds are poorly soluble in water and soluble in lipids, as described above. For this reason, these compounds readily cross biological membranes, being easily absorbed by the intestinal tract and the skin and mucous membranes [5].

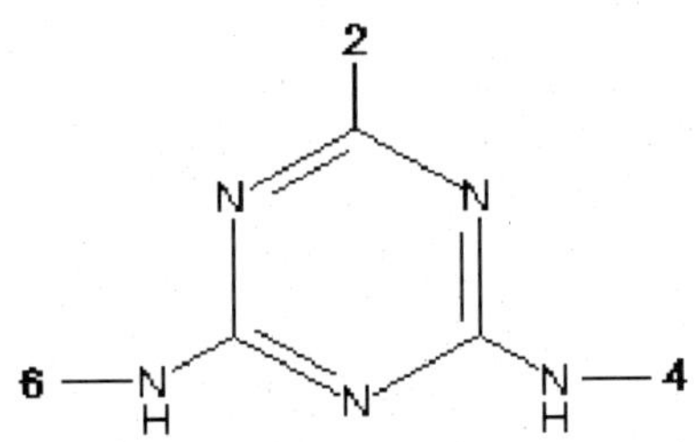

Substitution at ring position:

Atrazine: **2**: -Cl; **4**: $-C_2H_5$; **6**: $-CH(CH_3)_2$

Simazine: 2: -Cl; **4**: $-C_2H_5$; **6**: $-C_2H_5$

Ametryn: 2: $-SCH_3$; **4**: $-C_2H_5$; **6**: $-CH(CH_3)_2$

Figure 1. Chemical structure of the selected s-triazines.

For animals and humans, the main access rout of the triazine herbicides is the oral absorption [6]. Biotransformation starts in the gut and these xenobiotics are partially metabolized and the triazines and its metabolites are found in plasma. McMullin et al. [5] studied the plasma concentrations of atrazine and its metabolites after oral gavage and observed different retention times and concentrations in the plasma for each metabolite. Atrazine (150 mg/body weight) presented a 25 hours concentration peak with maximum concentration at ~3 hours (~7 μmol/L). Direct metabolites monodealkylated, ETHYL (2-

chloro-4ethyl-amino-6amino-1,3,5-triazine) presented a 20 hours peak concentration with maximum concentration at ~3 hours (2.8 μmol/L), and ISOPROPYL (2-chloro-4-amino-6-isoppropylamino-1,3,5 triazine) presented a 18 h concentration peak with maximum concentration at ~3 hours (0.8 μmol/L). The subsequent metabolism of these dealkylated intermediates produces the di-dealkylated metabolite, diaminochlorotriazine (DACT) [7] that presented a 70 hours peak in the plasma with maximal concentration at -18 hours. 64% of oral administered atrazine was systemically metabolized by the cytochrome P450 family in the intestine to the monodealkylated metabolites [5].

When a xenobiotic enters the cell, is induced a biotransformation process to avoid damage and to facilitate its excretion. In mammals, the biotransformation processes occurs mainly in the liver, but all cells can do this. Neurons, skin and kidney cells, among others, have active biotransformation systems [8]. As described above, the biotransformation of triazines that occurs in the absorptive cells of the intestine is very important mainly for atrazine.

In the cells the first line of biotransformation is the microsomal, located at smooth endoplasmic reticulum, performed by cytochrome P450 system superfamily (CYP). These systems catalyze oxidation of substrates by O_2. One atom enters the substrate and the other forms H_2O. A reducing agent is required in liver is the NADPH [8]. The overall reaction is described by the reaction below where SH represents the substrate and RH_2 the reducing agent:

$$SH + O_2 + RH_2 \longrightarrow SOH + R + H_2O$$

Overload of CYP activity induces leaking of electrons leading to O_2 univalent reduction producing superoxide radical ($O_2^{\cdot-}$). This radical can start a series of cascade of oxidative reactions mediated by reactive species which can lead to cellular damage. To resist oxidative attack, cells and biological fluids possess two antioxidant defense systems. The first is the enzymatic antioxidant system constituted, among others, by the activity of the enzymes superoxide dismutase (SOD), catalase and the enzymes of glutathione redox system. Reduced thiol protein groups, metabolites like urate, reduced glutathione (GSH) and birubin, and several dietary products (ascorbate, tocopherols, carotenoids, flavonoids etc.) are examples of second enzymatic system constituents [9].

Glutathione redox system deserves further explanation, because its importance to the cellular redox status maintenance. Changes in the cellular redox state, even very small ones are related to processes of cell damage and also the adaptive processes. Reduced glutathione (GSH) is a tripeptide synthesized by the enzyme system controlled by glutathione sintethase, mainly in liver [8, 10]. Cells can import GSH using the χ-glutamil acil transferase (χ-GT) transporter system of membranes. GSH is the substrate of the enzyme glutathione peroxidase (GPX) used to detoxify H_2O_2 and organic hidroperoxydes, oxidizing GSH to GSSG (oxidized glutathione). To maintain the high GSH/GSSG ratio (100:1), GSSG is reduced to 2GSH by the enzyme glutathione reductase (GR) that uses the reducing power of NADPH. Disturbs in the pentose phosphate pathway (that produces most of NADPH) induces lipid peroxidation in the erythrocyte membrane [10].

In the liver, the dominant metabolic pathway of atrazine and simazine in mammals is N-monodealkylation and isopropylhydroxilation, by the cytochrome P450 metabolic system.

Also reported is that atrazine and simazine are metabolized to N-deethylated, N-deisopropylated and isopropylhydroxylated products by the P450 system in rats [7].

Other important reactions of biotransformation are the addition to xenobiotics of GSH or glycolic acid to increase water solubility to facilitate excretion. Were found in blood and urine adults of chlorinated metabolites of chlorotriazines (atrazine and simazine) conjugated with GSH [5]. Literature presents increase of activity of the enzyme glutathione-s-transferase (GST), that add GSH to xenobiotics, in the liver in the presence of atrazine, and diminishment of antioxidant enzimes activity in peripheral susceptive tissues as testicular interstitial cells [11] leading to an oxidative situation.

Both biotransformation systems induced by the herbicides can induce or increase an oxidative stress situation. As described above the CYP activity overload increases oxidative species production, and the complexation reactions decreases the GSH/GSSG ratio because the use of GSH. During oxidative stress, the reactive species and/or it derivatives can attack lipids, proteins and nucleic acid molecules causing damages that cause metabolic alterations which may lead, in extreme conditions, to cellular death. Oxidative stress has been associated with the onset or progression of many diseases such atherosclerosis, cancer, psoriasis, Alzheimer, hypertension and heart and liver diseases and with important physiological processes such aging and physical exercise training [8, 12].

Other authors demonstrated triazine herbicide established oxidative stress in brain and kidney [13], brain, muscle and liver [14], liver [15, 16] erythrocytes [17, 18] of many animal models.

TRIAZINES AND CELLULAR DAMAGE

Catabolism of atrazine and other chlorotriazines, including simazine and propazine, occurs in animals by dealkylation, dechlorination, and conjugation [19]. After these processes, the metabolites will be capable of interact with cells through biomolecules affecting cellular, tissue and organisms metabolism. Baker et al. [20] demonstrated that triazines negatively affected amphibian survival and highlighted that understanding how different chemical classes of pesticides and fertilizers interact with amphibian populations can lead to new management practices and regulations. Cellular markers applied *in vivo* or *in vitro* can lead valuable information in many different levels of investigation using varied animals groups. Besides, a lot of research resulting from atrazine effects evaluations [21] and ametryn and simazyne effects are little known.

Pathological changes are powerful indicators of exposure to environmental stressors. However, studies concerning the effects of pesticide exposure on tissue histopathology are scarce. Common carp sub-chronically exposed to atrazine contamination were negatively affected at the tissue level. Atrazine exposure caused alterations to the brain and kidney structure of the common carp, as evidenced by the degeneration of Purkinje cells in the brain and hydropic degeneration of the kidney. Different degrees of granule cell loss in the hippocampus, reduction of Nissl bodies, degeneration of Purkinje cells, neuropil loss were observed. The kidney of common carp displaying different degrees of cloudy swelling of epithelial cells of renal tubules, necrosis in the tubular epithelium, contraction of the glomerulus and expansion of Bowman's space. So far, SOD, GSH-Px and CAT activities in the brain and kidney decreased after atrazine exposure [13].

Subchronic exposure of *Prochilodus lineatus* to 2, 10 and 25 μg/L of atrazine changes mechanisms of osmo and ionic regulation but does not result in significant changes in the Na^+/Cl^- ratio. Morphological responses exhibited by the gill cells explain, at least in part, the mechanisms of ion uptake maintenance and compensate for the possible effects of atrazine on the gills. Increase in plasma ion levels was observed that may have a cumulative effect leading to the significant increase observed in osmolality. These changes may be related to water shunts between plasma and interstitial fluid, suggesting altered water homeostasis and/or the changes in the concentration of metabolites in the blood [22].

Liu et al. [23] provided a partially understanding of the cytotoxic effects on fish cells caused by atrazine. These authors demonstrated that atrazine exhibited cytotoxic effects in cultured carp cells ZC7901 because of the induction of apoptosis.

Proteomic analysis was realized in adult female zebrafish (*Danio rerio*) liver exposed to atrazine. Several upregulated proteins were identified as 4-Hydroxyphenylpyruvate dioxygenase (HPPD) and heat shock protein (HSP) 10, which are responsible for stress response in zebrafish. In contrast, the main downregulated spots were identified as class III alcohol dehydrogenase, fatty acid-binding protein 7, and coatomer protein complex subunit zeta 1. These dates can aid to the development of new biomarkers that will be specific for pesticides allowing us to better understand the underlying mechanisms of atrazine-induced toxicity [24].

Atrazine sub-chronic exposure to 400 mg/kg/day concentration was able to induce hepatic rat oxidative stress, which resulted in higher levels of catalase and HSP90 but did not alter HSP70 levels or the levels of antioxidant enzymes (SOD and GST). Was also induced lipid peroxidation (LPO), hepatic degeneration accompanied by hepatocyte death, and the formation of micronuclei, confirming the cytotoxic and mutagenic potential of this herbicide [16].

Erythrocytes are highly susceptible to oxidative damage due to the presence of heme iron, polyunsaturated fatty acids (PUFA) and oxygen, which may initiate the reactions that induce oxidative changes in these cells. Some of the major alterations observed after atrazine treatment in rat erythrocytes were mild to moderate distortion in shape, significant ruptured membranes, echinocyte formation and central or peripheral protuberances. Authors suggested that these deformations might have occurred due to the underlying deformation of cytoskeleton in response to oxidative stress that induced an increase in lipid peroxidation (LPO) and changes in lipid composition of the membranes [17].

Similarly, Bhatti et al. [18] showed induction of erythrocyte LPO, changes the membrane content, and activities of antioxidant enzymes, suggesting that ROS may be involved in the toxic effects of atrazine. The changes in the cholesterol and phospholipid contents may be responsible for the changes in the activities of the membrane-bound acetylcholinesterase (AChE).

MCF-7 cells (derived from human breast cancer) were exposed to environmentally relevant concentrations of atrazine and were observed that this endocrine disrupting affected the proteomic level in these cells. Proteins belonged to various cellular compartments (nucleus, cytosol, membrane) and predominantly involved in transcription processes, stress regulation and structural components were underexpressed during the atrazine treatment indicating that atrazine treatment seems to decrease the activity of the cells [25].

Although the precise mechanism of action remains to be elucidated at certain tissue sites, studies have demonstrated that atrazine adversely affects the endocrine system and reproductive tissues in the rat. In a review presented by Sifakis et al. [26], pesticides can interfere with the hypothalamopituitary axis that regulates, through the production of the gonadotrophins FSH and LH, the function of Sertoli and Leydig cells, impairing spermatogenesis and steroidogenesis. Atrazine seems to have estrogenic and antiandrogenic properties and was suggested to reduce testicular testosterone and impair semen quality in male rats.

Sertoli-Germ Cells obtained of immature Wistar rats were submitted to concentrations of atrazine that corresponds to atrazine testicular tissue levels *in vivo* at the effect dose of 50 mg/kg/day. Atrazine-induced oxidative stress and decreased cell viability and increase of LPO and ROS production after a shorter period of culture than cell death, suggesting that atrazine-induced oxidative stress is the mechanism for reduction in cell viability. In addition, the increase in the activities for GPx and GR and their mRNA levels clearly show that the observed changes in enzyme activities are due to upregulated transcription of these genes. Increase in the antioxidant gene expression seems to be an adaptive response to oxidative stress, especially when atrazine has been reported to induce lipid peroxidation in several models [27].

According Pogrmic et al. [28], exposure to atrazine affects Leydig cell steroidogenesis via the inhibition of steroidogenesis gene expression, which is accompanied by decreased androgenesis. An interesting study developed by Pogrmic-Majkic's research group [29] to investigate direct effects of atrazine on rats Leydig cell steroidogenesis and the possible mechanisms of actions. They showed stimulatory effects of atrazine on cAMP accumulation and androgen production during the first 3 days of *in vivo* treatment (200 mg/kg body weight, by gavage) followed by a decline during further treatment, what indicates that atrazine has a transient stimulatory action on cAMP signaling pathway in Leydig cells and that further study would clarify duration of treatment when stimulatory atrazine action turns to inhibition.

Date yet no published obtained by our research group showed that an another triazine, ametryn, when orally administered for a chronic period was also capable to reduce sperm and cell Leydig number probably in consequence of oxidative stress induction, causing an impact on rat reproductive capacity. It is probably that ametryn have a similar atrazine mechanism of action.

Triazine Effects on Genetic Material

Excessive generation of intracellular ROS submits the organism to a process of oxidative stress and DNA damage [30, 31], leading to DNA base changes, single and double strand DNA breakage (SSB and DSB), and lesions in apurinic or apyrimidinic sites (AP sites) [32].

Oliveira-Brett and Silva [33] proposed that besides ROS induction, triazine herbicides can also interact with DNA directly through insertion mechanisms and formation of adducts between purine bases (Adenine and Guanine).

The agents that interact with DNA or its cellular components (spindle fibers) and enzymes (topoisomerase) are known to be genotoxic. The term genotoxicity refers to the nuclear changes caused by strand DNA breakage, abnormal DNA synthesis, and exchanges between sister chromatids. The genotoxic effects may be transient and susceptible to repairs, while mutagenic effects feature permanent changes in the content or structure of the genetic material of an organism [34].

There are different biological assays for determining the genotoxicity and mutagenicity of triazine herbicides in vertebrate organisms. However, the main assays are the comet and micronucleus assay (Mn), which can be performed *in vitro* or *in vivo*. The comet assay, also known as SCGE (single-cell gel electrophoresis) consists on DNA migration over a blade covered with agarose under electrophoretic conditions. The material is observed with a fluorescence microscope, and damaged cells show the appearance of a comet, including a head (nuclear region) and a tail containing the fragments of DNA. This assay is fast, inexpensive, and sensitive enough to assess the primary damages in the genetic material, and can be applied in different tissues and specific cells [35- 38].

Among the available assays for evaluating the xenobiotic mutagenicity, the Micronucleus assay (Mn) in bone marrow and peripheral blood of rats and mice is one of the most accepted tests in the field of genetic toxicology because it allows the systematic evaluation of the substance of interest, and the metabolic activation can be a differential parameter *in vivo* and *in vitro* studies [39, 40]. As a general rule, the formation of micronuclei is the result of lagging acentric fragments to be deleted from the nucleus during mitosis [41]. Micronuclei appear in daughter cells due to, unrepaired or incorrectly repaired damages, induced in the parental cells that are formed during mitosis (telophase) or meiosis [42].

Information on the genotoxic potential of triazine herbicides is still contradictory. Several test organisms have been exposed to different concentrations and during different exposure periods for elucidation of the biological responses caused by these compounds on the genetic material.

Fish represents a standard model for the study of aquatic ecosystems because they are exposed directly to chemicals and mutagens from agricultural production runoff, or indirectly through the ecosystem food chain [43].

Cavas [44] evaluated the genotoxicity of the Gesaprim® herbicide, which has atrazine is the active principle, in peripheral blood erythrocytes of the fish *Carassius auratus* L., 1758, (Pisces: Cyprinidae) using the Mn and comet assays. The fishes were exposed to concentrations of 5, 10 and 15 mg/L of Gesaprim® for a period of 2, 4 and 6 days. An increase in the frequency of micronuclei and DNA damage were observed in all treatments, showing a genotoxic potential of this herbicide. Concordant results were found by Campos Ventura et al. [45] in peripheral blood erythrocytes of the fish *Oreochromis niloticus*, treated for 72 h with concentrations of 6.5, 12.5 and 25 mg/L of atrazine. Genotoxic effects were also observed by Nwani et al. [46] in erythrocytes and gill cells of the fish *Channa punctatusi* after concentrations of 8.4 mg/L on the seventh day of exposure (Figure 2). These results support that triazine shows a genotoxic potential on aquatic organisms.

Genotoxic effects of triazine herbicides have been found *in vivo* tests in rodents, suggesting a possible genotoxic potential of these herbicides in mammals. Singh et al. [17] evaluated the atrazine genotoxicity in male Wistar rats. The animals were treated with 300mg/kg during a period of 7, 14 and 21 days, and shown a significant increase ($p < 0.001$) in the tail length of comets and a significant increase in the frequency of micronuclei in blood and liver cells.

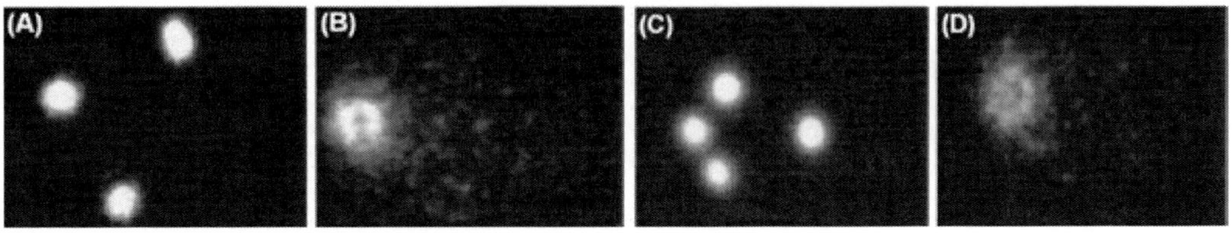

Figure 2. (A) Control, (B) atrazine exposed gill cells; (C) Control, and (D) atrazine exposed erytrocyte cells (adaptated of Nwani et al. [46]).

Campos-Pereira et al. [16] evaluated the bone marrow of the same model treated with 400 mg/kg body weight of atrazine for a period of 14 days, and observed an increase in the frequency of micronucleated polychromatic erythrocytes. However, Kligerman et al. [47], evaluated the genotoxicity of three triazine herbicides in the mouse bone marrow in different concentrations: Atrazine (125, 250 and 500 mg/Kg), simazine (2000 mg/kg) and cyamazine (100, 200, 300 and 400 mg/kg). They did not observe increases in micronuclei. Their results suggest that the triazine herbicides not have genotoxic potential in bone marrow Mn under the tested conditions. Some authors suggest that the toxic potential of some triazine herbicides can be reduced because of a conjugation with reduced glutathione (GSH) caused by the molecular structure of these herbicides and by the glutathione-S-transferase (GST) action [15].

Genotoxicity of the simazine herbicide was assessed through different techniques. The results were concordant *in vivo* and *in vitro* studies with no changes in the measured genetic parameters. The Mn in bone marrow and peripheral blood of rodents did not show significant changes, as well as tests of sister chromatid exchanges in Chinese hamster cells [48]. The information available in the literature suggests that simazine herbicide do not show genotoxic or mutagenic potential.

Ametryn herbicide has been poorly studied for biological effects in living organisms, but the genotoxic potential was verified by Maya-Flores et al. [49] and Calderon-Segura et al. [55]. Both groups studied the genotoxicity of this compound with direct application in cultured human lymphocytes, and with the application of extracts from the roots of *Vicia faba* treated with different concentrations of ametryn. The direct treatment performed with the herbicide did not induce sister chromatid exchanges, and no change in the comet assay was observed. However, genotoxic effects were observed in cells treated with the extracts of the roots of *V. faba* exposed to ametryn. These results indicate that this herbicide goes through a process of biological activation and produces active metabolites when metabolized by the plant. These metabolites are able to induce genetic damage in human lymphocytes cultures. The genotoxic effects using different assays systems are listed in table 1.

Triazine herbicides were extensively evaluated in various genotoxicity assays and different doses and concentrations were tested under different conditions. However, the results on its mutagenic potential are inconclusive [56].

Experimental studies have associated triazine herbicides to the development of cancer in rats. Sprague-Dawley rats developed mammary adenocarcinoma cells when exposed to atrazine, suggesting this herbicide could possibly be classified as a probable human carcinogen. However, more elaborate studies concluded that the mechanisms involved in tumor formation had a hormonal origin and were not associated with DNA damage. These results supported the classification of atrazine as "not likely to be carcinogenic in humans" by the United States Environmental Protection Agency (U.S. EPA) [57].

Table 1. Summary of positive results about genotoxicity of triazines

Test Organisms	DNA effect	Triazine	Dose	References
Japanese quail	Increased comet assay	Atrazine	500 mg/kg	Hussain et al., 2011 [50]
Peripheral blood leukocytes of mice	Increased comet assay	Atrazine Cyanazine	500 mg/kg 200/300 mg/kg	Tennant et al., 2001 [51]
Workers occupationally exposed	Increased comet assay Micronucleus test	Mixture of pesticides (atrazine, cyanazine)	——	Garaj-Vrhovac, Zeljezic, 2002 [52]
Multiple mouse tissues	Increased comet assay	Gesaprim® (atrazine)	540 mg/kg	Zeljezic et al., 2004 [53]
Human lymphocytes *(in vitro)*	Increased comet assay	Gesaprim® (atrazine)	0.47 e 4.7 μg/mL	Zeljezic et al., 2006 [54]
Human lymphocytes *(in vitro)*	Increased comet assay	Ametryn	50-500 mg/L	Calderon-Segura et al., 2007 [55]
Oreochromis niloticus (Fish)	Increased comet assay Micronucleus test	Atrazine	6.5, 12.5 and 25 μg/L	Campos Ventura et al., 2008 [45]
Wistar rats (liver and blood cells)	Increased comet assay Micronucleus test	Atrazine	300 mg/kg	Singh et al., 2008 [17]
Carassius auratus (Fish)	Increased comet assay Micronucleus test	Gesaprim® (atrazine)	5, 10 e 15 μg/L	Cavas, 2011 [44]
Channa punctatus (Fish)	Increased comet assay Micronucleus test	Rasayanzine	8.4 mg/L	Nwani et al., 2011[46]
Prochilodus lineatus (Fish)	Increased comet assay	Atrazine	10 μg/L	Santos and Martinez, 2012 [15]
Wistar Rats (bone marrow)	Micronucleus test	Atrazine	400 mg/kg	Campos-Pereira et al., 2012 [16]

Studies conducted with workers in contact with triazines show that different types of cancers are reported, however, none of the results are statistically significant. The epidemiological studies do not provide convincing scientific proof of a causal relationship between triazine herbicides and cancer in humans [57-60].

Another interesting aspect is available recent evidence that support the concept that epigenetics holds substantial potential for furthering our understanding of the molecular mechanisms of pesticides health effects, as well as for predicting health-related risks due to conditions of environmental exposure and individual susceptibility. Epigenetics effects are heritable changes in gene expression that occur without a change in the DNA sequence. Although no information about epigenetic effects of triazines, global DNA methylation levels have been reported to be inversely associated with blood levels of persistent organic pollutants (POPs), xenobiotics that accumulate in adipose tissue. So far, some pesticides that belong to the environmental endocrine disruptors (EDs) family, synthetic chemicals that resemble natural hormones, are also known to cause epigenetic perturbations [61].

Toxicity and Cell Death

A variety of environmental contaminants can induce apoptosis, necrosis, or autophagy, depending on both cell type and dose. Then, it is now apparent that multiple cell death programs can be activated during toxicity. It seems more likely that several death executing routines may be activated concomitantly within injured cells and that one or the other becomes predominant, depending on the stimulus and the metabolic state of the tissue. Although the predominance of one or the other death executing mechanism may be dictated by factors as different as energy requirement, signaling molecules or the intensity of a given insult, in many instances, the differentiation program within a given tissue dictates the way to die [62,63].

It is evident that chemical toxicity might be associated with multiple modes of cell death. There is the coexistence of different cell death modalities in pathological settings that involves an intricate cross talk of cellular signaling pathways. Depending on the type of lethal agent, the cell death process can be initiated in different intracellular compartments, and cross talk between these compartments appears essential for all cell death modalities. Importantly, depending on the nature and severity of the stimulus, and on the cell type, the hierarchy of interorganelle cross talk might result in different cell death modalities. Moreover, in some cases, suppression of the function of a particular intracellular compartment might switch one mode of cell death to another [63].

In this scenario, the classification of the cell death in tissues and organs of animals exposed to environmental chemical compounds become very difficult. Conversely, the presence or the absence of cell death in the histophatological analysis is an indicative of the level of stress triggers by herbicides in exposed animals and its detection is an important tool in ecotoxicological studies. On the basis of histopatological changes, such as alteration of tubular system of caudal kidney, the values of LOEC (lowest observed effect concentration) and NOEC (no observed effect concentration) for simazine were estimated in different developmental stages of common carp (*Cyprinus carpio*) [64].

Triazine herbicides can induce immunotoxicity. Atrazine, for example, is capable of inducing splenocytic apoptosis mediated by the Fas/FasL pathway in mice, which could be the potential mechanism underlying the immunotoxicity of atrazine [65]. Male mice which had been prenatally/lactationally exposed to atrazine had an increase in both T cell proliferation and cytolytic activity [66]. Atrazine-induced immunotoxicity is an example of cell death induced by extracellular stress signals that are sensed and propagated by specific transmembrane receptors. According to review performed by Galluzzi et al. [67], this kind of cell death is termed 'extrinsic apoptosis'.

Extrinsic apoptosis can be initiated by the binding of lethal ligands, such as FAS/CD95 ligand (FASL/CD95L), tumor necrosis factor α (TNFα) and TNF (ligand) superfamily, TNF-related apoptosis inducing ligand (TRAIL), to various death receptors (i.e., FAS/CD95, TNFR1 and TRAILR1–2) [67,68].

The first evidence that atrazine was able to induce apoptosis in fish cells was performed by Liu et al. [23]. In this study, the authors indicated the existence of a novel cytotoxic mechanism caused by atrazine that may improve our understanding of the complex relationship between contaminants and aquatic organisms. Apoptosis induced by atrazine was dose- and time-dependent and was involved in mitochondrial membrane potential (DeltaPsi(m)) disruption, elevation in intracellular $Ca^{2+,}$ generation of reactive oxygen species, and intracellular ATP depletion. According to review performed by Galluzzi et al. [67], this kind of cell death is termed 'intrinsic apoptosis'.

Intrinsic apoptosis can be triggered by a variety of intracellular stress conditions, including DNA damage, oxidative stress, cytosolic Ca^{2+} overload, mild excitotoxicity (related to glutamate receptor overstimulation in the nervous system), accumulation of unfolded proteins in the endoplasmic reticulum (ER) and many others. Although the signaling cascades that trigger intrinsic apoptosis are highly heterogeneous, including exogenous chemical compounds such as toxicants, as far as the initiating stimuli are concerned, they are all wired to a mitochondrion-centered control mechanism. Frequently, along with the propagation of the pro-apoptotic signaling cascade, anti-apoptotic mechanisms are also engaged, in an attempt to allow cells to cope with stress. In this scenario, both pro- and anti-apoptotic signals converge at mitochondrial membranes, which become permeabilized when the former predominate over the latter [69]. Thus, intrinsic apoptosis results from a bioenergetic and metabolic catastrophe coupled to multiple active executioner mechanisms.

Intrinsic apoptosis can be caspase-dependent and caspase-independent. An example of caspase-dependent apoptosis was described by Zaya et al. [70] that performed a study regarding the effects of atrazine on developing *Xenopus laevis* tadpoles. Livers from 400 μg/L exposed tadpoles had higher numbers of activated caspase-3 immunopositive cells suggesting increased rates of apoptosis. It is important to highlight that this kind of apoptosis was detected in tadpoles exposed to low levels of atrazine that are potentially found in puddles, vernal ponds and runoff soon after application (200 and 400 μg/L).

Conversely, caspase activation seems to have a prominent role in a limited number of instances of stress-induced intrinsic apoptosis in vitro, as demonstrated by the fact that – in contrast to extrinsic apoptosis – chemical and/or genetic inhibition of caspases rarely, if ever, confers long-term cytoprotective effects and truly prevents cell death. In this context, caspase inhibition only delays the execution of cell death, which eventually can even exhibit morphological features of necrosis [71-73].

In spite of the recent advances in our understanding of cell death mechanisms and associated signaling networks, much work remains to be done before we can fully appreciate the toxicological significance of these findings. Although it is clear that the activation of cell death pathways is responsible for acute toxicity of many chemical toxicants, their potential involvement in subacute or chronic toxicity caused by long-term exposure to other drugs or environmental pollutants remains to be investigated.

Conclusion

The major triazine effects on vertebrates are summarized at the figure below.

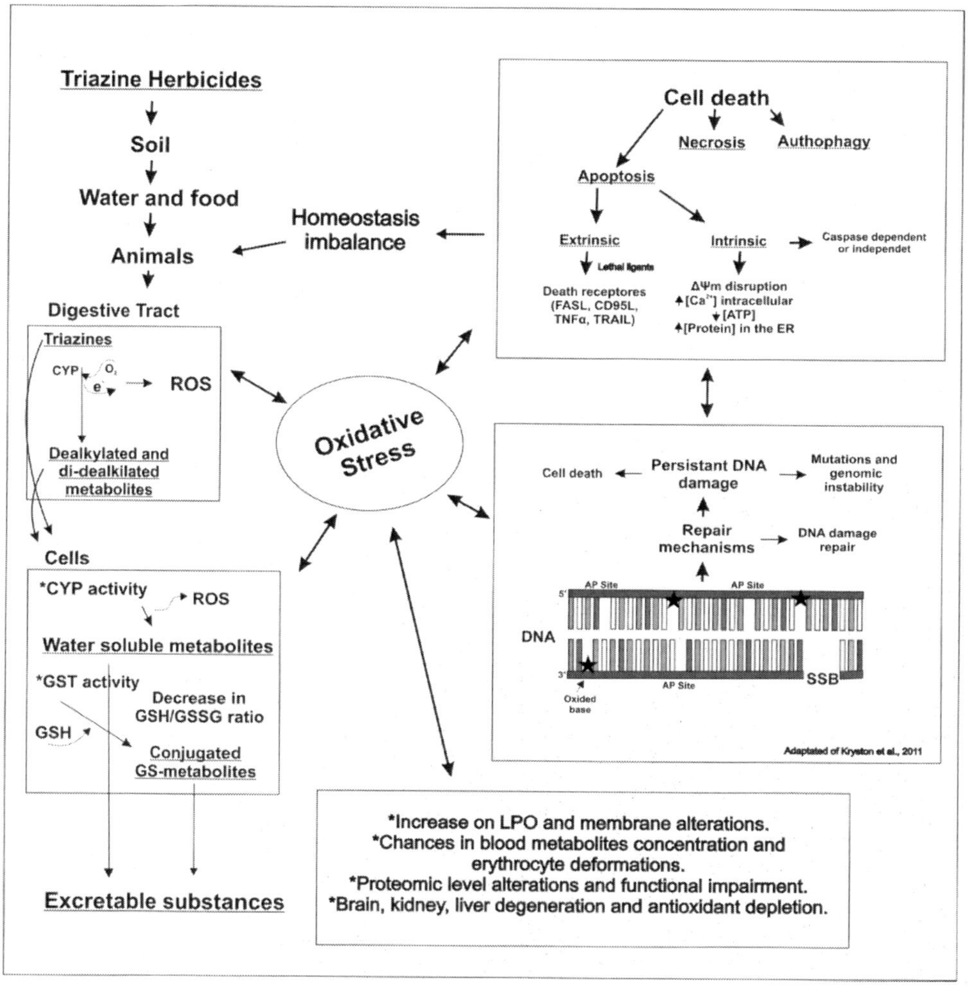

References

[1] Le Baron, H., McFarland, J., Burnside, O. C. (eds). 2008. The triazine herbicides, 1th. Ed. Elsevier.

[2] EPA-738-F-05-007. 2005. Prevention, pesticides and toxic substances September 2005. Avaliable from http://www.epa.gov/oppsrrd1/REDs/factsheets/ametryn_fact.pdf

[3] Kidd, H., James, D.R. (eds.). 1991. The Agrochemicals Handbook, 3th. Ed. Royal Society of Chemistry Information Services, Cambridge, UK.

[4] Agriculture Canada. 1982. Guide to the chemicals used in crop protection 7th. Ed. Publication no. 1093.

[5] McMullin, T.S., Hanneman, H., Cranmer, B. K., Tessari Andersen, M. E. 2007. Oral absorption and oxidative metabolism of atrazine in rats evaluated by physiological modeling approaches. *Toxicology* 240, 1-14.

[6] Singh, M., Sandhir, R., Kiran, R. 2011. Effects on antioxidant status of liver following atrazine exposure and its attenuation by vitamin E. *Exp. Toxicol. Pathol.* 63, 269–276.

[7] Hanioka, N., Jinno, H., Tanaka-Kagawa, T., Nishimura, T., Ando, M. 1999. *In vitro* metabolism of simazine, atrazine and propazine by hepatic cytochrome P450 enzymes of rat, mouse and guinea pig, and oestrogenic activity of chlorotriazines and their main metabolites. *Xenobiotica* 29, 1231-1226.

[8] Halliwell, B., Gutteridge, J. M. C. 2007. In: Free Radicals in Biology and Medicine. 3th ed. Oxford University Press.

[9] Gandra, P. G,. Alves, A. A., Macedo, D. V., Kubota, L. T. 2004. Determinação eletroquímica da capacidade antioxidante para avaliação do exercício físico. *Química Nova* 27, 980-985.

[10] Alves, A. A., Pereira da Silva, L., Macedo, D. V., Kubota, L. T. 2003. Amperometric sensor for glutathione reductase activity determination in erythrocyte hemolysate. *Anal. Biochem.* 323(1), 33-38.

[11] Pogrmic-Majkic, K. Kaisarevic, S., Fa, S., Dakic, V., Glisic, B., Hrubik, J., Kovacevic, R. 2012. Atrazine effects on antioxidant status and xenobiotic metabolizing enzymes after oral administration inperipubertal male rat. *Environ. Toxicol. Pharmacol.* 34, 495–501.

[12] Veloso, A. J.,Chow, A. M., Dhar, D., Tang, D. W. F. 2013. Ganesh H.V.S. Mikhaylichenko S. Brown I.R., Kerman K. Biological activity of sym-Triazines with acetylcholine-like substitutions as modulators of Alzheimer's disease. *ACS Chem. Neurosci.* 4, 924-929.

[13] Xing, H., Li,S., Wang, Z., Gao, X., Xu, S., Wang, S. 2012. Histopathological changes and antioxidant response in brain and kidney of common carp exposed to atrazine and chlorpyrifos. *Chemosphere* 88, 377-383.

[14] Stara, A., Machova, J., Velisek, J. 2012. Effect of chronic exposure to simazine on oxidative stress and antioxidant response in common carp (*Cyprinus carpio* L.) *Environ. Toxicol. Pharmacol.* 33, 334–343.

[15] Santos, T.G., Martinez, C.B.R. 2012. Atrazine promotes biochemical changes and DNA damage in a Neotropical fish species. *Chemosphere* 89, 1118–1125.

[16] Campos-Pereira, F. D., Olveira, C. A., Pigoso, A. A., Silva-Zacarin, E. C. M., Barbieri, R., Spatti, E. F., Marin-Morales, M. A., Severi-Aguiar, G. D. C., 2012. Early cytotoxic and genotoxic effects of atrazine on Wistar rat liver: A morphological, immunohistochemical, biochemical, and molecular study. *Ecotox. Environ. Saf.* 78, 170-177.

[17] Singh, M., Sandhir, R., Kiran, R. Atrazine-Induced Alterations in Rat Erythrocyte Membranes: Ameliorating Effect of Vitamin E. 2008. *J. Biochem. Molec. Toxicol.* 22(5), 363-369.

[18] Bhatti, J. S., Sidhu, I. P. S., Bhatti, G. K. 2011. Ameliorative action of melatonin on oxidative damage induced by atrazine toxicity in rat erythrocytes. *Mol. Cell. Biochem.* 353, 139–149.

[19] Stanko, J, P., Enoch, R. R., Rayner, J. L., Davis, C. C., Wolf, D. C., Malarkey, D. E., Fenton, S. E. 2010. Effects of Prenatal Exposure to a Low Dose Atrazine Metabolite

Mixture on Pubertal Timing and Prostate Development of Male Long Evans Rats. *Reprod. Toxicol.* 30(4), 540–549.

[20] Baker, N.J., Betsy A. Bancroft, B.A., Garcia, T.S., 2013. A meta-analysis of the effects of pesticides and fertilizers on survival and growth of amphibians. *Science of the Total Environment* 449, 150–156.

[21] Severi-Aguiar, G. D. C., Silva-Zacarin, E. C. M., 2011. Effects of herbicide atrazine in experimental animal models. In: Mohammed Naguib Abd El-Ghany Hasaneen (Ed.), Herbicides – properties, synthesis and control. Croatia: InTech, p. 285-296. (ISBN 979-953-307-729-0).

[22] Paulino, M. G., Sakuragui, M. M., Fernandes, M. N. 2012. Effects of atrazine on the gill cells and ionic balance in a neotropical fish, *Prochilodus lineatus*. *Chemosphere* 86, 1-7.

[23] Liu, X. M, Shao, J. Z., Xiang, L.X., Chen, X. Y. 2006. Cytotoxic Effects and Apoptosis Induction of Atrazine in a Grass Carp (*Ctenopharyngodon idellus*) Cell Line. *Environ. Toxicol.* 21, 80-90.

[24] Jin, Y., Zhang, X., Lu, D., Fu, Z. 2012. Proteomic Analysis of Hepatic Tissue in Adult Female Zebrafish (*Danio rerio*) Exposed to Atrazine. *Arch. Environ. Contam. Toxicol.* 62, 127–134.

[25] Lasserre, J. P., Fack, F., Revets, D., Planchon, S., Renaut, J., Hoffmann, L., Gutleb, A. C., Muller, C. P., Bohn, T. 2009. Effects of the Endocrine Disruptors Atrazine and PCB 153 on the Protein Expression of MCF-7 Human Cells. *J. Proteome Res.* 8, 5485-5496.

[26] Sifakis S., Mparmpas M., Soldin O. P., Tsatsakis A., 2011. Pesticide Exposure and Health Related Issues in Male and Female Reproductive System. In: Pesticides - Formulations, Effects, Fate. Margarita Stoytcheva (Ed.), pp. 495-526, InTech, Croatia, Retrieved from www.intechweb.org/books/show/title/pesticidesformulations-effects-fate, ISBN 978-953-307-532-7.

[27] Abarikwu, A. O., Pant, A. B., Farombi, E. O. 2012. Dietary Antioxidant, Quercetin, Protects Sertoli-Germ Cell Coculture from Atrazine-Induced Oxidative Damage. J. Biochem. Molec. Toxicol. 26(11), 477-485

[28] Pogrmic, K., Fa, S., Dakic, V., Kaisarevic, S., Kovacevic, R. 2009 Atrazine Oral Exposure of Peripubertal Male Rats Downregulates Steroidogenesis Gene Expression in Leydig Cells. *Toxicol. Sci.* 111(1), 189–197.

[29] Pogrmic-Majkic, K., Fa, S., Dakic, V., Kaisarevic, S., Kovacevic, R. 2010 Upregulation of Peripubertal Rat Leydig Cell Steroidogenesis Following 24 h *In Vitro* and *In Vivo* Exposure to Atrazine. *Toxicol. Sci.* 118(1), 52–60.

[30] Amado, L. L., Monserrat, J. M. 2010. Oxidative stress generation by microcystins in aquatic animals: why and how. *Environ. Int.* 36, 226-235.

[31] Kryston, T. B., Georgiev, A. B., Pissis, P., Georgakilas, A. G. 2011. Role of oxidative stress and DNA damage in human carcinogenesis. *Mutat Res.* 711, 193-201.

[32] Salmon, T. B., Evert, B. A, Song, B., Doetsch, P. W. 2004. Biological consequences of oxidative stress-induced DNA damage in *Saccharomyces cerevisiae*. *Nucleic. Acids. Res.* 32, 3712-3723.

[33] Oliveira-Brentt, A. M., da Silva, L. A. 2002. A DNA-electrochemical biosensor for screening environmental damage caused by s-triazine derivatives. *Anal. Bioanal. Chem.* 373, 717-723.

[34] Dearfield, K. L., Cimino, M. C., McCarroll, N. E., Mauer, I., Valcovic, L .R., U.S. Environmental Protection Agency. 2002. Genotoxicity risk assessment: a proposed classification strategy. *Mutat. Res.* 521, 121-135.

[35] Hartmann, A., Agurell, E., Beevers, C., Brendler-Schwaab, S., Burlinson, B., Clay, P., Collins, A., Smith, A., Speit, G., Thybaud, V., Tice, R. R. 2003. Recommendations for conducting the *in vivo* alkaline Comet assay. *Mutagenesis* 18, 45–51.

[36] Brendler-Schwaab, S., Hartmann, A., Pfuhler, S., Speit, G. 2005. The *in vivo* comet assay: use and status in genotoxicity testing. *Mutagenesis* 20, 245-254.

[37] Collins, A. R., Oscoz, A. A., Brunborg, G., Gaivao, I., Giovannelli, L., Kruszewski, M., Smith, C.C., Stetina, R. 2008. The comet assay: topical issues. *Mutagenesis* 23, 143–151.

[38] Speit, G., Vasquez, M., Hartmann, A. 2009. The comet assay as an indicator test for germ cell genotoxicity. *Mutat Res.* 681, 3-12.

[39] Tweats, D. J., Blakey, D., Heflich, R. H., Jacobs, A., Jacobsen, S. D., Morita, T., Nohmi, T. O'donovan, M.'R., Sasaki, Y. F., Sofunij, T., Tice, R. 2007. Report of the IWGT working group on strategies and interpretation of regulatory *in vivo* tests I. Increases in micronucleated bone marrow cells in rodents that do not indicate genotoxic hazards. *Mutat. Res.* 627, 78–91.

[40] Valadares, M. C., Castro, N. C., Cunha, L. C. 2007. *Synadenium umbellatum*: cytotoxicity and DNA damage to bone marrow cells from mice. *Rev. Bras. Cienc. Farm.* 43, 631-638.

[41] Kirsch-Volders, M., Plas, G., Elhajouji, A., Lukamowicz, M., Gonzalez, L., Vande Loock, K., Decordier, I. 2011. The *in vitro* MN assay in 2011: origin and fate, biological significance, protocols, high throughput methodologies and toxicological relevance. *Arch. Toxicol.* 85, 873-899.

[42] Ribeiro, L. R. 2003. Teste do micronúcleo em medula óssea de roedores *in vivo*. In: Ribeiro, L. R., Salvadori, D. M. F., Marques, E. K. (Org.). Mutagênese Ambiental. Canoas: Ulbra, p.355. (ISBN 85-7528-067-8).

[43] Lakra, W. S., Nagpure, N. S. 2009. Genotoxicological studies in fishes: a review. *Indiano J. Anim. Sci.* 79, 93-98.

[44] Cavas, T. 2011. *In vivo* genotoxicity evaluation of atrazine and atrazine-based herbicide on fish Carassius auratus using the micronucleus test and the comet assay. *Food Chem. Toxicol.* 49, 1431-1435.

[45] Campos Ventura, B., De-Angelis, D. F., Marin-Morales, M. A. 2008. Mutagenic and genotoxic effects of the atrazine herbicide in *Oreochromis niloticus* (Perciformes, Cichlidae) detected by the micronuclei test and the comet assay. *Pesticide Biochem. Physiol.* 90, 42–51.

[46] Nwani, C. D., Lakra, W. S., Nagpure, N. S., Kumar, R., Kushwaha, B., Srivastava, K. 2011. Toxicity of the Herbicide Atrazine: Effects on Lipid Peroxidation and Activities of Antioxidant Enzymes in the Freshwater Fish *Channa Punctatus* (Bloch) *Int. J. Environ. Res. Public Health.* 7, 3298-3312.

[47] Kligerman, A. D., Doerr, C. L., Tennant, A. H., Peng, B. 2000. Cytogenetic studies of three triazine herbicides. II. *In vivo* micronucleus studies in mouse bone marrow. *Mutat. Res.* 471, 107-112.

[48] Suárez, S., Rubio, A., Sueiro, R. A., Garrido, J. 2003. Sister chromatid exchanges and micronuclei analysis in lymphocytes of men exposed to simazine through drinking water. *Mut. Res.* 537, 141–149.

[49] Flores-Maya, S., Gómez-Arroyo, S., Calderón-Segura, M. E., Villalobos-Pietrini, R., Waliszewski, S. M., de la Cruz, L. G. 2005. Promutagen activation of triazine herbicides metribuzin and ametryn through Vicia faba metabolism inducing sister chromatid exchanges in human lymphocytes in vitro and in *V. faba* root tip meristems. Toxicol. *In Vitro* 19, 243-251.

[50] Hussain, R., Mahmood, F., Khan, M. Z., Khan, A., Muhammad, F. 2011. Pathological and genotoxic effects of atrazine in male Japanese quail (*Coturnix japonica*). *Ecotoxicology,* 20, 1–8.

[51] Tennant, A. H., Peng, B., Kligerman, A. D. 2001. Genotoxicity studies of three triazine herbicides: *in vivo* studies using the alkaline single cell gel (SCG) assay. *Mutat. Res.* 493, 1-10.

[52] Garaj-Vrhovac, V., Zeljezic, D. 2002. Assessment of genoma damage in a population of Croatian workers employed in pesticide production by chromosomal aberration analysis, micronucleus assay and Comet assay. *J. App. Toxicol.* 22, 249–255.

[53] Zeljezic, D., Garaj-Vrhovac, V. 2004. Genotoxicity evaluation of pesticide formulations containing alachlor and atrazine in multiple mouse tissues (blood, kidney, liver, bone marrow, spleen) by comet assay. *Neoplasma* 51, 198-203.

[54] Zeljezic, D., Garaj-Vrhovac, V., Perkovic, P. 2006. Evaluation of DNA damage induced by atrazine and atrazine-based herbicide in human lymphocytes *in vitro* using a comet and DNA diffusion assay. *Toxicol. In Vitro* 20, 923-935.

[55] Calderón-Segura, M. E., Gómez-Arroyo, S., Molina-Alvarez, B., Villalobos-Pietrini, R., Calderón-Ezquerro, C., Cortés-Eslava, J., Valencia-Quintana, P. R., López-González, L., Zúñiga-Reyes, R., Sánchez-Rincón, J. 2007. Metabolic activation of herbicide products by *Vicia faba* detected in human peripheral lymphocytes using alkaline single cell gel electrophoresis. *Toxicol. In Vitro* 21, 1143-1154.

[56] Jowa, L., Howd, R. 2011. Should Atrazine and Related Chlorotriazines Be Considered Carcinogenic for Human Health Risk Assessment? *J. Environ. Sci. Health C Carcinog. Ecotoxicol. Rev.* 29, 91-144.

[57] Sathiakumar, N., MacLennan, P. A., Mandel, J., Delzell, E. 2011. A review of epidemiologic studies of triazine herbicides and cancer. *Crit. Rev. Toxicol.* 41, 1–34.

[58] McLennan, P. A., Delzell, E., Sathiakumar, N., Myers, S. L., Cheng, H., Grizzle, W., Chen, V., and, Wu, X. C. 2002. Cancer incidence among triazine herbicide manufacturing workers. *J. Occup. Environ. Med.* 44, 1048–1058.

[59] Gammon, D. W., Aldous, C. N., Carr, W.C. Jr, Sanborn, J. R., Pfeifer, K. F. 2005. A risk assessment of atrazine use in California: human health and ecological aspects. *Pest. Manag. Sci.* 61, 331-355.

[60] Freeman, L. E. B., Rusiecki, J. A., Hoppin, J. A., Lubin, J. H., Koutros, S., Andreotti, G., Zahm, S. H., Hines, C. J., Coble, J. B., Barone-Adesi, F., Sloan, J., Sandler, D. P., Blair, A., Alavanja, M. C. R. 2011. Atrazine and cancer incidence among pesticide applicators in the agricultural health study (1994-2007). *Environ. Health Perspect.* 119, 1253 -1259.

[61] Collotta, M., Bertazzi, P.A., Bollati, V., 2013. Epigenetics and pesticides. *Toxicol.,* 307, 35-41.

[62] Orrenius, S., Zhivotovsky, B. 2006. The future of toxicology–does it matter how cells die? *Chem. Res. Toxicol.* 19, 729–733.

[63] Orrenius, S., Nicotera, P., Zhivotovsky, B. 2011. Cell Death Mechanisms and Their Implications in Toxicology. *Toxicol. Sci.* 119(1), 3–19.

[64] Velisek, J., Stara, A., Machova, J., Dvorak, P., Zuskova, E., Svobodova, Z. 2012. Effects of low-concentrations of simazine on early life stages of common carp (*Cyprinus carpio* L.). *Neuro. Endocrinol. Lett.* 33(Suppl 3), 90-5.

[65] Xiaofeng Zhang, X., Wang, M., Gao, S., Ren, R., Zheng, J., Zhang, Y. 2011. Atrazine-induced apoptosis of splenocytes in BALB/C mice. *BMC Med.* 9, 117.

[66] Rowe, A. M., Kathleen, M., Brundage, K.M., Schafer, R., Barnett, J.B. 2007. Immunomodulatory effects of maternal atrazine exposure on male Balb/c mice. *Toxicol. Appl. Pharmacol.* 214, 69-77.

[67] Galluzzi, L. et al. 2012. Molecular definitions of cell death subroutines: recommendations of the Nomenclature Committee on Cell Death 2012. *Cell Death and Differentiation* 19, 107–120.

[68] Wajant H. 2002. The Fas signaling pathway: more than a paradigm. Science 296, 1635–1636.

[69] Kroemer, G., Galluzzi, L., Brenner, C. 2007. Mitochondrial membrane permeabilization in cell death. *Physiol. Rev.* 87, 99–163.

[70] Zaya, R. M., Amini, Z., Whitaker, A. S., Kohler, S. L., Ide, C. F. 2011. Atrazine exposure affects growth, body condition and liver health in *Xenopus laevis* tadpoles. *Aquat. Toxicol.* 104 (3-4), 243-53.

[71] Lemaire, C., Andreau, K., Souvannavong, V., Adam, A. 1998. Inhibition of caspase activity induces a switch from apoptosis to necrosis. *FEBS* Lett. 425, 266–270.

[72] Chipuk, J. E., Green, D. R. 2005. Do inducers of apoptosis trigger caspase-independent cell death? *Nat. Rev. Mol. Cell Biol.* 6, 268–275.

[73] Kroemer, G., Martin, S. J. 2005. Caspase-independent cell death. *Nat. Med.* 11, 725–730.

In: Handbook on Herbicides
Editors: Daiki Kobayashi and Eito Watanabe
ISBN: 978-1-62948-053-4

Chapter 7

PHOTOLYSIS EXPERIMENTS ON ALLOXYDIM HERBICIDE AND BIOLOGICAL RESPONSE OF ITS TRANSFORMATION PRODUCT

P. Sandín-España[1,*], B. Sevilla-Morán[1], M. Villarroya-Ferruz[1], J. L. Alonso-Prados[1] and I. Santín-Montanyá[2]

[1]DTEVPF - Unit of Plant Protection Products, Instituto Nacional de Investigación y Tecnología Agraria y Alimentaria (INIA), Madrid, Spain
[2]Plant Protection Department, Instituto Nacional de Investigación y Tecnología Agraria y Alimentaria (INIA), Madrid, Spain

ABSTRACT

The photochemical degradation of alloxydim herbicide was investigated in different organic solvents simulating natural constituents of plant cuticles. Methanol and n-hexane were used as surrogates for primary alcohol groups and saturated hydrocarbon chains, respectively. Alloxydim was rapidly photodegraded in both solvents, showing a half-life of 21.94 ± 0.51 min in methanol and 7.5 ± 0.75 min in n-hexane.

During the photolysis, two transformation products were detected and their identification was achieved using liquid chromatography coupled to quadrupole time-of-flight mass spectrometry with electrospray ionization (LC-ESI-Qtof). The main photolysis reaction was the reduction of the oxime group leading to the formation of imine compound, whereas the isomerization of C-N bond was observed to a lesser extent.

Imine by-product was stable and it was isolated in order to study its phytotoxicity in a succeeding cereal crop (wheat) and a grass weed (Bromus). Alloxydim herbicide was more toxic than its main by-product to the tested species, in both bioassays tested, seed germination and plant growth. For instance, in seed germination bioassay, EC50 for alloxydim ranged from 0.33 to 0.88 ppm while for its by-product ranged from 100 to 277 ppm. Roots of all species were more sensitive than coleoptile to both compounds.

* Corresponding author: sandin@inia.es.

INTRODUCTION

After application, plant surfaces, especially leaf surfaces, are the first reaction compartment for an herbicide molecule. Furthermore, spray drift after herbicide application is an indirect route by which an herbicide also reaches plant surfaces. Once on the plant surfaces, different degradation pathways are possible such as hydrolysis, oxidation or photolysis.

Photodegradation is found to be one of the most important factors in dissipation process for pesticides on plant surfaces [1-4].

Many herbicides are hydrophobic molecules, and thus the larger lipid- covered surface of leaves (cuticles) form an ideal sink for accumulation of these compounds [5]. Wax chemistry, especially of epicuticular waxes, has been investigated for many plant species [6]. Waxes are basically classified into even- and odd-carbon-numbered straight chain homologues and cyclic compounds [7]. The first class consists of acids, aldehydes, primary alcohols and alkyl esters whereas the odd-numbered homologues are hydrocarbons, secondary alcohols, ketones and β-diketones. Considering this wax chemistry, photodegradation of pesticides can be studied in simple organic solvents as surrogates of waxes [8-10].

Degradation reactions lead to the formation of new compounds with unknown properties. In some cases, herbicide degradates can either be more persistent [11] and/or toxic [12, 13] than their parent compound. Moreover, in some cases it has been suggested that herbicidal activity were due to the stable transformation products. Thus, obtaining data on parent compounds and their primary degradation products is critical form understanding the fate of herbicides in the field. However identify degradation products is an analytical challenge since the majority of by-products are new compounds, for which analytical standards are not available.

Alloxydim (Figure 4) was developed by BASF AG for post-emergence control of grass weeds and volunteer cereals in sugar beet, vegetables and broad-leaved crops and it is applied at doses between 0.5 - 1.0 kg a.s./ha. Alloxydim is a selective systemic herbicide from the cyclohexanedione oxime class. This family of compounds has been developed during the last thirty years as post-emergence herbicides that inhibit acetyl-CoA carboxylase [14, 15], the enzyme that catalyses fatty acid synthesis. The cyclohexanedione ring and the ethoxyimine group linked to ring position 3 seem to be essential for the phytotoxic activity, whereas substitutions in position 6 may vary considerably without large differences on herbicidal activity [15].

Regarding the photochemical behaviour of alloxydim it has been mainly investigated in aqueous samples [16, 17]. However, to the best of our knowledge, this is the first time that the photodegradation of alloxydim in models resembling plant leaves has been studied.

In our research group a photodegradation study of alloxydim was performed under simulate solar irradiation (Suntest apparatus) at different irradiation intensities in aqueous media. The results demonstrated that degradation rate in direct photolysis was higher as the irradiation intensities was increased. Moreover, indirect photolysis of the tested herbicide was investigated under the presence of humic acids (HA), nitrate and Fe (III) ions. It was observed that HA inhibited the photolysis kinetic whereas rate constants measured in the presence of nitrate ions indicated no effect on degradation. On the contrary, Fe (III) ions enhanced the photodegradation rate of alloxydim. Half-lives ranged from 165.78 to 4.63 min for different intensities in direct photolysis and from 104.81 to 1.14 min for indirect photolysis.

Furthermore, we have investigated the photochemical degradation of the herbicide alloxydim in different aqueous media such as ground, river, mineral and ultrapure water under natural and simulated solar irradiation [17]. The study revealed that alloxydim half-lives depend on the composition of the water and the radiation source showing a low persistence in natural waters under sunlight and simulated solar irradiation photolysis.

As for degradation products (DPs), two DPs have been identified in these studies [16, 17]. Both DPs are more polar than alloxydim which increase their solubility and could make them easily leach to groundwater and potentially contaminate drinking water sources. However, isomer *Z* is formed temporarily and it is not stable. The major DP has been isolated by means of solid phase extraction and the use of NMR techniques made possible to elucidate its structure as a mixture of imine tautomers which were not described in literature till now. This by-product seems stable and could accumulate in the environment.

Hashimoto et al. (1979) [18] observed that the main photoproducts of alloxydim on soybean plants were imine and two isomeric oxazoles of alloxydim. Similar results were obtained by Soeda et al. (1979) [19], in acetone solution, on TLC plate and on sugar beet plants. Ono et al. (1984) [20] observed the photodegradation of alloxydim in soil under natural sunlight to imine, two isomeric oxazoles of alloxydim, and other by-products.

As mentioned before, some degradation products can be more toxic than the parent compound. Regarding cyclohexanediones chemical class, few studies have been performed to evaluate the toxicity of their degradation products. In this sense, we have study the toxicity of irradiated solutions by means of a bioassay following the standard microtest procedure based on the decreasing of light emission by the marine bacterium *Vibrio fischeri*. The results indicated that the toxicity of the photoproducts was higher than the toxicity of the parent compound.

In this sense, bioassays offer several advantages (short, quick and simple methods) as the detection of very low phytotoxic residues and the detection of its bioavailability. The laboratory bioassay is the first step to investigate the involvement of herbicides and their by-products on plant growth and represent an important tool in studies of crop selectivity, herbicide resistance development and herbicide resistant weeds detection [21, 52]. Besides, the development of bioassays using sensitive species is necessary to detect the presence of phytotoxic levels of herbicide and by-products in growth medium [22, 23].

Among the different types of diagnostic test, soil-free assays (e.g. Petri dishes, pots and *in vitro* tests) are the most attractive for early detection of herbicide phytotoxicity. Seed germination and plant growth bioassays are the most common techniques used to evaluate herbicide phytotoxicity [24]. Both types of bioassays have been widely accepted as main parameters to monitor growth responses to different chemicals. Numerous studies also include plant growth bioassays but the effect on germination is often not separated from the effect on growth [25]. A large number of studies have been carried out with different plant species [22, 26-29]. Itävaara et al. (1997) [30] proposed a lettuce seed germination test in Petri dishes as a potential herbicide toxicity test and observed that there was a low sensibility. Fuentes et al. (2004) [31] observed that seed germination was considered as a less sensitive method compare to root length when used as a bioassay for the evaluation of phytotoxicity. According to findings by Kapustka (1997) [32] and Araujo & Monteiro (2005) [33], the seed germination bioassay could be relatively low sensitive to many toxic substances, because many chemicals may not be absorbed by seeds and the embryonic plant draws its nutritional requirements internally from seed stored materials and is effectively isolated from the environment.

In general, seed germination bioassays are conducted in petri dishes by placing seeds of receiver species on substrata (often filter paper) moistened with the herbicide solution. The petri dishes are placed in an incubator under controlled light and dark periods are regularly checked for their germination, usually up to seven days. Data taken in the end is used to calculate percentage germination, which is mostly served to validate phytotoxicity in ecosystems or agro-systems. But this has received great criticism because some researchers have proved that percentage germination do not provide sufficient information on the potential activity of the herbicide [34]. It has been found difficult to interpret data from seed germination bioassays about the most effective herbicide or by-product and the rates that can be selected for greenhouse and field studies. To overcome this dilemma many investigators have used different indices to show chemical effects on germination [35-37]. These indices have been classified in maximum percentage germination, germination capacity, germination progress and shape of germination curve [38].

Plant growth bioassays under hydroponic conditions promote herbicide activity because allow the maximum herbicide bioavailability to the plant, all the roots were confined within the solution with the plant at maximum water uptake, which may overestimate the risk of plant damage. The roots are responsible for absorption and accumulation of substances so the root lengths will be more affected by the concentration of the herbicides [33, 39]. Overestimating the risk may be useful as it offers a safety margin for extrapolating plant sensitivity data derived using hydroponic culture to a field situation [40].

Therefore, bioassays are, on one hand, useful tools to screen herbicide phytotoxicity and provide additional information to assess the potential risk for sensitive rotational crops or non-target plants, by means of dose-response curves, and on the other hand, can complement, from a biological point of view, the information given by analytical methods.

In this sense, our group studied the phytotoxicity of alloxydim and its main metabolite obtained in its degradation with chlorine with hydroponic bioassays on wheat [41]. Results showed that after seven days of treatment the most sensitive biological parameter for alloxydim was root length, causing in the root growth of plants a 40% of significant reduction at the dose of 0.3 mg L^{-1} and 94% of reduction at the highest dose. However, the effect of metabolite on root growth only occurred at the highest metabolite dose (10 mg L^{-1}), causing a 32% of reduction in root growth.

In this work, three main objectives can be highlighted; first, to study the photodegradation of alloxydim in plant surfaces simulating the plant cuticle with organic solvent. Once the kinetic of photolysis is performed, the second step is to identify and isolate the main degradation products generated in this process. The third objective is to evaluate the individual toxicity of alloxydim and its imine by-product in a succeeding cereal crop (wheat) and in a grass weed (Bromus). Plant sensitivity was determined using the doses-response curves of root, coleoptyle and shoot growth. The doses for 50% inhibition of root, coleoptile and shoot growth (EC50) were used to estimate the relative sensitivity of the different species.

EXPERIMENTAL

Reagents and Solutions

Alloxydim (methyl (E)-(RS)-3-[1-(allyloxyimino)butyl]-4-hydroxy-6,6-dimethyl-2-oxocyclohex-3-enecarboxylate) was acquired from Dr. Ehrenstorfer GmbH (Augsburg, Germany) as the sodium salt (98% purity) and was used without further purification.

Ultrapure water and formic acid (p.a.) were obtained from a Millipore system (Milli-Q-50 18mV) and Merck (Damstadt, Germany), respectively. Acetonitrile (HPLC far UV grade), methanol (HPLC grade) and n-hexane (HPLC grade) were supplied by Labscan (Stillorgan, Co., Dublin, Ireland). Commercial mineral water (FontVella) was purchased from a local supermarket.

The plant growth bioassays were conducted with alloxydim and its imine by-product dissolved in Hewitt nutrient solution, at pH 6.7, prepared with mineral water [38].

Irradiation Experiments

To perform experiments to simulate the plant cuticle, stock solution of alloxydim (50 mg L^{-1}) in methanol was prepared by dissolving directly the appropriate amount of the herbicide in the respective solvent. Because of the low solubility of alloxydim in n-hexane, their stock solutions were prepared at a lower concentration (10 mg L^{-1}) as follows: alloxydim (0.51 mg) was dissolved firstly in the minimum amount of methanol and then the solvent was evaporated by means of a gentle air stream to dryness. The residue was dissolved in ethyl acetate (3 mL) and n-hexane was added to bring a final volume of 50 mL. All stock solutions of alloxydim were stored at 4°C in the dark and were used to prepare more dilute standard solutions (5 mg L^{-1}) by addition of the corresponding solvent.

Photodegradation experiments were carried out in a Suntest CPS+ apparatus from Atlas (Linsengericht, Germany) equipped with a xenon arc lamp (1500 W) and a special glass filter restricting the transmission of wavelength below 290 nm. All photodegradation experiments were performed at an irradiation intensity of 750 W m^{-2}. A Suncool chiller was used to maintain a mean internal temperature of 25 ± 1°C.

Non-aqueous solutions of alloxydim in methanol and n-hexane were exposed to simulated solar radiation in round capped quartz cuvettes with magnetic stirring. At selected time intervals, aliquots were withdrawn and subsequently analysed by HPLC-DAD in order to follow the reaction kinetics. Before chromatographic analysis, the solvent of samples in n-hexane was removed using a vacuum centrifuge (*Concentrator plus*, Eppendorf AG, Hamburg, Germany) during 5 minutes and the residue was dissolved in methanol, whereas samples in methanol were injected directly.

Concurrently with irradiation experiments, dark control experiments were performed in order to evaluate the stability of alloxydim dissolved in non-aqueous solutions. All experiments were carried out until complete herbicide disappearance. Three replicates were carried out for each photodegradation experiment and the numerical results presented correspond to the mathematical average of these three independent analyses.

Chromatographic Analysis

Photodegradation kinetics were performed on a HPLC system (series 1100; Agilent Technologies, Palo Alto, CA, USA) equipped with a photodiode array detector (DAD). The analytical column used was a Waters Nova-Pak® C18 column (4 μm particle size, 3.9mm×150 mm) with a ODS precolumn and were maintained at 25°C. The mobile-phase was a mixture of water acidified with 0.1% of formic acid (A) and acetonitrile (B).

A gradient method was used in order to follow the decay of alloxydim as well as to study the by-products formation during irradiation experiments. The percentage of mobile phase B was as follows: 0-1.2 min, 50%; 1.2-2 min, 50-60%; 2-3 min, 60-70% and the separation finished after eight minutes. The flow rate was 1 mL min^{-1} and the injection volume was 20 μL.

For the identification of by-products, mass spectrometry experiments were performed on a HPLC system (series 1100; Agilent Technologies, Palo Alto, CA, USA) coupled to a hybrid Qtof mass spectrometer (QStar Pulsar I, Applied Biosystems). Just before the separation, an external calibration in the mass spectrometer was performed with a mixture of phosphazenes. The experiments were performed in positive ion mode. The instrumental parameters were set as follows—mass range analysed: 50–1200; ion spray voltage (IS): 5000 V; ion source gas pressure (GS1): 65 psi; ion source gas 2 (GS2): 65 psi; curtain gas pressure (Cur): 20 psi; declustering potential (DP): 70 V; focusing potential (FP): 250 V; declustering potential 2: 15 V. In MS/MS experiments the collision energy (CE) for each ion selected was kept at 22 eV. The column, precolumn, mobile phase and gradient employed were the same as described above except the flow rate that was 0.7 mL min^{-1}.

Isolation of Alloxydim By-product

To obtain imine by-product, concentrated aqueous solutions of alloxydim (500 mg L^{-1}) were irradiated during 110 hours. SPE was used to extract and isolate this by-product. Irradiated samples were passed through Isolute ENV+ cartridges (500 mg, 3 mL) under vacuum using a vacuum manifold. Extraction was performed under gravity with an elution volume of 3x2 mL of methanol. The eluate was evaporated to dryness under a gentle air stream and stored at 4ºC in the dark.

Bioassay Experiments

The study was conducted with two winter wheat, *Triticum aestivum* L. (cv. Nita), *Triticum durum* L. (cv. Pavon) and one grass weed (*Bromus diandrus* L.). The analytical standard alloxydim-sodium and imine by-product were the active substances employed in this research. The doses assayed with alloxydim were from 0.5 to 5 mg L^{-1}. The doses assayed with imine were from 20 to 800 mg L^{-1}. Controls with nutritive solution were included in each assay.

Seed Germination Bioassay

For the seed germination assay we used the method according to Warman (1999) [42] and Gariglio et al. (2002) [43]. The assays were conducted in a growth chamber with 16 h of light (illumination 100 μE m^{-2} s^{-1}) at 22 ± 1°C and 8 h of darkness at 16 ± 1°C.

Twenty seeds of all species assayed were placed in a Petri dish (9 cm diameter) on double-layered Whatman filter paper No.1. The filter paper was moistened with 15 mL at different concentrations of herbicide solution (alloxydim doses used in this work were 0-0.5-1-2-3-4-5 mg L^{-1} and imine doses used by wheat were 0-20-40-100-400-800 mg L^{-1} and by Bromus were 0-150-300-500-700-800 mg L^{-1}) and with mineral water in the case of control treatment. Seeds were regularly checked for moisture. Germination of seeds was recorded five days after seeding by wheat and nine days after seeding by Bromus.

Plant Growth Bioassay

The assays were conducted in a growth chamber with 16 h of light (illumination 100 μE m^{-2} s^{-1}) at 24 ± 1°C and 8 h of darkness at 15 ± 1°C.

All seeds were pre-germinated in petri dishes of 9 cm of diameter, on filter paper discs soaked in deionised water for 48 h. When the root appeared and the coleoptile had 2 or 3 mm, five seedlings per pot were placed on a plastic grid, which was placed into a 220 ml plastic container wrapped in black cardboard. There were 5 replications per dose. Each container was filled with respective herbicide dose, dissolved in 175 ml of Hewitt nutrient solution.

Alloxydim and imine doses used in this work were 0-0.1-0.2-0.3-0.4-0.5-0.6-0.7-0.8-0.9-1.0-1.5 mg L^{-1} for all species. The plastic containers were randomly placed in a growth chamber and after 2 and 4 days, the containers were refilled with Hewitt nutrient solution so as to keep the solution at the grid level and cover the plant roots. Plants were removed after 7 days and the root length was determined from the point of root initiation to the tip of the root. Shoots were also measured after 7 days of treatment. The experimental design was a randomized complete block with five replications for each concentration and five seed per glass.

Data Analysis

Photolysis rates of alloxydim were described by a first order kinetic, given by the following equation:

$$C_t = C_0 e^{-k t} \qquad \text{Equation 1}$$

where C_0 is the initial concentration of alloxydim, C_t is the concentration at irradiation time t and k is the rate constant of the photodegradation process.

The half-lifes ($t_{1/2}$) for the photolysis process were also determined. This parameter is defined as the time taken for alloxydim concentration to fall to half of its initial value and it is related to the rate constant, k, by means of the equation $t_{1/2} = \ln 2/k$.

Non-linear regression was used to fit log-logistic (equation 2) [44] and Gompertz models (equation 3) [45] to the root length data over dose for each species.

The equations of these curves are:

$$Y = C + ((D-C)/(1+\exp(b.\ln(X)-\ln(EC50+1)))) \quad \text{Equation 2}$$

$$Y = \exp[\ln(A).\exp(-rX)] \quad \text{Equation 3}$$

where Y was the root length (cm) and X the herbicide dose (mg a.i. L^{-1}). In equation 2, D was the upper asymptote (maximum root growth of plants), C the lower asymptote (minimum root growth of plants), b the slope of the curve around the EC50, and EC50 the dose giving 50% root length equation. In equation 3, A was the upper asymptote and r the slope of the linearized function. All these parameters were adjusted afterward by nonlinear least squares iterative method using both equations.

Additionally, a lack-of-fit F test was used to determine whether the models adequately described the data according test method described by Seefeldt et al. (1995) [44]. The EC50, the dose giving 50% root growth reduction, is calculated by Gompertz model using the following expression:

$$EC50 = [\ln[\ln(A/2)/(\ln(A)]]/-r \quad \text{Equation 4}$$

All the statistical analyses were carried out using Statgraphics® (Statgraphics Plus for Windows 1996) [46].

Results and Discussion

Photodegradation Kinetics

UV absorption spectra of alloxydim in methanol and n-hexane showed a strong absorption above 290 nm. Therefore, it can be expected that direct photodegradation of this herbicide occurs under natural conditions, since the absorbance spectrum of alloxydim and the UV emission spectrum of sunlight overlap in the region of 290-325 nm (data not showed).

Photodegradation experiments were conducted under controlled conditions in laboratory in order to avoid any variation caused by geographical situation, seasonal or meteorological conditions. In this sense, the device described on experimental section is an appropriate source of simulated sunlight, since it provides a wavelength distribution close to natural sunlight and a constant irradiance.

Photolysis of alloxydim was studied in two different non-aqueous media in order to simulate several functional groups present in plant cuticle constituents. Thus, methanol was selected as the simplest model solvents for primary alcohol groups present in cutin acids, free fatty alcohols or sterols, whereas n-hexane was used as models of saturated hydrocarbons of alkanes, lipids and sterols.

Figure 1 shows the photodegradation kinetics of alloxydim dissolved in methanol and n-hexane, respectively.

Under the irradiation conditions studied, alloxydim underwent a rapid photolysis. The data fitted well to the first-order rate law and good correlation coefficients were achieved (> 0.94). Thus, the herbicide was completely degraded in non-aqueous media after 1.3-2 h of irradiation, showing a half-life of 21.94 ± 0.51 min in methanol and 7.5 ± 0.75 min in n-hexane. It was observed that the concentration of alloxydim remained constant in dark control experiments during the irradiation time, which allowed to discard reactions no photoinitiated (thermolysis, hydrolysis ...). Moreover, the phototransformation of alloxydim was faster in non-aqueous media than in aqueous media. In a previous work the photolysis rate constant of alloxydim in mineral water was $13.65 \cdot 10^{-3}$ min^{-1} [17], while in this study the rate constant in methanol was $31.68 \cdot 10^{-3}$ min^{-1}. It indicates that photolysis rate of alloxydim is dependent on the composition of the reaction media. Therefore, it can be expected that alloxydim suffers photolysis on plant surfaces and the rate of this degradation process can depend on the functional groups present in plant cuticle.

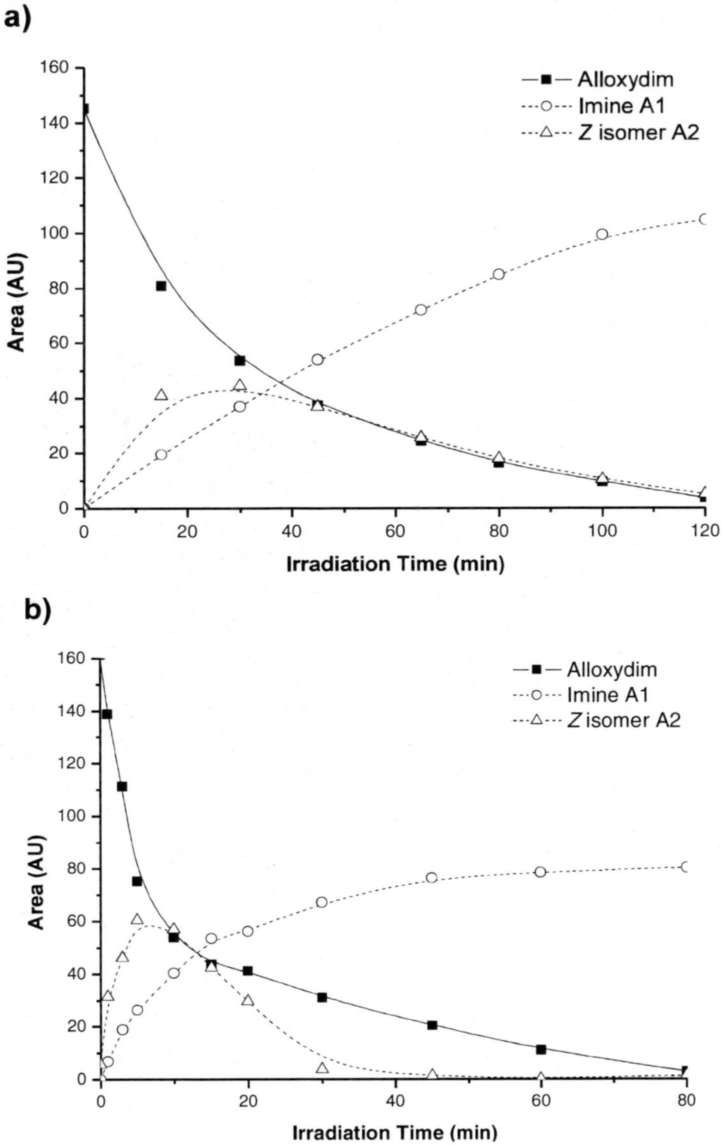

Figure 1. Evolution of transformation products during the irradiation under simulated sunlight of alloxydim dissolved in methanol (a) and n-hexane (b).

By-product Identification

During the irradiation of alloxydim two different photoproducts (A1 and A2) were detected (Figure 2). Both by-products presented at shorter retention time than that of alloxydim, showing a more polar character. Thus, their mobility to aqueous media or soil could be favoured, being these compounds potential contaminants of these compartments.

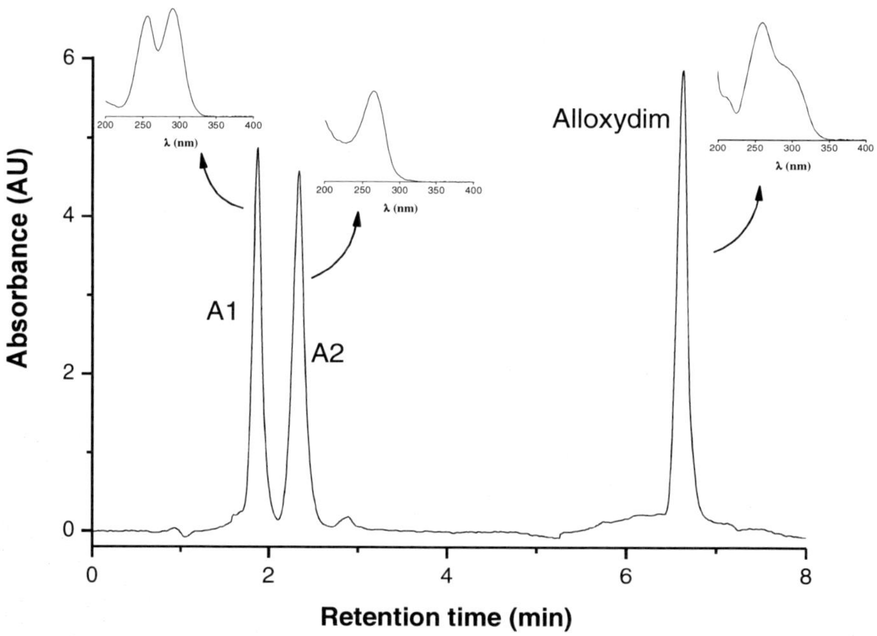

Figure 2. HPLC–DAD chromatogram of photolised solution of alloxydim in methanol exposed to simulated sunlight. Inserts are the UV spectrum of alloxydim and their photoproducts under chromatographic conditions.

Photoproducts A1 and A2 were separated and identified by means of HPLC-ESI-Qtof. On the basis of its mass spectra (Figure 3) and the accurate mass measurements, the most probable elemental composition for photoproduct A2 turned out to be identical with the starting compound alloxydim. Moreover, the fragmentation patterns of A2 by-product and alloxydim are essentially equivalent (Figure 3b). The main fragments from its molecular ion $[M+H]^+$ (m/z 324) were due to the loss of oxime moiety as 2-propenol molecule (m/z 266), the loss of methanol of the methyl ester (m/z 234) and the combined loss of a methanol molecule with one or two CO molecules (m/z 206 and m/z 178, respectively). These results allowed to identify A2 compound as the *Z*-isomer of alloxydim at the oxime ether double bond. Several authors have stated that some *E*-isomers of cyclohexanedione oxime herbicides may equilibrate with the Z-isomer in polar solvents [47-49] or in chlorinated water [22, 50]. Moreover, it has been reported that isomerization can be induced by light [16, 48] and temperature [51].

Photoproduct A1 showed an $[M+H]^+$ ion at *m/z* 268 and its main fragments observed appear at m/z 236, 208, 194 and 180 (Figure 3a). These fragments are the same that the ones obtained in a previous study of the photodegradation of alloxydin in natural waters [16, 17].

There are several possible formulae for A1 on the basis of the accurate mass measurement, but only the formula $C_{14}H_{22}NO_4$ is consistent with previously remarks. The fragment at m/z 236 corresponds to the loss of methanol molecule. From this fragment, a loss of one or two CO molecules leads to the formation of ions at m/z 208 and 180, respectively. By the other hand from the ion 236, via ketene (CH_2=CO) elimination the ion at m/z 194 could be formed. Attending to the fragmentation pattern and to the elemental composition, the structure of photoproduct A1 was assigned as the corresponding imine compound of alloxydim.

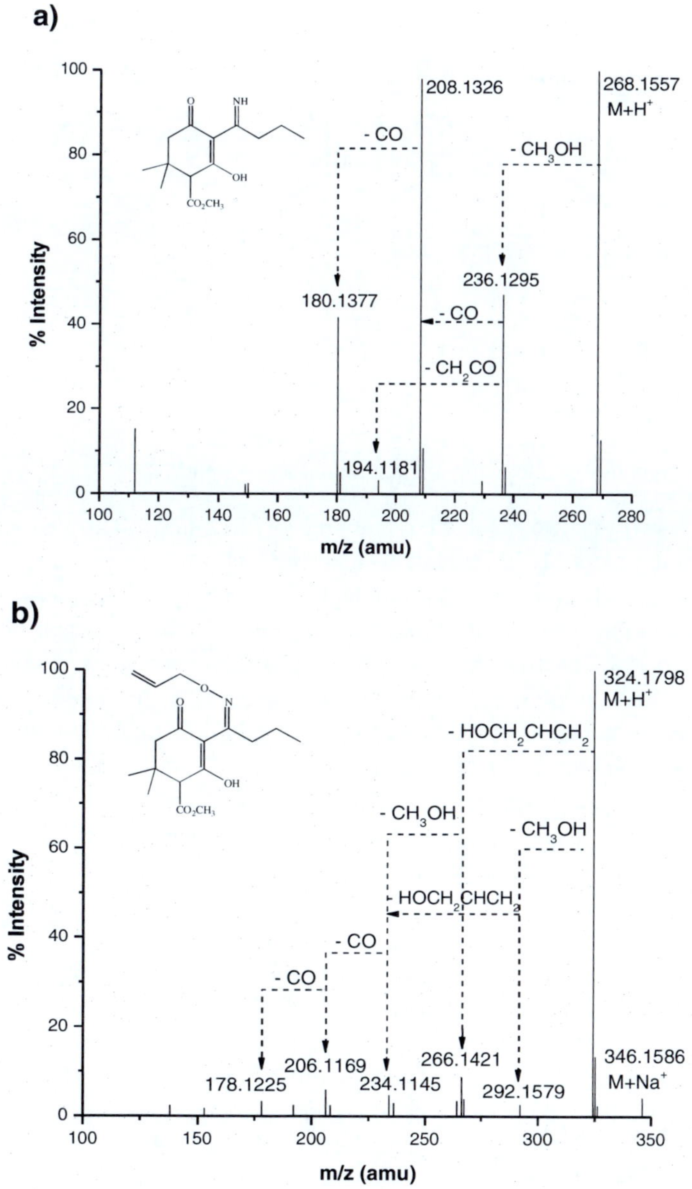

Figure 3. Mass spectra of alloxydim photoproduct A1 (a) and A2 (b).

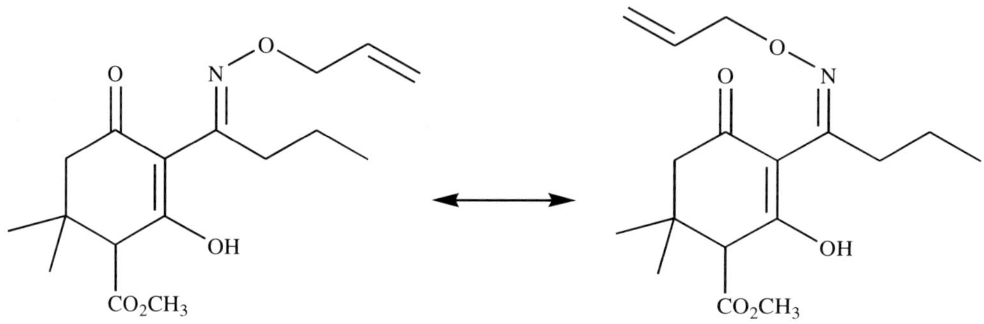

Alloxydim (*E* isomer) **A2 (Z isomer)**

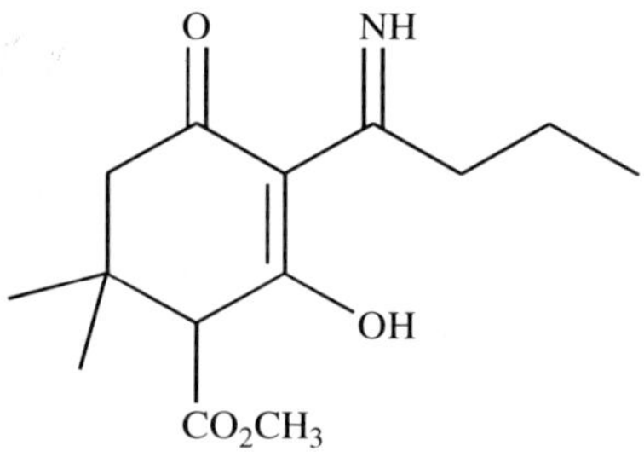

A1

Figure 4. Chemical structures of alloxydim and photoproducts A1 and A2.

In our previous works, we detected A1 as one of the main photoproducts of herbicide alloxydim in natural waters [16, 17]. However, to the best of our knowledge, A1 derivative has not been previously observed in any photodegradation study on plant leaves.

The kinetic evolution of these by-products during the photolysis was followed by means of HPLC-DAD and their profiles were similar both organic solvents (Figure 1). Thus, in methanol and n-hexane samples, the imine A1 was detected as the major product and their maximum amounts were reached at the end of the irradiation time. This by-product seems stable and could accumulate in the environment. However, the available information about this compound is still scarce. Hence, it would be necessary to ascertain if residues of this compound represent any danger to the environment. In this sense, Sandín et al. (2013) [17] stated that the irradiated solutions of alloxydim presented toxicity higher to the bacteria *Vibrio fischeri* than that of parent and this toxicity was mainly attributed to the appearance of imine A1 in the photolysed solutions.

Regarding to by-product A2, it is formed in smaller amounts and the maximum levels of this minor product were reached at the first stages of photolysis and then went down slowly until its complete disappearance (Figure 1). Although it has been reported that *Z* isomer (A2) is unstable in aqueous media and quickly reverts to the initial *E* isomer (Figure 4), it could be expected that *Z* isomer also undergoes photolysis and also gives rise to the formation of imine A1 [17].

Although kinetic profiles of A1 and A2 formation by-products in the two solvents are similar, their distribution depends on the photo-reaction media (Figure 1). Thus, in n-hexane an increase of 36% in the amount of *Z* isomer A2 was observed in comparison with photolysis in methanol. In contrast, the formation of imine A1 in methanol was more favoured than in n-hexane, raising about 30% in methanol compared to n-hexane. Similar results were obtained when alloxydim was subjected to photodegradation in natural waters [16, 17]. Their photoproducts were the same in different conditions, but their proportion showed a dependence on the constitution of the irradiated media. For instance, in the presence of Fe (III) ions alloxydim was readily photodegraded to A1 and A2, but A1 is formed in smaller amounts than in the presence of other natural substances as humic acids or nitrate ions, while by-product A2 is produced in higher amounts [16]. In the same way, in river water photoproducts AP1 and AP2 of alloxydim were detected at lower concentration than in ultrapure water [17].

Therefore, it can be expected that photoproducts under field conditions were the same but their profile depended on functional groups present in plant cuticle.

Bioassay Experiments

Seed Germination Bioassay

Our results showed clear differences in the sensitivity to alloxydim herbicide and imine by-product on winter wheat (Nita and Pavon) and grass weed Bromus. The alloxydim herbicide was more phytotoxic than imine by-product which caused less injury in seed germination of all species (Figures 5 and 6). Root and coleoptile growths of all species assayed were significantly affected by alloxydim (Figure 5). The EC50 values calculated by two wheat root length, and Bromus were similar. However, EC50 values by coleoptile length were bigger than those of root length indicating that roots of all species are more sensitive to the herbicide tested in this study than coleoptile, especially in *Triticum aestivum* L., cv. Pavon (Table 1).

Table 1. EC50 values of alloxydim tested by root and coleoptile length, in seed germination bioassay

Specie	Bioassay	EC50 (ppm)	R^2 (%)
T. durum cv. Nita	Root length	0,518	89,44%
	Coleoptile length	1,959	78,59%
T. aestivum cv. Pavon	Root length	0,878	89,48%
	Coleoptile length	53,01	73,83%
B. diandrus L.	Root length	0,329	82,67%
	Coleoptile length	3,338	33,92%

We could observe phytotoxicity effects in tested species at highly doses of imine (Figure 6). The parameters measured in wheat, root and coleoptile length, were significantly affected by imine at 100 and 400 mg L^{-1} respectively. The EC50 values of coleoptile length calculated in wheat were bigger than those of root length indicating that roots are more sensitive than coleoptile. However, with grass weed tested the imine had similar EC50 value in root and coleoptile length (277 mg L^{-1}) (Table 2).

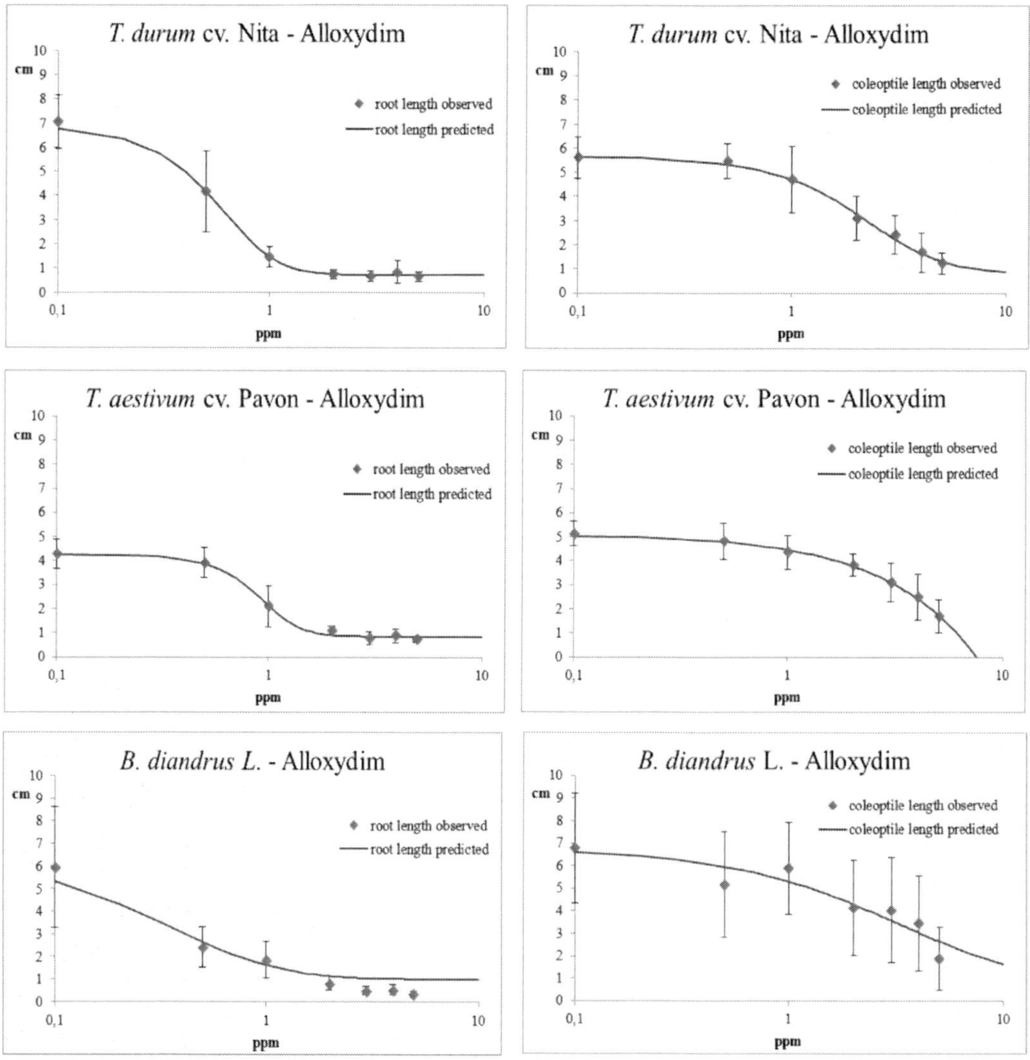

Figure 5. Dose-response curves to alloxydim, in seed germination bioassay, in root length and coleoptile length of wheat (Nita and Pavon) and Bromus.

The results obtained confirm that wheat plants and grass weed (Bromus spp.) employed could be susceptible species capable to detect the presence of cyclohexanodione herbicides and by-products. Overall, there is no one species or endpoint that is consistently the most sensitive for all species or all chemicals in all soils, and differences in bioavailability among compounds may confound comparison of test results [53]. The foregoing results suggest that the use of alloxydim can produce damage on seed germination and seedling of cereal succeeding crops, neighboring cereal crops and on non- target plants. We also observed that high doses of imine by-product showed injury in seed germination and seedling of all tested species (Figures 7 and 8).

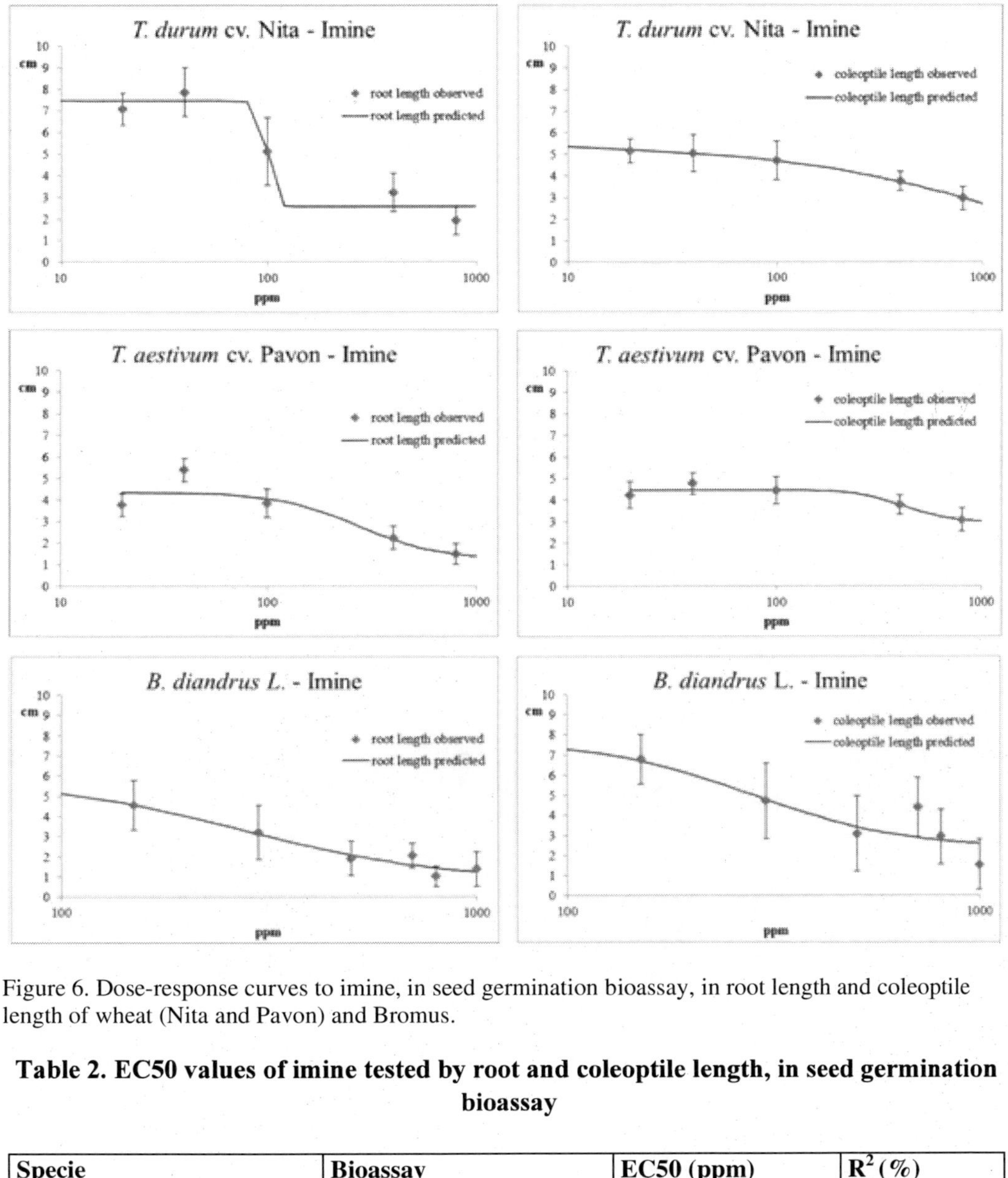

Figure 6. Dose-response curves to imine, in seed germination bioassay, in root length and coleoptile length of wheat (Nita and Pavon) and Bromus.

Table 2. EC50 values of imine tested by root and coleoptile length, in seed germination bioassay

Specie	**Bioassay**	**EC50 (ppm)**	**R^2 (%)**
T. durum cv. Nita	Root length	100,28	80,86%
	Coleoptile length	442,60	65,74%
T. aestivum cv. Pavon	Root length	270,00	70,41%
	Coleoptile length	416,65	49,05%
B. diandrus L.	Root length	277,42	64,51%
	Coleoptile length	277,91	57,40%

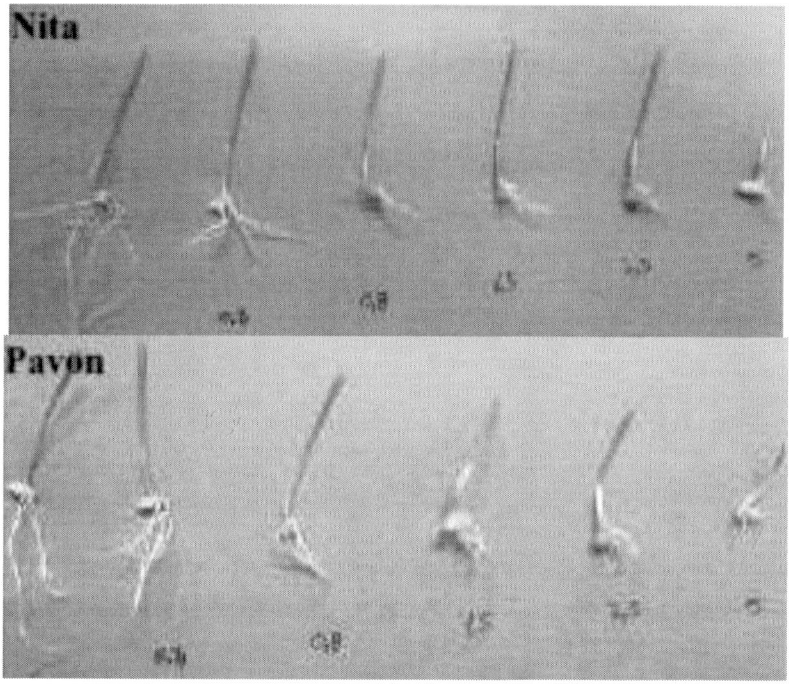

Figure 7. Response of seed germination of wheat, cv. Nita and Pavon, at 5 days after treatment with several doses of alloxydim (Control-0.4-0.8-1.5-2.5-5 mg L^{-1}).

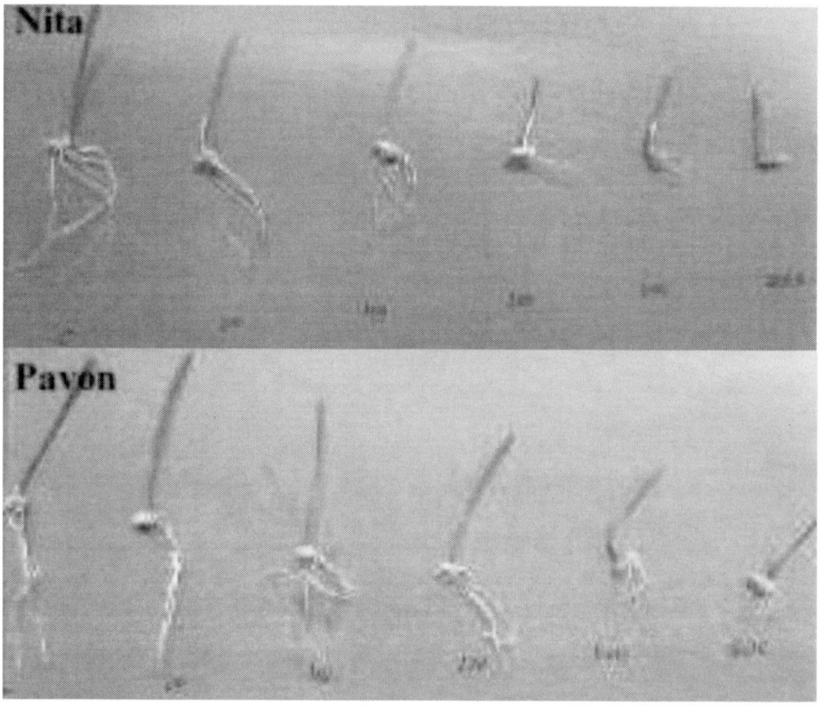

Figure 8. Response of seed germination of wheat, cv. Nita and Pavon, at 5 days after treatment with several doses of imine (Control-20-40-100-400-800 mg L^{-1}).

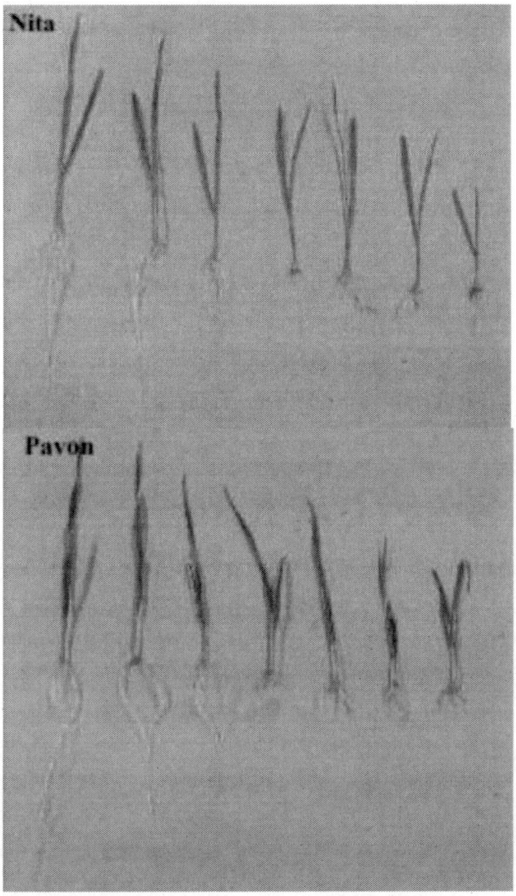

Figure 9. Response of wheat, cv. Nita and Pavon, at 7 days after treatment with several doses of alloxydim (Control-0.1-0.2-0.3-0.4-0.5-0.6 mg L^{-1}).

Plant Growth Bioassay

We have studied the phytotoxicity of alloxydim and its imine by-product on plant growth, with hydroponic bioassays, on winter wheat and Bromus. Alloxydim compound demonstrated a strong post-emergent toxicity on cereal crops and grass weed under controlled conditions (Figures 9 and 10).

Results showed that after seven days of treatment the most sensitive biological parameter for alloxydim was root length (Figure 11). Root system control presented normal growth, while those from injured plants were increasingly deformed (main root twisted and lack of secondary roots).

Wheat root growth was increasingly affected with alloxydim doses assayed from 0.1 to 1 mg L^{-1}, causing in the root growth of wheat plants a 50% of reduction (EC50) at the dose of 0.14 mg L^{-1} and 0.22 mg L^{-1}. Bromus root growth was also increasingly affected with the same doses, and the EC50 value obtained was 0.57 mg L^{-1} (Table 3).

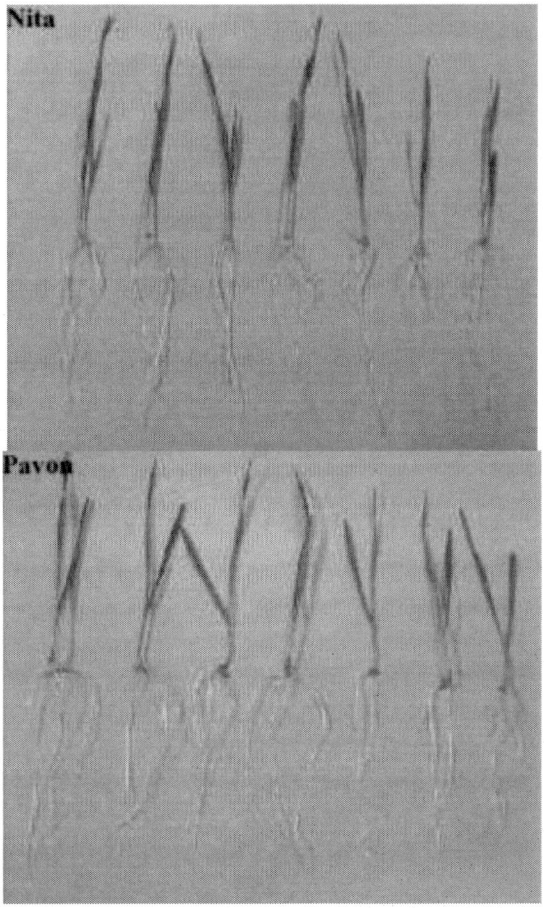

Figure 10. Response of wheat, cv. Nita and Pavon, at 7 days after treatment with several doses of imine (Control-0.1-0.2-0.3-0.4-0.5-0.6 mg L^{-1}).

We observed also a phytotoxicity effect of alloxydim on shoot elongation of tested species (Figure 12). Shoot growth was increasingly affected with alloxydim doses assayed from 0.1 to 1 mg L^{-1}, causing in the shoot growth of wheat plants a 50% of reduction (EC50) at the dose of 0.42 mg L^{-1} and 0.58 mg L^{-1}. Bromus shoot growth was also increasingly affected with the same doses, and the EC50 value obtained was 1.08 mg L^{-1} (Table 3).

In general, when were assayed and data analysed by equation regression (EC50), the comparison of observed with predicted effects showed good agreement. This phenomenon is a strong argument for inclusion of bioassays in routine monitoring applications to measure toxicity of all chemicals on plants.

Based on the responses of root and shoot growth inhibition and injury produced by alloxydim on tested species, the same doses were chosen to study the imine effects. But imine on root and shoot growth of tested species did not produced any significant effect at the dosages assayed. The regression equations did not adjust at observed data and we could not calculate the EC50 values for each parameter (Figure 12).

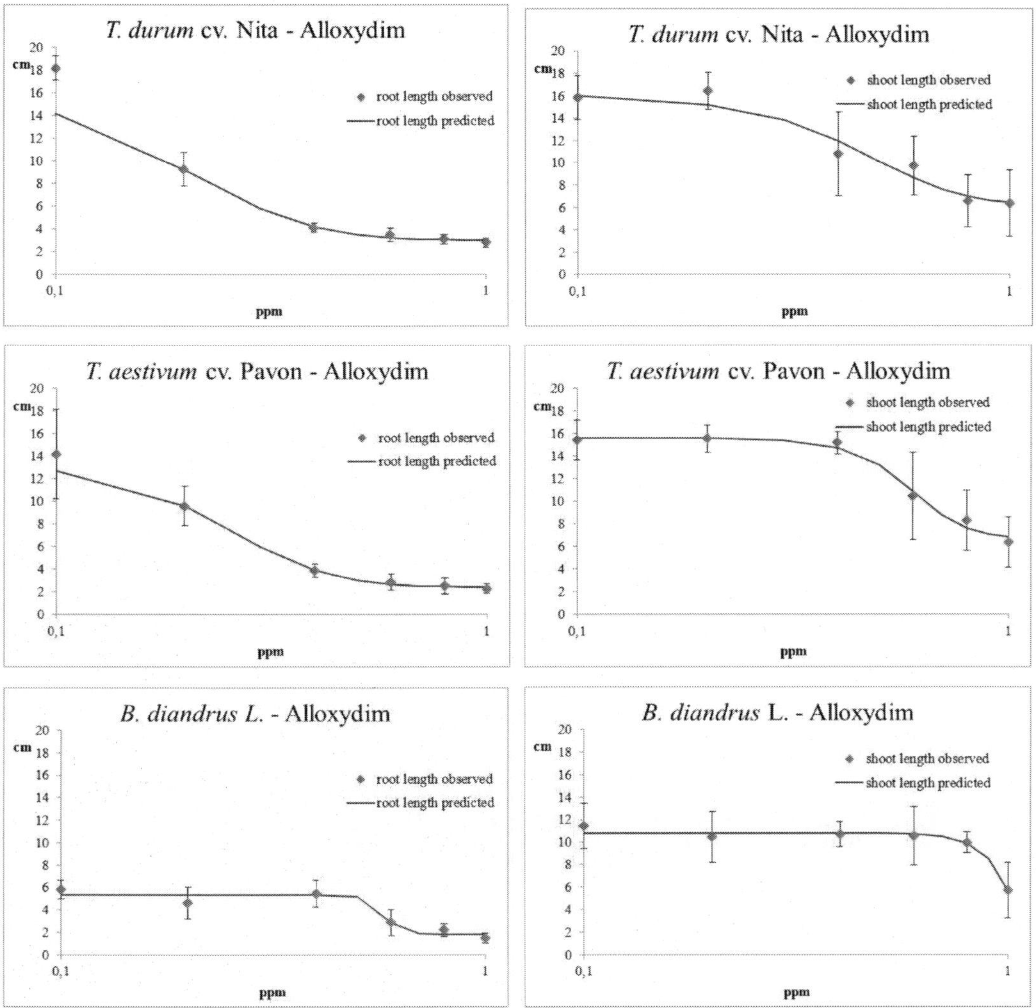

Figure 11. Dose-response curves to alloxydim, in plant growth bioassay, in root length and shoot length of wheat (Nita and Pavon) and Bromus.

Table 3. EC50 values of alloxydim tested by root and shoot length, in plant growth bioassay

Specie	Bioassay	EC50 (ppm)	R^2 (%)
T. durum cv. Nita	Root length	0,145	97,93%
	Shoot length	0,428	68,17%
T. aestivum cv. Pavon	Root length	0,226	89,97%
	Shoot length	0,588	72,73%
B. diandrus L.	Root length	0,578	72,01%
	Shoot length	1,081	51,21%

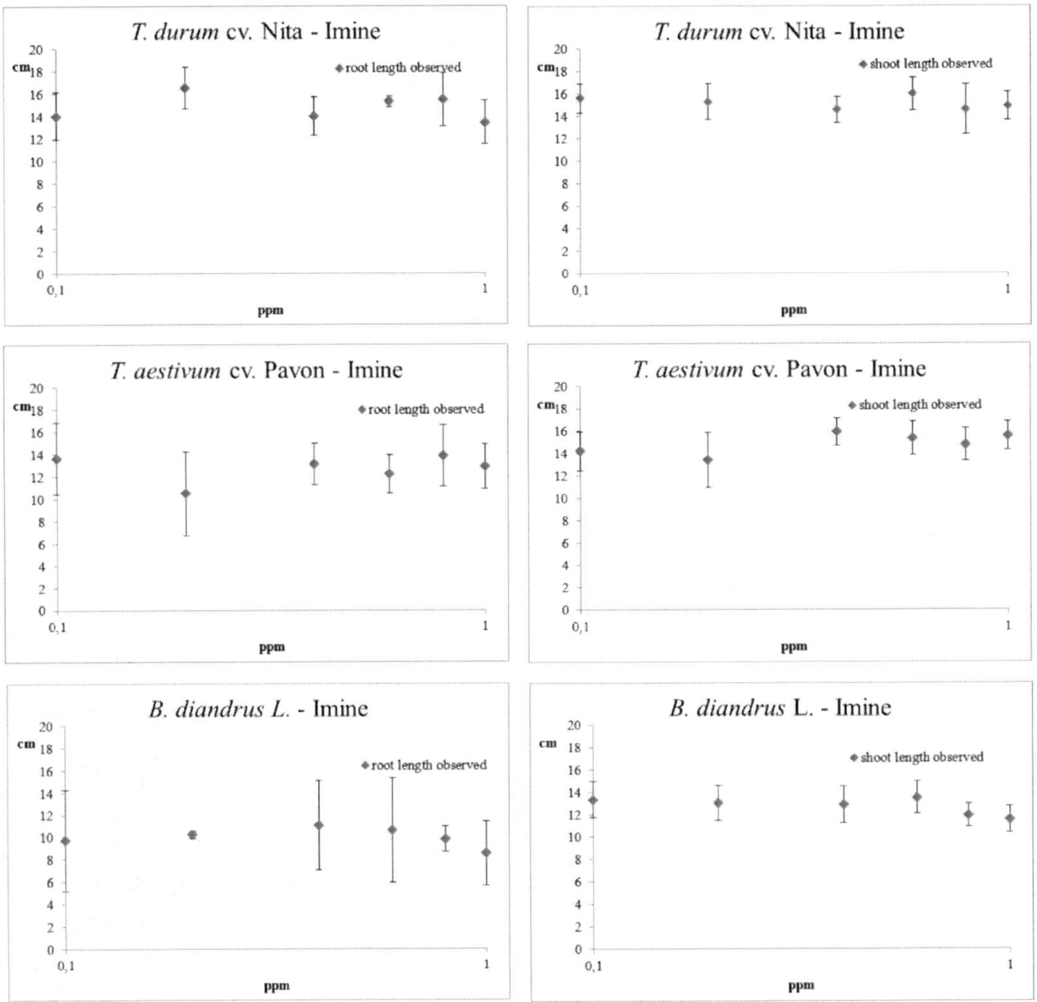

Figure 12. Dose-response curves to imine, in plant growth bioassay, in root length and shoot length of wheat (Nita and Pavon) and Bromus.

Therefore, the estimated parameters of the dose-response curves for individual species confirmed the trend that grass in vegetation are more affected by alloxydim doses than imine by-product in the same vegetation.

CONCLUSION

Once applied in the field, most of the initial concentration does not enter immediately into the target specie, but remains in plant leaves, soil and water to which direct applications are made. In these compartments, active substances are subject to different agents capable of transforming by different processes into one or more transformation products. On plant surfaces, different possible degradation pathways such photolysis can take place. This reaction can lead to the formation of degradation products that can be more persistent and/or toxic than their parent compound. Thus, information on degradation products is necessary to understand the environmental fate of pesticides and to established important degradation pathways.

In this sense, it is also important not only study the phytotoxicity of the herbicide but also of their degradation by-products. Bioassays or biological tests provide practical information based on the observation of the response of the plant to different chemical products. Bioassays offer several advantages as the detection of very low phytotoxic residues and the detection of its bioavailability. In this sense, bioassays can be used as a valuable tool to detect the possible biological activity or the risk of phytotoxicity of herbicides degradation products.

The results obtained in this work clearly indicate that alloxydim herbicide undergoes a rapid photodegradation in solvents that simulate plant surfaces. The predominant reaction is the N-O cleavage of the oxime ether bond yielding to an imine compound. To a lower extend, another by-product is generated as a result of the isomerization to Z isomer.

Alloxydim is by far more toxic than its by-product to winter wheat (Nita, Pavon) and grass weed (Bromus), in both bioassays tested, seed germination and plant growth. Conversely, previous results showed that this by-product was more toxic to bacteria *Vibrio fisheri*. On the basis of these results, more toxicity and persistence studies would be necessary to ascertain if residues of this compound represent any danger to the environment. Furthermore, further researches are needed to transfer the results from laboratory to natural field conditions.

The findings of this work highlight the importance of study not only the herbicide behaviour, but also the fate of the new by-products generated which could even be more toxic and persistent than the parent compound.

REFERENCES

[1] Breithaupt, D. E., Schwack W. *Chemosphere* 2000, 41, 1401-1406.

[2] Schwack, W., Andlauer W., Armbruster W. *Pestic Sci* 1994, 40, 279 - 284.

[3] Schynowski, F., Schwack W. *Chemosphere* 1996, 33, 2255-2562.

[4] Schwack, W. *Toxicol Environ Chem* 1987, 14, 63-72.

[5] Katagi, T. *Rev Environ Contam Toxicol* 2004, 182, 1-195.

[6] Bianchi, G. Chapter 5: Plant Waxes. In: *Waxes: Chemistry, Molecular Biology and Functions*; Hamilton, R. J., Ed.; The Oil Press Ltd.: Dundee, 1995; 6, pp 175-222.

[7] Baker, E. A. Chemistry and Morphology of Plant Epicuticular Waxes. In: *The Plant Cuticles*; Cutler, D. F., Alvin K. L., Price C. E., Eds.; Academic Press Inc.: London, 1982; pp 139-165.

[8] Kopf, G., Schwack W. *J Pesticide Sci* 1995, 43, 303-309.

[9] Schwack, W., Walker J. F., Bourgeois B. *J Agric Food Chem* 1995, 43, 3088-3092.

[10] Schippers, N., Schwack W. *J Agric Food Chem* 2008, 56, 8023-8029.

[11] Durand, G., Barceló D., Albaigés J., Mansour M. *Chromatographia* 1990, 29, 120-124.

[12] Tixier, C., Meunier L., Bonnemoy F., Boule P. *Int J Photoen* 2000, 2, 1-8.

[13] Osano, O., Admiraal W., Klamer H. J. C., Pastor D., Bleeker E. A. J. *Environ Pollut* 2002, 119, 195-202.

[14] Incledon, B. J., Hall J. C. *Pestic Biochem Physiol* 1997, 57, 255-271.
[15] Lichtenthaler, H. K. *Z Naturforsch C* 1990, 45, 521-528.
[16] Sevilla-Morán, B., Sandín-España P., Vicente-Arana M. J., Alonso-Prados J. L., García-Baudín J. M. *J Photochem Photobiol A* 2008, 198, 162-168.
[17] Sandín-España, P., Sevilla-Morán B., Calvo L., Mateo-Miranda M., Alonso-Prados J. L. *Microchem J* 2013, 106, 212-219.
[18] Hashimoto, Y., Ishihara K., Soeda Y. *J Pesticide Sci* 1979, 4, 299-304.
[19] Soeda, Y., Ishihara K., Iwataki I., Kamimura H. *J Pesticide Sci* 1979, 4, 121-128.
[20] Ono, S., Shiotani H., Ishihara K., Tokieda M., Soeda Y. *J Pesticide Sci* 1984, 9, 471-480.
[21] Sandín-España, P., Loureiro I., Escorial C., Chueca C., Santín-Montanya I. Chapter 20: *The Bioassay Technique in the Study of the Herbicide Effects.* In: Herbicides, Theory and Applications; Soloneski, S., Larramendy M. L., Eds., InTech: Rijeka, 2011; pp 431-454.
[22] Sandín-España, P., Llanos S., Magrans J. O., Alonso-Prados J. L., García-Baudín J. M. *Weed Res* 2003, 43, 451-457.
[23] Santín-Montanya, I., Sandín-España P., García Baudín J. M., Coll-Morales J. *Chemosphere* 2007, 66, 1315-1322.
[24] Kapanen, A., Itavaara M. *Ecotox Environ Safe* 2001, 49, 1-6.
[25] Kato-Noguchi, H., Kosemura S., Yamamura S., Mizutani J., Hasegawa K. *J Chem Ecol* 1994, 20, 309-314.
[26] Hernández-Sevillano, E., Villarroya M., Alonso-Prados J. L., García-Baudín J. M. *Weed Technol* 2001, 15, 447-452.
[27] Alonso-Prados, J. L., Hernández-Sevillano E., Llanos S., Villarroya M., García-Baudín J. M. *Crop Prot* 2002, 21, 1061-1066.
[28] Blacklow, W. M., Pheloung P. C. *Aust J Agric Res* 1991, 42, 1205-1216.
[29] Günther, P., Pestemer W., Rahman A., Nordmeyer H. *Weed Res* 1993, 33.
[30] Itävaara, M., Vikman M., Venelampi O. *Compost Sci Util* 1997, 5, 84-92.
[31] Fuentes, A., Lloréns M., Sáez J., Aguilar M. I., Ortuño J. F., Meseguer V. F. *J Hazard Mater* 2004, 108, 161-169.
[32] Kapustka, L. A. Chapter 17: *Selection of Phytotoxicity Tests for Use in Ecological Risk Assessments.* In: Plants for Environmental Studies; Wang, W., Gorsuch J. W., Hughes J. S., Eds.; CRC Press LLC.: Boca Raton, 1997; pp 515-548.
[33] Araújo, A. S. F., Monteiro R. T. R. *Sci Agric* 2005, 62, 286-290.
[34] Inderjit, Dakshini K. M. M. *Bot Rev* 1995, 61, 28-44.
[35] Bewley, J. D., Black M. *Seeds: Physiology of Development and Germination*; Plenum Press: New York, 1985; pp 367.
[36] Bradbeer, J. W. *Seed Dormancy and Germination*; Blackie and Son: Glasgow, 1988; pp 146.
[37] Orchard, T. *Seed Sci Tech* 1977, 5, 61-69.
[38] Chiapusio, G., Sánchez A. M., Reigosa M. J., González L., Pellissier F. *J Chem Ecol* 1997, 23, 2445-2453.
[39] Jettner, R. J., Walker S. R., Churchett J. D., Blamey F. P. C., Adkins S. W., Bell K. *Weed Res* 1999, 39, 287-295.
[40] Santín-Montanyá, I., Alonso-Prados J. L., Villarroya M., García-Baudín J. M. *J Environ Sci Health B* 2006, 41, 781-793.

[41] Sandín-España, P., Santín I., Magrans J. O., Alonso-Prados J. L., García-Baudín J. M. *Agron Sustain Dev* 2005, 25, 331-334.
[42] Warman, P. R. *Compost Sci Util* 1999, 7, 33-37.
[43] Gariglio, N. F., Buyatti M. A., Pilatti R. A., Russia D. E. G., Acosta M. R. *N Z J Crop Hort Sci* 2002, 30, 135-139.
[44] Seefeldt S. S., Jensen J. E., Patrick Fuerst E. *Weed Technol* 1995, 9, 218-225.
[45] Streibig, J. C., Rudemo M., Jensen J. E. Chapter 3: *Dose Response Curves and Statistical Models.* In: Herbicide Bioassays; Streibig, J. C., Kudsk P., Eds.; CRC Press LLC: Boca Raton, 1993; pp 29-55.
[46] Statgraphics®. Statgraphics Plus for Windows (1996). *Advanced Regression Module Manual.* Version 2. Rockville, Md: Manugistics, pp. 307.
[47] Falb, L. N., Bridges D. C., Smith A. E. *J Agric Food Chem* 1990, 38, 875-878.
[48] Sevilla-Morán, B., Alonso-Prados J. L., García-Baudín J. M., Sandín-España P. *J Agric Food Chem* 2010, 58, 3068-3076.
[49] Sandín-España, P., González-Blázquez J. J., Magrans J. O., García-Baudín J. M. *Chromatographia* 2002, 55, 681-686.
[50] Sandín-España, P., Magrans J. O., García-Baudín J. M. *Chromatographia* 2005, 62, 133-137.
[51] Curtin, D. Y., Grubbs E. J., McCarty C. G. *J Am Chem Soc* 1966, 88, 2775-2786.
[52] Sandín-España, P., Sevilla-Morán B., Alonso-Prados J. L., Santín-Montanya I. Chapter 5: *Chemical Behaviour and Herbicidal Activity of Cyclohexanedione Oxime Herbicides.* In: Herbicides - Properties, Synthesis and Control of Weeds; Naguib, M.; Hasaneen A. E.-G., Eds.; InTech: Rijeka, 2012; pp 75-102.
[53] Clark, J., Ortego L. S., Fairbrother A. *Chemosphere* 2004, 57, 1599-1612.

In: Handbook on Herbicides
Editors: Daiki Kobayashi and Eito Watanabe
ISBN: 978-1-62948-053-4

Chapter 8

Glyphosate-Resistant Weeds in Southern Europe: Current Status, Control Strategies and Future Challenges

***Demosthenis Chachalis*[1,*] *and Ilias S. Travlos*[2]**
[1]Benaki Phytopathological Institute, Laboratory of Weed Science, Athens, Greece
[2]Agricultural University of Athens, Laboratory of Agronomy, Athens, Greece

Abstract

Glyphosate has been extensively used for more than three decades and it is clearly considered to be the most common herbicide worldwide. However, its use is now in danger because of the evolution and spread of glyphosate-resistant weeds. Adoption of glyphosate-resistant crops has certainly increased the reliance on glyphosate. Unfortunately, repeated use of glyphosate along with the absence of other proactive methods has greatly increased the risk of glyphosate resistance, even in the absence of glyphosate-resistant crops. That is also the case in southern Europe, where many herbicides are no longer registered, making weed control even more difficult and depense on glyphosate more intense. As a result, totally 5 weed species have been reported to have developed resistance to glyphosate in Spain, Greece, Italy, France and Portugal. Even if the number of weed species is rather low (being only about 20% of the global total), the problem is very serious and ongoing since most of these cases from southern Europe have arisen the last decade and the number of glyphosate-resistant biotypes keeps increasing at a worrying rate. This review summarizes the current status of glyphosate resistant weeds reported in these countries, as well as several methods to avoid future spread of glyphosate resistance. Moreover, the importance of diversity and good agricultural practices is highlighted, since they seem to be the only way for the maintenance of sustainability. Increased awareness of weed resistance by farmers, agronomists and other stakeholders is clearly among the first requirements of a proactive approach.

* Corresponding author: Demosthenis Chachalis. Benaki Phytopathological Institute, Laboratory of Weed Science, 8 S. Delta Str., 14561 Athens, Greece. E-mail: d.chachalis@bpi.gr.

1. INTRODUCTION

Although glyphosate is considered a non-selective herbicide, recently characterized as a once-in-a-century herbicide (Duke and Powles, 2008) its effectiveness is rather limited on some weed species due to their natural tolerance and selectivity. On the contrary, glyphosate resistance is considered to be the capacity of plants to grow and reproduce following its application, resulting in a steady increase of numbers of resistant individuals in the initial population. Development of glyphosate-resistant weeds in conventional perennial crops in Europe is based on two key principles, set long ago by Harper 1956, as follows: 1) the intensity of the selection pressure, and 2) the reproduction of the surviving resistant individuals.

For glyphosate, the most important factors that affect the shifting of populations of a weed species from the susceptible to the resistant end are the following: 1) the intensity of the selection pressure that is mainly influenced by the glyphosate rate and application frequency, 2) the frequency of the resistant individuals after exposure of the initial population to the herbicide, 3) the genetic base of resistance (dominant or semi-dominant genes), and 4) the type of the weed species (annual or perennial, self or cross pollinated).

2. CURRENT STATUS, CONTROL STRATEGIES AND FUTURE CHALLENGES

Crops and Current Crop Practices

The majority of European glyphosate resistance problems is focused on Southern European countries (France, Greece, Italy, Spain and Portugal) on perennial crops (orchards and vineyards) (Heap, 2013). In addition, there have been a couple of reports for glyphosate resistance cases in two central European countries (Czech Republic and Poland) on non-crop areas. It is now well known that a number of factors have been implicated on the development of glyphosate resistant weed species in Europe (Chachalis et al., 2013). The most important contributing factors to such a development of glyphosate resistance are:

1 The long history (more than 30 years) of sole glyphosate application,
2 The frequent glyphosate application (more than once a year),
3 No rotation of herbicides with different mode of action,
4 Use of sub-optimal glyphosate application rates,
5 Spray of glyphosate with local-made and inproper “spraying” equipment,
6 Limited use of tillage or cover crops
7 Little use of integrated weed management approaches

It is now recognized that most perennial crops in Europe are considered as a “high risk” situation for glyphosate weed resistance development (Brants, 2012). During the last years and especially after the recent withdrawals of many herbicides, there have been many reports from Greek growers that *Conyza* spp. has become increasingly difficult to control with several herbicides, especially in no-tillage or minimum-tillage systems (Travlos et al., 2009).

In Greece, in kiwi in northern regions, in pomes in central regions, in stone fruits in northern and central regions, in olives (Peloponnese, western and central regions), in citrus (Peloponnese) and grapevines (Peloponnese, central regions) are considered as high risk cases (Chachalis and Travlos, 2012). The situation is similar in Portugal, with olives and vineyards (in coastal and mainland regions) being considered as high risk crops (Calha et al., 2011; Calha and Portugal, 2012). In Italy, glyphosate resistance is focused in north-west regions in vineyards and hazelnuts (Piedmont area) and the olives in southern Italy in Apulia and Calabria region (Collavo and Sattin, 2012). In France, the grapevines are considered as a high risk crop in many regions (Cottet and Favier, 2012). In Spain, olives, citrus and vineyards (from many regions in coastal and mainland regions) are also under serious threat (De Prado, 2012).

Resistant Weeds and Perennial Crops

In perennial crops, modification of the canopy closure as a means of suppression of glyphosate resistant weeds is not possible, as in other annual cropping systems (e.g. cereals). It is known that *Lolium* spp. is highly affected by the early canopy closure, achieved by narrow rows and higher seeding rates, tactics used by many Australian wheat farmers resulting in reduced weed growth and numbers of seeds returned to the seedbank (Walsh and Powels, 2007). Regarding *Conyza* spp., shading tends to favour its growth as evidenced by the very high numbers of plants immediately above the tree canopy.

Although timing of spraying varying by perennial crop, region, and even from grower to grower within a region and a given crop, there are some general similar principles. In olives, the fields must be clean during the late autumn and early winter time for proper olive collection. In many regions (in Greece and elsewhere), it is quite common that farmers tend to avoid spraying herbicides in autumn but rather employ mowing as a means of weed control. Reasons for such a tactic, is their reluctance to use synthetic herbicides very near the olive collection. In these cases, herbicides are mainly applied in early spring time as a primary means of weed control for the spring and summer time period, aiming for clean fields and subsequent savings of irrigation water that is a rather limiting factor in many Mediterranean regions.

In grapevines, most of the herbicide applications are performed in winter time when the vines are in the dormant state prior to the initiation of the new growth. Similar timing of herbicide application applies to most other orchards (citrus, pomes, stone fruits, kiwi etc.).

Tillage operation is greatly varied by perennial crop, region, and even from grower to grower within a region and a given crop. Generally in olives, tillage tends to be avoided close to the trees in order to avoid transmission of phytoparasistic problems (e.g. *Verticilium* spp.). In grapevines, tillage is quite common between rows as a primarily means of weed control in this area of the field whereas herbicides are applied in the line close to the vines. As tillage (type or intensity) is declined in orchards, from moldboard to chisel and no-till, shifts on weed species have been documented favoring particular species such as members of the *Conyza* spp. family (Davis et al., 2008).

In addition, tillage cannot be performed in grapevines in hilly areas with slopes, and in a number of those regions the glyphosate-resistant weed problems are more pronounced, such as in Aigialeia region in Greece (Chachalis and Travlos, 2012).

Similar approaches are employed in pomes, stone fruits and other orchards. Conservation-tillage practices that have been established to prevent soil erosion and promote water conservation, as part of anti-desertification national plans, restrict the wider application of tillage for weed management. In addition, the guidelines derived from the European Common Agricultural Policy (CAP) not only discourage it but in several cases prohibit tillage operations on vulnerable lands.

In perennial crops, establishing a cover crop between the plant rows is quite common. This cover crop acts as a physical barrier, reducing the light availability, and suppresses weed germination and emergence as part of an effective weed management tool (Norsworthy et al., 2012). In addition, use of mulches (living or nonliving) such as straw, stubbles, hay is not common in perennial crops but rather suitable for high-value vegetable crops.

Mowing in perennial crops is used as another mechanical option for control of weed growth especially prior to seed set. However, mowing alone has a rather limited efficacy on both current weed glyphosate-resistant species. In *Lolium* spp. mowing cannot remove the growing point as in other annual weed species (Meiss et al., 2008). In *Conyza* spp. mowing has got only a temporarily effect of weed growth since a vigour re-growth is documented and new seed is set in the plant (Travlos and Chachalis, 2013). In fact, sustainable herbicide-resistance management requires a longer-term perspective. Reducing seed numbers in the soil seedbank reduces the number of future plants and consequently resistance risk (Neve et al., 2011).

In Southern Europe, there has been a variation on the recommended rates of glyphosate application, targeting mainly the species of the *Conyza* family, based on the formulation and the registration of the product. In France, glyphosate rates vary from 900 to 1680 g. a.i./ha, in Spain from 1080 to 2160 g. a.i./ha (Sansom et al., 2013), and in Greece from 720 to 2160 g. a.i./ha (Chachalis and Travlos, 2012).

Weed Species That Have Developed Glyphosate Resistance in Southern Europe

From the time of the first European case of glyphosate resistance (*C. bonariensis*, in Spain, in 2004) (Urbano et al., 2007), there has been an increasing number of reports with such weed species. By July 2013, according to I. Heap International Survey of Herbicide Resistant Weeds Database, five (5) weed species have been documented to develop resistance to glyphosate in Southern European countries (Table 1). Those weed species are: three members of the *Conyza* family (*C. canadensis*, *C. bonariensis*, and *C. albida*), and two members of the *Lolium* family (*L. multiflorum* and *L. rigidum*).

In addition to the previous five weed species presented on the Table 1, there have been reports of a resistant population of *Eleusine indica* collected from a citrus orchard in Huelva, South Spain (González-Torralva et al., 2010), and a resistant population of *Lolium perene* from Douro region, in North Portugal (Calha and Portugal, 2012).

Regarding *Conyza spp.*, three are the most important species across mediteranean countries, namely hairy fleabane (*C. bonariensis* L.), horseweed (*C. canadensis* L. Cronq.), and tall fleabane (*C. albida* Willd. ex Spreng = *C. sumatrensis* (Retz.) E. Walker) and they are commonly found in perennial crops such as alfalfa, vineyards, orchards and along field margins, roadsides and other noncrop areas (Saavedra et al., 2002; Travlos et al., 2009).

Table 1. Weed species that have developed resistance to glyphosate in Southern European countries and the year of resistance documentation

Weed Species	Country	Year documented
Conyza canadensis	Greece	2012
	Italy	2011
	Spain	2006
Conyza bonariensis	Greece	2010
	Portugal	2010
	Spain	2004
Conyza albida	Greece	2012
	Spain	2009
Lolium rigidum	France	2005
	Italy	2007
	Spain	2006
Lolium multiflorum	Spain	2006

From Heap, 2013.

In the case of Greece, hairy fleabane and horseweed are the most common and abundant of the three species, while *C. albida* was more recently introduced in the country (Travlos and Chachalis, 2010; Travlos and Chachalis, 2013). These cosmopolitan species of North American origin, although present in the region for many years have recently become much more prevalent pests and major weeds for many crops and arable lands (Prieur-Richard et al., 2000; Travlos et al., 2009; Travlos and Chachalis, 2010).

Conyza can be prolific seed producers, with a single plant capable of producing thousands of nondormant seeds (Kempen and Graf, 1981), rapidly and widely dispersed by the wind (Shields et al., 2006). *Conyza* spp. has become one of the most common and problematic weeds in agronomic crops (Weaver, 2001) since it adapts to periodically plant-free, undisturbed soil and establish with an absence of tillage in crop production (Brown and Whitwell, 1988; Buhler, 1992). Therefore, the opportunistic nature of *Conyza* in undisturbed areas makes them well-suited for becoming established in agricultural fields (alfalfa, orchards, vineyards, vegetables etc) and surrounding areas, especially in no-tillage crop production and other systems using conservation tillage (Brown and Whitwell, 1988; Bruce and Kells, 1990). Taking into account that over the last two decades, the percentage of crop hectares using conservation tillage systems has been significantly increased, *Conyza* spp. has grown importance.

Concerning *Lolium spp.*, Italian ryegrass (*Lolium perenne* ssp. *multiflorum*), also called annual ryegrass, is an upright annual grass that may also behave like a biennial or short-lived perennial which grows vigorously in winter and early spring. Italian ryegrass is found in roadsides and crop fields. Italian ryegrass and a related species, perennial ryegrass (*Lolium perenne L.*), can cross and produce offspring which are difficult to identify as either species (Justice et al., 1994). Ryegrasses are cultivated for turf and forage, as cover crops, while both species are very competitive and problematic weeds.

Italian ryegrass is a widely used forage grass in temperate regions of the world and also is a competitive weed in orchards and crops in the United States (Hoskins et al., 2005; Tucker et al., 2006).

The control of Italian ryegrass in orchards is frequently based on the intensive use of glyphosate. The annual graminaceous weed *Lolium rigidum* Gaud. (annual ryegrass) is the most widespread and troublesome weed of Australian agriculture, infesting c. 6 million ha (Alemseged et al., 2001). Its wide distribution, frequent high densities and high degree of genetic variability, together with a long history of intense selection with herbicides, have resulted in evolution of resistance to herbicides with many different modes of action in Australia.

The first confirmed cases of glyphosate resistance were in *L. rigidum* biotypes from Australia (Powles et al., 1998; Pratley et al., 1999). Subsequently, other resistant *L. rigidum* populations have been found elsewhere (in Australia and South Africa). Resistance in the closely related *Lolium multiflorum* Lam. (Italian ryegrass) has been documented in Chile (Perez and Kogan, 2003). In some cases where detailed application histories have been available, resistance has evolved in *L. rigidum* following c. 15 applications of glyphosate (Neve, 2003).

The ability of rigid ryegrass to accumulate resistance mechanisms and acquire cross-resistance is attributed to its widespread distribution within cropping regions, prolific seed set, cross-fertilization, as well as significant genetic variability and phenotypic plasticity. The first evidence of evolved resistance to glyphosate in a weed species was reported by Powles et al. (1998).

Determination of the Level of Glyphosate Resistance

The screening procedure for glyphosate resistance in weed species populations is given on Figure 1. Initial screening utilizing seedling assays (such as in Petri-dishes, growing media) should be performed as a rapid test to detect resistance that is manifested at the seedling stage. In addition, at a whole plant level, a discriminating dose (DS) could be applied to distinguish susceptible (S-) from resistant (R-) biotypes. In principle, the DS should be high enough to remove the S- individuals and low enough to allow the R- ones. Given that the level of glyphosate resistance is frequently between 3 to 10-fold, selection of the appropriate DS is not always easy. Various DS have been used: in *Conyza bonariensis*, 0.24 Kg a.i./ha (Urbano et al., 2007), in *C. canadensis* 0.8 Kg a.i./ha (Trainer et al., 2005), 1.72 Kg a.i./ha (Davis et al., 2008), and in *C. canadensis*, *C. bonariensis* 1.44 Kg a.i./ha (Chachalis and Travlos, 2012).

Screening of the populations is an essential procedure to reveal their inheritable variability in response to the molecule and document the evolution of glyphosate resistance in various weeds (Smith and Hallett, 2006). In this procedure, understanding the mechanism(s) of resistance and the genetics of the resistant traits will provide clear information on the potential for evolution and spread of the glyphosate resistant weeds. For example, in *C. canadensis*, the resistance was based on a single incompletely dominant locus on a nuclear gene, explaining the pervasive and rapid spread of such populations (Zelaya et al., 2004). The non-target-site based physiological mechanisms responsible for reduced translocation may involve vacuolar sequestration of glyphosate or reduced uptake and exclusion and have been reported especially in GR weed biotypes of *C. canadensis, C. bonariensis*, *L. rigidum* and *L. multiflorum* (Powles and Preston, 2006; Shaner, 2009; Ge et al., 2010; Beckie, 2011; De Prado, 2012; Rojano-Delgado et al., 2012b).

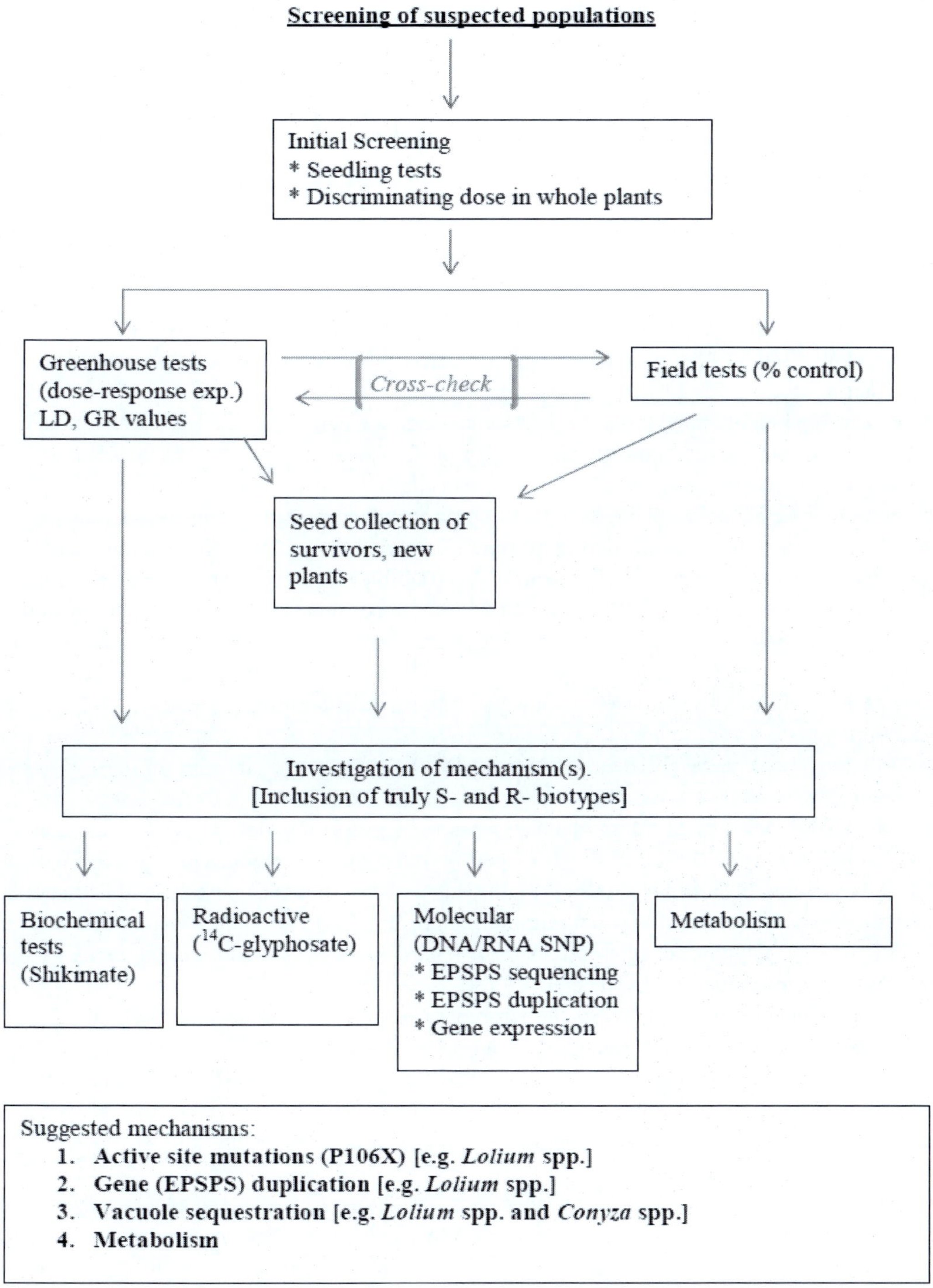

Figure 1. The screening procedure to test weed species populations for glyphosate resistance.

It is also clear that in terms of the overall procedure, several other tests and experiments are rather essential. Dose-response experiments, shikimate test and several quick tests play their own crucial role and have been extensively used in many relative studies (Rojano-Delgado et al., 2012a). Moreover, the role of field scouting and reporting and also an extension system is also very important (Powles, 2008; Travlos et al., 2009; Beckie, 2011).

Pro-Active Management Strategy

The core objective of the pro-active strategy is to mitigate resistance evolution by implementing a comprehensive weed management strategy in order to minimize the risks for development of glyphosate resistance. The pro-active management strategy should focus on some major areas, with proper use of glyphosate being one of the most crucial ones. Respecting label recommendations for herbicide usage are part of the on-farm GAP and strict adherence is needed to maintain sustainable use of glyphosate (Brants, 2012). More than that special attention should be given to the decision of the right on-farm dose, avoiding the use of sub-lethal glyphosate doses. Prior to the widespread development of glyphosate resistance, in the US (Mortensen et al., 2000) and in Europe (Ketel and Lotz, 1998; Kempenaar and Lotz, 2004), there was a strong support favoring reduced herbicide rates as means environmentally sound practices. This approach was in parallel with the concept of weed economic thresholds that were very popular in that time (O'Donovan, 1996). It is now known that a weed population under selection from repeated and sub-lethal doses, risks accumulation of minor genes (with only slight increase in fitness) that provide significant levels of resistance (Norsworthy et al., 2012).

Decision about the right dose of glyphosate is not always straight forward for two reasons; firstly there is a wide variation of glyphosate observed response and secondly the glyphosate label has a wide range of listed rates for control of different species and size of plants. Johnson et al. (2009) calculated a specific index for the glyphosate dose needed for control based on the weed according to the label. This index is based on the equation [(minimum rate / maximum size at minimum rate) + (maximum rate / maximum size at maximum rate)]. Glyphosate rates fluctuate in regard to both size and development by the nearly nine fold difference (Johnson et al., 2009).

Application of the maximum registered glyphosate dose will ensure minimum possibility of the weed population for incremental enrichment of its gene pool with minor resistance alleles, as clearly described as a "*creeping resistance*" by Gressel (1995). In addition, in some weed species (such as in *Lolium* spp.), individuals that had survived sub-lethal glyphosate doses had initiated a very rapid population response accelerating the shift from S- to R-populations (Busi and Powles, 2009).

Glyphosate application should be done at the right weed growth stages to ensure the lowest possible natural tolerance of the plant. When weeds exceed the appropriate growth stage or height, it is equivalent to apply glyphosate at sub-lethal doses with consequences previously described.

For *Conyza* spp., the appropriate growth stage is up until the small rosette stage (5-8 fully expended leaves, 15-20 cm height) (De Prado, 2007). Although weed size is critical for ensuring maximum glyphosate efficacy, very often under real field conditions there has been a quite variable size of weeds.

This is particular true in *Conyza* spp. whereas plants of many growth stages are present [newly emerged plants (seedling stage), fully developed (large rosette stage) plants from the previous year, and overwintered plants (small rosette)] (Loux et al., 2004). This situation is more exacerbated in many southern European regions with mild winter or high temperatures in early spring (Chachalis and Travlos, 2012) and therefore there is a portion of individuals with the possibility to survive glyphosate application.

On-farm glyphosate applications should ensure its maximum efficacy. Treatments under drought stress conditions results in plants with closed stomata and generally reduced systemic activity (De Ruiter and Meinen, 1998; Chachalis and Reddy, 2005). In addition, avoidance to use of hard water for filling up the spraying tank and ensuring rainfastness of the herbicide are also important factors.

Finally, the most important factor leading to evolution of herbicide resistance is overreliance on a single herbicide without using other weed management options (Heap, 2013). It is now known that *Lolium* spp. populations had developed resistance in orchards where glyphosate had been used for at least 14 years in the US (Perez-Jones et al. 2005; Simarmata et al. 2005) and in Australia (Powles et al., 1997). In *Conyza canadensis*, confirmation of resistance was done only after 3 years (Van Gessel, 2001). Presentation of all options available regarding the use of herbicides with other mode of action will be done in the subsequent sections. As reported by Beckie (2011): "the main herbicide strategy for proactively or reactively managing GR weeds is to supplement glyphosate with herbicides with alternative modes of action and with soil-residual activity." These herbicides can be applied in sequences or mixtures in order to ensure a residual activity, which is very crucial for the effective control of weeds such as *Conyza* spp. The time of application should be also taken into account, while practices such as incorporation or irrigation may be also required for maximizing efficacy (Chachalis and Travlos, 2012). Often inspection of equipment is also essential for right spraying, since in many cases the spraying pressure is very high and the procedure totally wrong (Travlos et al., 2009). Weeds in field margins and roadsides should be also managed, since they can serve as a corridor for the introduction and movement of new or GR weeds (Boutin and Jobin, 1998). As stated by Noordijk et al. (2011), a detrimental effect of managing field borders can be a reduction in farmland biodiversity. However, fortunately this does not seem to be a major problem in most Mediterranean countries, whereas biodiversity is already high or threatened only in specific vulnerable zones (Travlos et al., 2009). Among "best management practices" (as stated by Norworthy et al., 2012), effective and often scouting is a vital component of any successful weed management program, including one aimed at managing GR cases.

Re-Active Management Strategy

Proactive herbicide resistance management is generally not practised worldwide. One of the most common reasons cited by growers for not implementing preventive herbicide-resistant weed management practices is the future availability of a herbicide with a new mode of action (Scott et al., 2009; Beckie, 2011). As reported by Llewellyn et al. (2002), producers hope that new herbicides capable of overcoming resistance issues will continue to enter the market. However, the reality is that herbicide discovery is significantly lower and therefore other alternatives should be implemented (Ruegg et al., 2007).

The core objective of the re-active strategy is to restore weed control after resistance has evolved by implementing a strategic plan to counteract glyphosate-resistant weeds in an economic, effective, environmental-friendly and sustainable way. Unfortunately, most re-active management examples come from outside Europe (Beckie, 2011). However, several lessons should be learnt and may offer solutions like the one described as the "double knock" treatment which incorporates an application of a second herbicide mode of action, other than glyphosate, shortly after the glyphosate preseeding application (Neve et al., 2003; Weersink et al., 2005). Moreover, the use of mixtures with glyphosate (sequencing within or across growing seasons) and other cultural practices (e.g. tillage) should be also involved in an integrated GR weed management system.

However, it should be noted that early adoption of diverse resistance management practices is rather necessary since costs of control after herbicide resistance has evolved are often higher than the cost of a program for reducing the risk of resistance in the first place (Mueller et al., 2005; Orson, 1999). Moreover, complexity of weed management programs consumes management time, attention, and labor and therefore it should be avoided (Pardo et al., 2010).

Conclusion

Repeated use of glyphosate along with the absence of other proactive methods has clearly increased the risk of glyphosate resistance, even in the absence of glyphosate-resistant crops. Glyphosate resistance in weeds is a clear and present economic problem and it can be characterized as one of the major agronomic "technical" problems that the countries of Southern Europe face. The problem is ongoing since most of these cases from southern Europe have arisen the last decade and the number of glyphosate-resistant biotypes keeps increasing at a worrying rate. Diversity and good agricultural practices are crucial, since they seem to be the only way for the maintenance of sustainability. Increased awareness by farmers, agronomists and other stakeholders is among the first steps of a long-term management. The simplicity and cost effectiveness of alternative weed management programs is essential if glyphosate-resistance management is to be viable. Growers clearly do not wish to abandon glyphosate, even when dealing with GR weeds, because of its low cost and high efficacy on a wide range of weed species. However, weed management ought to be case specific and not only relying on one technique or one herbicide, especially in the case of highly diverse agricultural systems of southern European countries. Proactive approaches should play a pivotal role in order to ensure the profitability and sustainability of the special agroecosystems of Mediterranean countries.

References

Alemseged, Y., Jones, R. E. and Medd, R. W. (2001) A farmer survey of weed management and herbicide resistance problems of winter crops in Australia. *Plant Protection Quarterly* 16: 21–25.

Beckie, H. J. (2011) Herbicide-resistant weed management: focus on glyphosate. *Pest. Management Science* 67: 1037–1048.

Boutin, C. and Jobin, B. (1998) Intensity of agricultural practices and effects on adjacent habitats. *Ecol. Appl.* 8: 544–557.

Brants, I. (2012). Dissemination of weed resistance management recommendations. Pp. 49-52 in Proceedings of Workshop on "*European status and solutions for glyphosate resistance*". Universidad de Cordoba, 3-4 May, 2012, Cordoba, Spain.

Brown, S. M. and Whitwell, T. (1988) Influence of tillage on horseweed (*Conyza canadensis*). *Weed Technology* 2: 269–270.

Bruce, J. A. and Kells, J. J. (1990) Horseweed (*Conyza canadensis*) control in notillage soybeans (*Glycine max*) with preplant and preemergence herbicides. *Weed Technology* 4: 642–647.

Buhler, D. D. (1992) Population dynamics and control of annual weeds in corn (*Zea mays*) as influenced by tillage systems. *Weed Science* 40: 241–248.

Busi, R. and Powles, S. B. (2009) Evolution of glyphosate resistance in a *Lolium rigidum* population by glyphosate selection at sublethal doses. *Heredity* 103: 318–325.

Calha, I. M. and Portugal, J. (2012). Glyphosate resistance in Portugal: status and potential solutions. Pp. 27-28 in Proceedings of International Workshop on "*European status and solutions for glyphosate resistance*". Universidad de Cordoba, 3-4 May, 2012, Cordoba, Spain.

Calha, I. M., Portugal, J. and de Prado, R. (2011) Resistência aos herbicidas. Jornadas Resistência Adquirida a Herbicidas em Culturas Perenes, Beja (Portugal).

Chachalis, D. and Travlos, I. S. (2012). Glyphosate resistance status and potential solutions in Greece. Pp. 25-26 in Proceedings of International Workshop on "*European status and solutions for glyphosate resistance*". Universidad de Cordoba, 3-4 May, 2012, Cordoba, Spain.

Chachalis, D. and Reddy, K. N. (2005). Factors affecting sprouting and glyphosate translocation in rootstocks of redvine (*Brunnichia ovata*) and trumpetcreeper (*Campsis radicans*). *Weed Technology* 19: 141–147.

Chachalis, D., Tani, E., Travlos, I., Costa, J., and Sammons, R. D. (2013). Glyphosate resistant *Conyza* spp. in perennial crops in Greece: mechanisms of resistance and control strategies. *In: Proceedings of the 2013 Conference of Weed Science Society of America*, Abstract no.166, 4-7 February 2013, Baltimore, Maryland.

Collavo, A. and Sattin, M. (2012). Glyphosate resistance in Italy: status and potential solutions. Pp. 29-30 in Proceedings of International Workshop on "*European status and solutions for glyphosate resistance*". Universidad de Cordoba, 3-4 May, 2012, Cordoba, Spain.

Cottet, C. and Favier, T. (2012). Glyphosate resistance in French vineyard: current situation and perspectives. Pp. 31-32 in Proceedings of International Workshop on "*European status and solutions for glyphosate resistance*". Universidad de Cordoba, 3-4 May, 2012, Cordoba, Spain.

Davis, V. M., Gibson, K. D. and Johnson, W. G., (2008) A field survey to determine distribution and frequency of glyphosate-resistant horseweed (*Conyza canadensis*) in Indiana. *Weed Technology* 22: 331–338.

De Prado, R. (2007) Agronomic, biological and molecular bases of *Lolium* spp. and *Conyza* spp. biotype resistants to glyphosate in Spain. Project collaboration between University of Córdoba and Monsanto.

De Prado, R. (2012). Glyphosate impact on Mediterranean agriculture: expectations and solutions. Pp. 11-14 in Proceedings of International Workshop on "*European status and solutions for glyphosate resistance*". Universidad de Cordoba, 3-4 May, 2012, Cordoba, Spain.

De Ruiter, H. and Meinen, E. (1998) Influence of water stress and surfactant on the efficacy, absorption and translocation of glyphosate. *Weed Science* 46: 289–296.

Duke, S. O. and Powles, S. B. (2008) Glyphosate: a once-in-a-century herbicide. *Pest. Management Science* 64: 319–325.

Ge, X., d'Avignon, D. A., Ackerman, J. J. H., and Sammons, R. D. (2010) Rapid vacuolar sequestration: the horseweed glyphosate resistance mechanism. *Pest. Management Science* 66: 345–348.

González-Torralva, F., Cruz-Hipolito, H., Bastida, F., Mülleder, N., Smeda, R. J., De Prado, R. (2010) Differential susceptibility to glyphosate among the *Conyza* weed species in Spain. *Journal of Agricultural and Food Chemistry* 58: 4361–4366.

Gressel, J. (1995) Creeping resistances: the outcome of using marginally effective or reduced rates of herbicides. Pages 587–590 in *Proceedings of the Brighton Crop Protection Conference–Weeds*. Farnham, UK: British Crop Protection Council.

Harper, J. C. (1956) The evolution of weeds in relation to herbicides. *Proceedings of British Weed Control Conference* 3: 179–188.

Heap, I. (2013) The International Survey of Herbicide Resistant Weeds. Available: http://www.weedscience.org. Accessed 12 June 2013.

Hoskins, A., Young, B., Krausz, R., and Russin, J. (2005) Control of Italian ryegrass (*Lolium multiflorum*) in winter wheat. *Weed Technology* 19: 261–265.

Johnson, W. G., Owen, M. D. K., Kruger, G. R., Young, B. G., Shaw, D. R., Wilson, R. G., Wilcut, J. W., Jordan, D. L., and Weller, S. C. (2009) US farmer awareness of glyphosate-resistant weeds and resistance management strategies. *Weed Technology* 23: 308–312.

Justice, G. G., Peeper, T. F., Solie, J. B., and Epplin, F. M. (1994) Net returns from Italian ryegrass (*Lolium multiflorum*) control in winter-wheat (*Triticum aestivum*). *Weed Technology* 8: 317–323.

Kempen, H. M. and Graf, J. (1981) Weed seed production. *Proceedings of Western Society of Weed Science* 34: 78–81.

Kempenaar, C. and Lotz, L. A. P. (2004) Reduction of herbicide use and emission by new weed control methods and strategies. *Water Science Technology* 49: 135–138.

Ketel, D. H. and Lotz, L. A. P. (1998) Influence of allocation and detoxification of metribuzin in *Chenopodium album* on the reliability of prediction of the minimum lethal herbicide dose rate. *Weed Research* 38: 267–274.

Llewellyn, R. S., Lindner, R. K., Pannell, D. J., and Powles, S. B. (2002). Resistance and the herbicide resource: perceptions of Western Australian grain growers. *Crop Protection* 21: 1067–1075.

Loux, M., Stachler, J., Johnson, B., Nice, G., Davis, V., and Nordby, D. (2004): Biology and Management of Horseweed. Extension Leaflet No. 323. Available at http://www.btny.purdue.edu/weedscience/marestail/ID-323%20 HorseWeed.pdf

Meiss, H., Munier-Jolain, N., Henriot, F., and Caneill, J. (2008) Effects of biomass, age and functional traits on regrowth of arable weeds after cutting. *Journal of Plant Diseases and Plant Protection* 21(SI XXI): 493–499.

Mortensen, D. A., Bastiaans, L. and Sattin, M. (2000) The role of ecology in the development of weed management systems: an outlook. *Weed Research* 40: 49–62.

Mueller, T. C., Mitchell, P. D., Young, B. G., and Culpepper, A. S. (2005) Proactive versus reactive management of glyphosate-resistant or tolerant weeds. *Weed Technology* 19: 924–933.

Neve, P., Diggle, A. J., Smith, F. P., and Powles, S. B. (2003). Simulating evolution of glyphosate resistance in *Lolium rigidum* I: population biology of a rare resistance trait. *Weed Research* 43: 404–417.

Neve, P., Norsworthy, J. K., Smith, K. L., and Zelaya, I. A. (2011) Modelling evolution and management of glyphosate resistance in *Amaranthus palmeri*. *Weed Research* 51: 99–112.

Noordijk, J., Musters, C. J. M., van Dijk, J., and de Snoo, G. R. (2011) Vegetation development in sown field margins and on adjacent ditch banks. *Plant Ecology* 212: 157–167.

Norsworthy, J. K., Ward, S. M., Shaw, D. R., Llewellyn, R. S., Nichols, R. L., Webster, T. M., Bradley, K. W., Frisvold, G., Powles, S. B., Burgos, N. R., Witt, W. W., and Barett, M. (2012). Reducing the risks of herbicide resistance: best management practices and recommendations. *Weed Science* 31–62.

O'Donovan, J. T. (1996) Weed economic thresholds: Useful agronomic tool or pipe dream? *Phytoprotection* 77: 13–28.

Orson, J. H. (1999) The cost to the producer of herbicide resistance. *Weed Technology* 13: 607–611.

Pardo, G., Riravololona, M. and Munier-Jolain. N. M. (2010) Using a farming system model to evaluate cropping system prototypes: are labour constraints and economic performances hampering the adoption of Integrated Weed Management? *European Journal of Agronomy* 33: 24–32.

Perez, A. and Kogan, M. (2003) Glyphosate resistance in *Lolium multiflorum* in Chilean orchards. *Weed Research* 43: 12–19.

Perez-Jones, A., Park, K. W., Colquhoun, J., Mallory-Smith, C., and Shaner, D. (2005) Identification of glyphosate-resistant Italian ryegrass (*Lolium multiflorum*) in Oregon. *Weed Science* 53: 775–779.

Powles, S. B. (2008) Evolved glyphosate-resistant weeds around the world: lessons to be learnt. *Pest Management Science* 64: 360–365.

Powles, S. B., Lorraine-Colwill D. F., Dellow J. J., and Preston, C. (1998). Evolved resistance to glyphosate in rigid ryegrass (*Lolium rigidum*) in Australia. *Weed Science* 46: 604–607.

Powles, S. B. and Preston, C. (2006) Evolved glyphosate resistance in plants: biochemical and genetic basis of resistance. *Weed Technol.* 20: 282–289.

Powles S. A. B., Preston, C., Jutsum, A. R., and Bryan, I. B. (1997) Herbicide resistance: Impact and management. *Advances in Agronomy* 58: 57–93.

Pratley, J., Urwin, N., Stanton, R., Baines, P., Broster, J., Cullis, K., Schafer, D., Bohn, J., and Kruger, R. (1999). Resistance to glyphosate in *Lolium rigidum*. I. Bioevaluation. *Weed Science* 47: 405–411.

Prieur-Richard, A. H., Lavorel, S., Grigulis, K., and Dos-Santos, A. (2000) Plant community diversity and invasibility by exotics: invasion of Mediterranean old fields by *Conyza bonariensis* and *Conyza canadensis*. *Ecology Letters* 3: 412–422.

Rojano-Delgado, A. M., Algaba, I. M., Rios, A., and De Prado, R. (2012a). New biomarker to detect glyphosate resistance in weeds. Pp. 125-126 in Proceedings of International Workshop on "*European status and solutions for glyphosate resistance*". Universidad de Cordoba, 3-4 May, 2012, Cordoba, Spain.

Rojano-Delgado, A. M., Carvalho, L., Algaba, I. M., and De Prado, R. (2012b). Glyphosate metabolism in *Digitaria insularis* biotypes. Pp. 123-124 in Proceedings of International Workshop on "*European status and solutions for glyphosate resistance*". Universidad de Cordoba, 3-4 May, 2012, Cordoba, Spain.

Ruegg, W. T., Quadranti, M. and Zoschke, A. (2007) Herbicide research and development: challenges and opportunities. *Weed Research* 47: 271–275.

Saavedra, M. M. and Pastor, M. (2002) *Sistemas de cultivo en olivar (manejo de malas hierbas y herbicidas)*. Ed. Agrícola Española, Madrid, España. 429 p.

Sansom, M., Saborido, A. A. and Dubois, M. (2013) Control of *Conyza* spp. with glyphosate – A reviewe of the situation in Europe. *Plant Protection Science* 1: 44–53.

Scott, B. A., VanGessel, M. J. and White-Hansen, S. (2009) Herbicide-resistant weeds in the United States and their impact on extension. *Weed Technology* 23: 599–603.

Shaner, D. L. (2009) Role of translocation as a mechanism of resistance to glyphosate. *Weed Science* 57: 118–123.

Shields, E. J., Dauer, J. T., VanGessel, M. J., and Neumann, G. (2006) Horsewwed (*Conyza canadensis*) seed collected in the planetary boundary layer. *Weed Science* 54: 1063–1067.

Simarmata, M., Bughrara, S. and Penner, D. (2005) Inheritance of glyphosate resistance in rigid ryegrass (*Lolium rigidum*) from California. *Weed Science* 53: 615–619.

Smith, D. A. and Hallett, S. G. (2006). Variable response of common waterhemp (*Amaranthus rudis*) populations and individuals to glyphosate. *Weed Technology* 20 (2): 466–471.

Trainer, G. D., Loux, M. M., Harrison, S. K., and Regnier, E. (2005) Response of horseweed biotypes to foliar applications of cloransulam-methyl and glyphosate. *Weed Technol.* 19: 231–236.

Travlos, I. S. and Chachalis, D. (2010) Glyphosate-resistant hairy fleabane (*Conyza bonariensis*) is reported in Greece. *Weed Technology* 24: 569–573.

Travlos, I. S., Chachalis, D. and Economou, G. (2009) Characters for the in situ recognition of some *Conyza* species and glyphosate resistant populations from Greece. *Proceedings of the 2nd International Conference on Novel Sustainable Weed Management in Arid and Semi-arid Agro-ecosystems*, EWRS, Santorini, Greece, p 63.

Travlos, I. S. and Chachalis, D. (2013). Relative competitiveness of glyphosate-resistant and glyphosate-susceptible populations of hairy fleabane, *Conyza bonariensis*. *Journal of Pest Science* 86: 345–351.

Travlos, I. S., Costa, J., Brants, I., and Chachalis, D. (2013) Strategies for the effective management of glyphosate-resistant *Conyza* spp. in Greece and ecological fitness parameters of the resistant populations. *Proceedings of 16th EWRS Symposium,* 24-27 June 2013, Samsun, Turkey.

Tucker, K., Morgan, G., Senseman, S., Miller, T., and Baumann, P. (2006) Identification, distribution and control of Italian ryegrass (*Lolium multiflorum*) ecotypes with varying levels of sensitivity to triasulfuron in Texas. *Weed Technology* 20: 745–750.

Urbano, J. M., Borrego, A., Torres, V., Leon, J. M., Jimenez, C., Dinelli, G., and Barnes, J. (2007) Glyphosate-resistant hairy fleabane (*Conyza bonariensis*) in Spain. *Weed Technology* 21: 396–401.

VanGessel, M. J. (2001) Glyphosate-resistant horseweed from Delaware. *Weed Science* 49: 703–705.

Walsh, M. J. and Powles, S. B. (2007) Management strategies for herbicide resistant weed populations in Australian dryland crop production systems. *Weed Technology* 21: 332–338.

Weaver, S. E. (2001) The biology of Canadian weeds. 115. *Conyza canadensis*. *Canadian Journal of Plant Science* 81: 867–875.

Weersink, A., Llewellyn, R. S. and Pannell, D. J. (2005) Economics of preemptive management to avoid weed resistance to glyphosate in Australia. *Crop Protection* 24: 659–665.

Zelaya, I. A., Owen, M. D. K. and VanGessel, M. J. (2004) Inheritance of evolved glyphosate resistance in *Conyza canadensis* (L.) Cronq. *Theor. Appl. Genet.* 110: 58–70.

In: Handbook on Herbicides
Editors: Daiki Kobayashi and Eito Watanabe

ISBN: 978-1-62948-053-4

Chapter 9

HERBICIDE PARAQUAT GENOTOXICITY-ENHANCEMENT BY THE PHENOLIC ANTIOXIDANTS DL-α - TOCOPHEROL AND 2,6-DI-TERT-BUTYL-P-CRESOL

Hideki Hanada*
Institute for Amphibian Biology, Graduate School of Science,
Hiroshima University, Hiroshima, Japan

ABSTRACT

This study investigates the enhancement mechanism of toxicity of the herbicide 1,1'-dimetyl-4,4'-bipyridium dichloride (paraquat, PQ) in combination with dl-α-tocopherol (vitamin E, α-TH) or vitamin E synthetic analog 2,6-di-tert-butyl-p-cresol (butylated hydroxytoluene, BHT). Acute structural chromosomal damage induced by PQ + α-TH and PQ + BHT was investigated cytogenetically using *Pelophylax* (*P.*) *nigromaculatus* and *Rana* (*R.*) *ornativentris* leukocytes *in vitro*. Chemical PQ-cation reduction by BHT and sodium nitrite was examined by analysis of the results obtained from PQ cation reduction test as well. In a preliminary study, nitrite (one of nitric oxide degradation products) production test was conducted using the tails of *R. rugosa* tadpoles because of investigation of the possible involvement of nitrite in the chromosomal damage-enhancement-mechanism. From these results obtained so far, the following process is proposed for PQ toxicity-enhancement: an increase in PQ monocation radical formation initiated by vitamin E and BHT → reactive oxygen species (ROS) generation → nitric oxide synthase (NOS) activation → nitric oxide (NO) generation → nitrite production → further enhancement of PQ monocation radical formation by nitrite → ROS generation further increased → an acute increase in structural chromosomal damage.

---

* Address to: Institute for Amphibian Biology, Graduate School of Science, Hiroshima University, Higashihiroshima, 739-8526, Japan Tel: +81-82-424-7485 Fax: +81-82-424-0739 E-mail: hanada@hiroshima-u.ac.jp.

Keywords: CAT, Structural chromosomal damage, Free BHT radical, Mn(III)TMpyP, Paraquat, α-TH, Tocopheroxyl radical

INTRODUCTION

Chemical pollutants released into the environment threaten the health of human and wildlife. One of the toxic chemical substances is paraquat (PQ) (Wang et al. 1987; Jee et al. 1995; Śpiewak 2001; Anderson and Scerri 2003; Valavanidis et al. 2006). PQ-induced adverse effects on human health are classified into two types through patients' conditions, acute toxicity or chronic toxicity. Former acute toxicity is fatal toxicity accompanied by lung injury (main injury), kidney injury, liver injury and myocardial dysfunction based on symptom analysis of unfortunately PQ-exposed workers and children (Wesseling et al. 1997, 2001). The later is roles as carcinogen that can initiate skin malignant tumor growth (Wang et al. 1987; Śpiewak 2001), and as causative agent of central nervous system dysfunction (Parkinson's disease) (Wesseling et al. 2001). The both of PQ toxicity increase the risk of adverse effects on human health.

In the effects of PQ toxicity on anuran embryos, PQ is seen to have a high level of teratogenic toxicity and fatal toxicity (Dial and Bauer 1984; Dial and Bauer Dial 1987; Bauer Dial and Dial 1995; Osano et al. 2002). Even low level of PQ dissolved in water is absorbed by aquatic plants, and then *R. berlandieri* tadpoles fed on the PQ-contaminated plants result in fatality and tail abnormalities by bioaccumulation, suggesting the difficulty to decompose PQ in plant body and animal body (Bauer Dial and Dial 1995). Recently, it has been reported that PQ-induced abnormal *Chaunus arenarum* embryos increases the activity of endogenous protective enzymes Mn-superoxide dismutase (SOD) against ROS (Mussi and Calcaterra 2010). PQ is thought to be bioaccumulative potential, and to cause lethal symptom and morphological abnormality thorough ROS generation. These adverse effects of PQ on health of human and anurans show tissue non-specific toxicity and intensity of the toxicity (Wesseling et al. 2001).

PQ cation (PQ^{2+}) is enzymatically reduced to the blue-colored monocation radical ($PQ^{+}\bullet$) by nicotinamide adenine dinucleotide phosphate (NADPH) in combination with NADPH-cytochrome *c* reductase (Gage 1968; Dodge and Harris 1970; Bus et al. 1974; Bus et al. 1976), suggesting that NADPH functions as an electron donor to PQ. PQ monocation radical ($PQ^{+}\bullet$) formed by electron derived from NADPH reacted with molecular oxygen, and then generates superoxide, and is thereafter converted into hydrogen peroxide by SOD (Gage 1968; Bus et al. 1974; Sofuni and Ishidate 1988). Excessive accumulation of hydrogen peroxide generates more highly toxic hydroxyl radical by Haber-Weiss reaction and Fenton reaction (Nicotera et al. 1985; Tanaka and Amano 1989). Hydroxyl radical is thought to induce lipid peroxidation leading to chromosomal aberrations (Bus et al. 1976; Nicotera et al. 1985; Tanaka and Amano 1989). Another point of view of the mechanism of PQ toxicity is NO generation. PQ is seen to induce NO through activation of NO synthase (NOS), and NO generation plays a key role in causing *in vivo* cytotoxicity (Berisha et al. 1994a, 1994b; Ahmad et al. 2008).

It is crucial to investigate the mechanism involved in suppression of adverse PQ influence using the radical scavengers (Mn(III)tetrakis(1-methyl-4-pyridyl)porphyrin and catalase) and physiologically active substance such as acetyl-L-carnitine (acetylated form of L-carnitine). Mn(III)tetrakis(1-methyl-4-pyridyl)porphyrin (Mn(III)TMpyP) has SOD-like action that is conversion of superoxide into hydrogen peroxide (Pasternack et al. 1981). Catalase (CAT) is an important endogenous enzyme to catalyze the decomposition of hydrogen peroxide into water and oxygen. Suppressive effect of these radical scavengers and ALC on PQ-induced oxidative structural chromosomal damage was therefore estimated.

Phenolic substances including α-TH and BHT have antioxidant actions to inhibit generation of lipidic free radicals induced by autoxidation of unsaturated fatty acids (USFA) (Březina et al. 1990; Frankel 1991; Witter 2005; Valavanidis et al. 2006; Pokorny 2008). The antioxidant actions of phenolic substances to inhibit formation of lipid peroxy radicals (LOO•) and formation of alkoxy radicals (LO•) initiated by lipid radicals (L•) is shown below.

Antioxidant mechanism of phenolic antioxidants

1. USFA autoxidation chain reaction

L-H + various stimulations (irradiation, heat treatment, reaction with metal ion and reaction with another free radicals) → (– H•) → L•
L• + O_2 → LOO•
LOO• + L –H → LOOH + L•
LOOH + Metal ion (an example, Fe^{2+}) → LO • + Metal ion (Fe^{3+}) + OH^{-}•

2. Hydrogen transfer reaction by phenolic antioxidant

A-H + LO• → A• + LOH
A-H + LOO• →A• + LOOH

3. Termination reaction by free antioxidant radical

A• + A• →A-A (Antioxidant dimer)
A• + LO• → A-LO (Lipid-antioxidant copolymer)
A• + LOO• →A-LOO (Peroxidic copolymer)

(A-H, phenolic antioxidant; A•, free antioxidant radical; - H•, hydrogen transfer reaction; L-H, unsaturated fatty acids; L•, lipid radical; LO•, alkoxy radical; LOO•, peroxy radical; LOH, hydroxy acid; LOOH, lipid hydroperoxides; OH^{-}•, hydroxyl radical)

Inhibitory mechanism of lipidic free radical generation by phenolic antioxidant is as follows; hydrogen transfer reaction that produces hydroxy acid and lipid hydroperoxides; termination reaction that forms lipid-antioxidant copolymers and peroxidic copolymers. Free antioxidant radical generated at the USFA autoxidation also reacts with another free antioxidant radical, resulting in forming antioxidant dimer by termination reaction. Thus, phenolic antioxidants suppress generation of lipidic free radicals, which cause adverse effects on health of human and wildlife.

α-TH is a representative one of phenolic antioxidants and an essential nutrient for bodies of human and wildlife. Pro-oxidative α-TH action in the presence of the initiators on human lipoprotein has been reported, however (Bowry et al. 1992; Kontush et al. 1996; Upston et al.

1999). D-α-tocopheryl succinate (most effective vitamin E form) increases the chromosomal damage frequency in Hela cells and ovarian carcinoma cells *in vitro* and does not prevent γ-irradiation-induced chromosomal damage in the same cancer cells, however, further enhances the γ-irradiation-induced chromosomal damage in spite of the result that d-α-tocopheryl succinate does not induce chromosomal damage in the human normal cells *in vitro* and the γ-irradiated normal human cells (Kumar et al. 2002). On the other hand, disrupted antioxidant function of BHT has been little known until today (Bomhard et al. 1992; OECD SIDS, CAS N°: 128-37-0). In this study, the possible mechanism of PQ toxicity enhanced by α-TH and BHT is described.

STRUCTURAL CHROMOSOMAL DAMAGE

Upper figure of Figure 1 shows normal karyotype of *P. nigromaculatus* consisted of 5 pairs of large chromosomes (No.1 ~ 5 pairs) and 8 pairs of small chromosomes (No.6 ~ 13). No.11 chromosome has second constriction examined in center portion of long arms, and the others do not have.

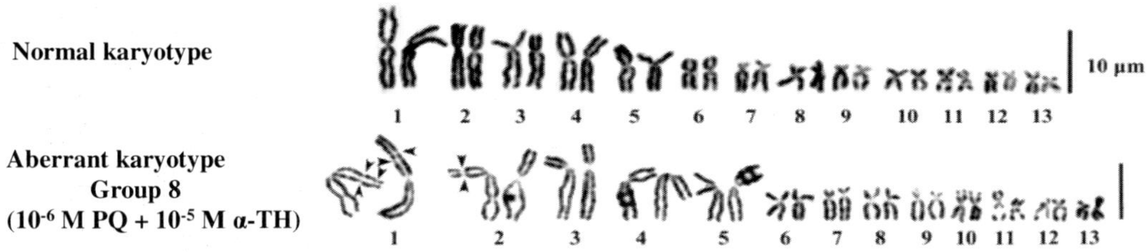

Figure 1. Normal (untreated control) and aberrant *P. nigromaculatus* karyotype (group 8, 10^{-6} M PQ + 10^{-5} M α-TH treatment). Animal maintenance: *P. nigromaculatus* specimens were used in the present investigation. Specimens were derived from standard strains maintained in the Institute for Amphibian Biology, Graduate School of Science, Hiroshima University. Specimens of *P. nigromaculatus* were fed on live crickets and raised at room temperature (26-27°C) (Kashiwagi et al. 2005). Culture: Blood samples were collected from frogs anesthetized with diethyl ether and then incubated at 25°C in 1 mL of 70 % Hank's balanced salt solution (pH 7.2) containing 630 mg L^{-1} lactose, 24.0 mg L^{-1} L-gultamine, 0.2 % (v/v) MEM amino acids solution 50×, 0.1% (v/v) MEM vitamin solution 100×, 11.8 mg L^{-1} succinate, 5 ~ 10 % fetal bovine serum (inactivated at 80 °C for 5 min), 2 IU mL^{-1} heparin sodium, 1 µl mL^{-1} PHAM, 100 Units mL^{-1} penicillin and 100 µg mL^{-1} streptomycin for 5 days (Hanada 2002). Each volume of blood per 1 mL of the basal medium was approximately 20 µL. Chromosome preparation: Chromosome preparations were made according to the method of Hanada (2002). Briefly, after treating with colchicine (final concentration 0.5 µg mL^{-1}) for 4 h, culture mediums were removed. And then blood samples were transferred to each centrifuge tube containing 1 mL of 0.075 M potassium chloride solution and incubated for 20 min at 25°C. After fixation in 200 µL of 1 : 3 acetic acid/methanol fixative, blood samples were centrifuged at 266 ×*g* for 5 min. 1 mL of new fixative was poured into centrifuge tubes after supernatant removal. Obtained cell suspensions of blood samples were stored at -20°C until use. Chromosome preparations made using air-drying method were stained with 4 % Giemsa solution in sodium phosphate buffer solution at pH 6.8 and observed using light microscopy Axioscope 2 *plus* (ZEISS Co. Ltd.). Photographs of metaphase chromosomes were recorded using the digital camera Nikon D80. Arrowheads show breakpoints.

Karyotypical abnormality in the bottom figure (10^{-6} M PQ + 10^{-5} M α-TH treatment, see 'legend of Figure 8' below for further details) is characterized by the 5 chromosomal breaks on short arms of No.1 chromosomes and the 2 breaks on short arm of No. 2 chromosome. Chromosomal damage mainly observed in this study was chromatid break and isochromatid break.

PQ GENOTOXICITY

PQ increased the frequency of leukocytes including aberrant chromosome(s) as PQ concentration increase (Figure 2). PQ seems to have cytogenetic toxicity in cultured *R. ornativentris* leukocytes. In Figure 3, 10^{-6} M PQ only induced increase in the frequency of leukocytes including structurally aberrant chromosomes by 23 %. This cytogenetic toxicity of PQ was significantly inhibited by the radical scavengers 10 to 100 μg mL^{-1} Mn(III)TMpyP and 1 to 10 U mL^{-1} CAT (Figure 3). Furthermore, 10^{-6} M PQ plus 1 to 10 μg mL^{-1} Mn(III)TMpyP plus 0.1 to 1 U mL^{-1} CAT decreased the induction of structural chromosomal damage significantly by the control level, suggesting that PQ induces structural chromosomal damage through superoxide generation and hydrogen peroxide generation (Figure 4).

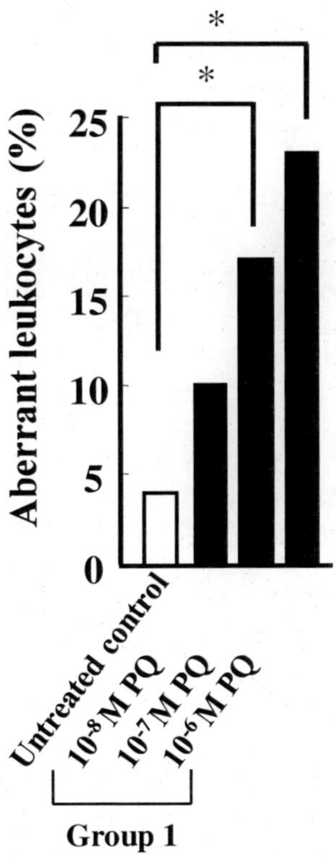

Figure 2. Genotoxic effect of PQ on cultured *R. ornativentris* leukocytes. Experimental conditions were the same as described in Figure 1. Group1 leukocytes were incubated in solution containing 10^{-8} to 10^{-6} M PQ for 6 h. Control leukocytes were incubated in medium to which PQ was not added. Data were analyzed by Chi-square test. *P*-values below 0.05 are considered significant. *Significantly greater than corresponding values for untreated control leukocytes.

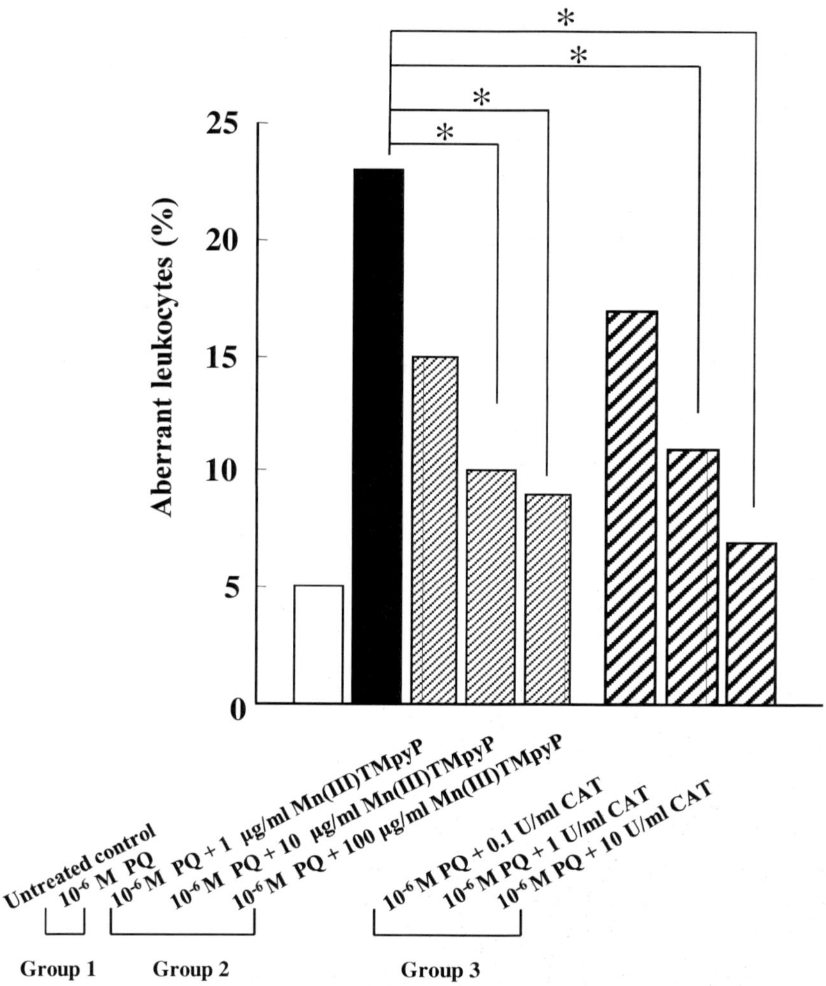

Figure 3. Suppressive effect of Mn(III)TMpyP and CAT on PQ-induced structural chromosomal damage in cultured *R. ornativentris* leukocytes. Experimental conditions were the same as described in Figure 1. Group 1 leukocytes were incubated in medium containing 10^{-6} M PQ for 6 h, and group 2 leukocytes were incubated in medium containing 10^{-6} M PQ and 1 to 100 μg mL^{-1} Mn(III)TMpyP for 6 h. Group 3 leukocytes were incubated in medium containing 10^{-6} M PQ and 0.1 to 10 U mL^{-1} CAT for 6 h. Control leukocytes were not exposed to either PQ, Mn(III)TMpyP or CAT. Data were analyzed by Chi-square test. *P*-values below 0.05 are considered significant. *Significantly less than corresponding values for PQ only (group 1).

Acetyl-L-carnitite (ALC) is thought to enhance β-oxidation, and the β-oxidation enhancement suppresses accumulation of free fatty acids leading to inhibiting lipid peroxidation and mitochondrial permeability transition pore opening, resulting in inhibition of apoptosis (Di Lisa et al. 1985; Paradies et al. 1999; Kashiwagi et al. 2001; Nishimura et al. 2008; Hanada et al. 2013). Suppressive effect of ALC on PQ-treated *R. ornativentris* leukocytes *in vitro* was shown in Figure 5.

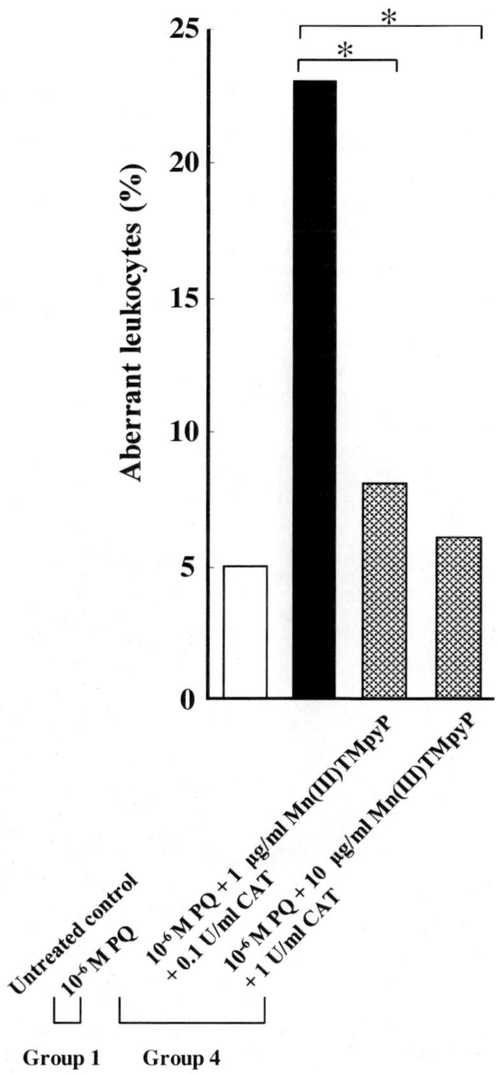

Figure 4. Suppressive effect of Mn(III)TMpyP plus CAT on PQ-induced structural chromosomal damage in cultured *R. ornativentris* leukocytes. Experimental conditions were the same as described in Figure 1. Group 1 leukocytes were incubated in medium containing 10^{-6} M PQ for 6 h. Group 4 leukocytes were incubated in medium containing 10^{-6} M PQ, 1 to 10 μg mL^{-1} Mn(III)TMpyP and 0.1 to 1 U mL^{-1} CAT for 6 h. Untreated control was not exposed to either PQ, Mn(III)TMpyP or CAT. Data were analyzed by Chi-square test. *P*-values below 0.05 are considered significant. *Significantly less than corresponding values for PQ only (group 1).

10^{-6} M PQ only induced the increase in the frequency of leukocytes including structurally aberrant chromosomes by 23 %. 10^{-6} M PQ plus 10^{-3} M ALC decreased the induction of structural chromosomal damage significantly, however. The results show that ALC decreases an amount of free fatty acids in cultured *R. ornativentris* leukocytes by β-oxidation, and thereby protecting the leukocytes against lipid peroxidation and PQ-induced structural chromosomal damage.

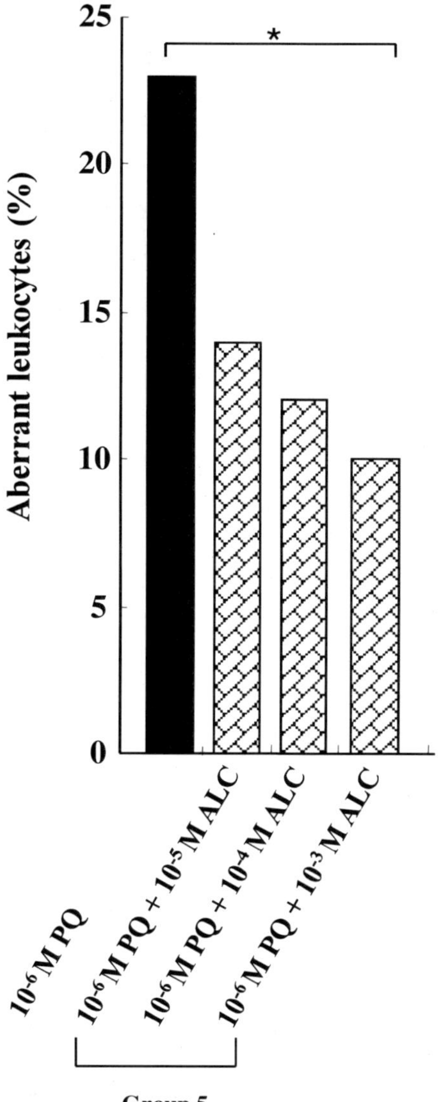

Figure 5. Suppressive effect of acetyl-L-carnitine (ALC) on PQ-induced structural chromosomal damage in cultured *R. ornativentris* leukocytes. Experimental conditions were the same as described in Figure 1. Group 1 leukocytes were incubated in medium containing 10^{-6} M PQ for 6 h. Group 5 leukocytes were incubated in medium containing 10^{-6} M PQ and 10^{-5} to 10^{-3} M ALC for 6 h. Data were analyzed by Chi-square test. *P*-values below 0.05 are considered significant. *Significantly less than corresponding values for PQ only (group 1).

Cultured *P. nigromaculatus* leukocytes were exposed to PQ in combination with NADPH as shown in Figure 6 (positive control). 10^{-6} M PQ induced the increase in the frequency of leukocytes including structurally aberrant chromosomes by 26 % (Figure 6). 10^{-6} M PQ plus 10^{-4} M NADPH increased the frequency of the aberrant leukocytes significantly by 48 %, as expected (Figure 6). These results suggest that NADPH functions as an electron donor to PQ.

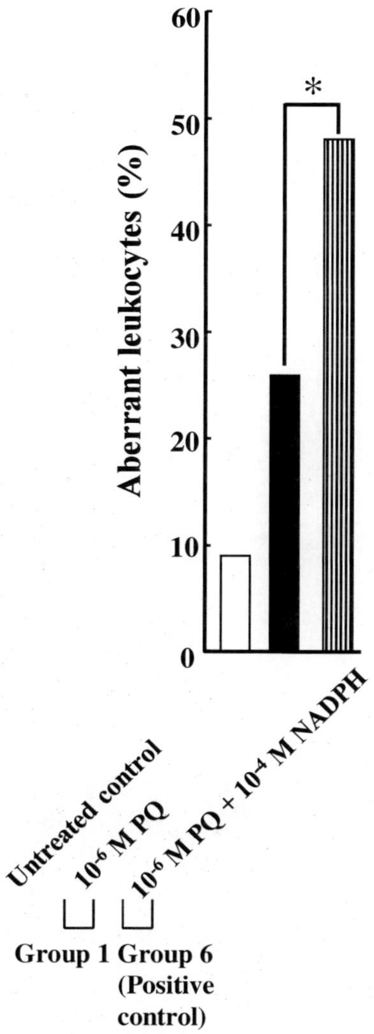

Figure 6. Genotoxic effect of PQ plus NADPH on cultured *P. nigromaculatus* leukocytes. Experimental conditions were the same as described in Figure 1. Group 1 leukocytes were incubated in medium containing 10^{-6} M PQ for 6 h, and group 6 (positive control) leukocytes were incubated in medium containing 10^{-6} M PQ plus 10^{-4} M NADPH for 6 h. Control leukocytes were incubated in medium and not exposed either to NADPH or PQ. Data were analyzed by Chi-square test. *P*-values below 0.05 are considered significant. *Significantly greater than corresponding values for PQ only (group 1).

Genotoxic Effects of Vitamin E (Alpha-TH) on Cultured *P. Nigromaculatus* Leukocytes

10^{-7} to 10^{-5} M α-TH only had no genotoxic effect on cultured leukocytes derived from individual of *P. nigromaculatus*, as shown in Figure 7.

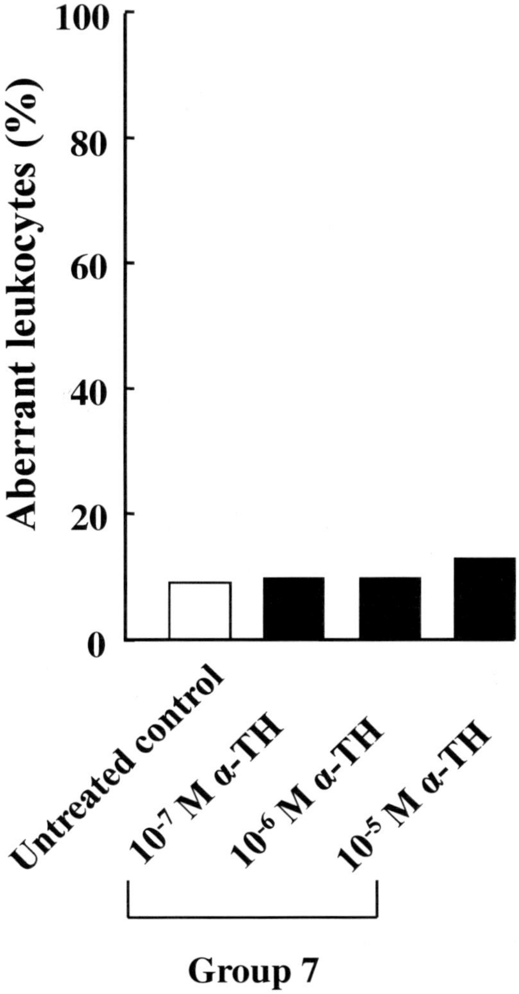

Figure 7. Genotoxic effect of α-TH on cultured *P. nigromaculatus* leukocytes. Experimental conditions were the same as described in Figure 1. Group 7 leukocytes were incubated in medium containing 10^{-7} to 10^{-5} M α-TH for 6 h. Control leukocytes were incubated in medium and not exposed to α-TH.

But, 10^{-5} M α-TH in combination with 10^{-6} M PQ increased the frequency of the aberrant leukocytes significantly by 48 %, more than 10^{-6} M PQ only (Figure 8). The excessive structural chromosomal damage induced by 10^{-6} M PQ plus 10^{-5} M α-TH was decreased by control level because of dual inhibition of Mn(III)TMpyP plus CAT (Figure 8). The results show that α-TH enhances the genotoxic action of PQ through ROS generation such as superoxide and hydrogen peroxide.

α-TH is representative phenolic antioxidant. PQ-induced chromosomal damage should be theoretically inhibited by tocopheroxyl radical generated after hydrogen transfer reaction of protective α-TH actions against lipidic free radicals, however, α-TH enhanced the PQ genotoxicity. These results suggest that extra electron of tocopheroxyl radical from α-TH is chemically transferred to PQ.

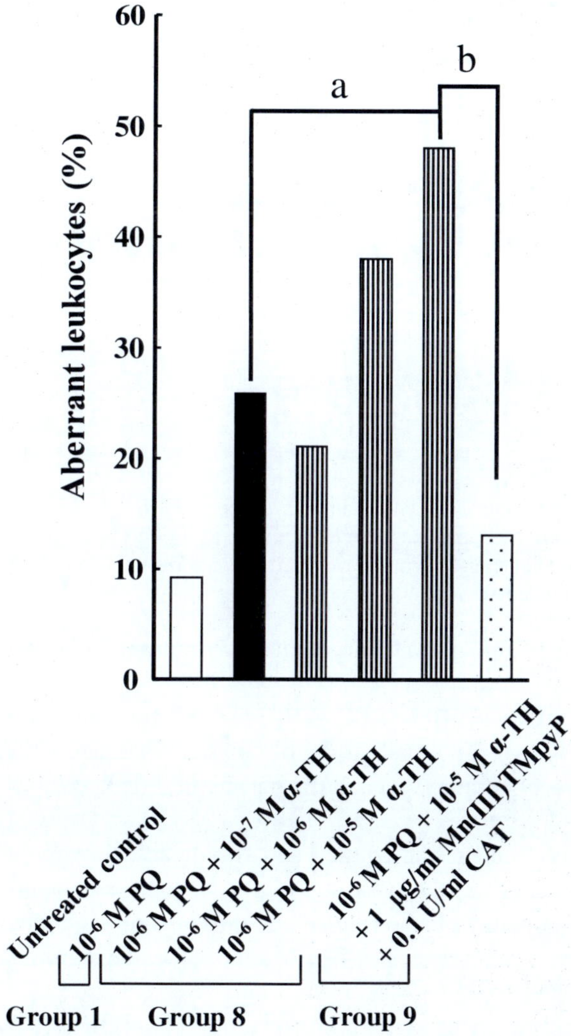

Figure 8. Genotoxic effect of PQ plus α-TH on cultured *P. nigromaculatus* leukocytes. Experimental conditions were the same as described in Figure 1. Group1 leukocytes were incubated in medium containing 10^{-6} M PQ for 6 h, and group 8 leukocytes were incubated in medium containing 10^{-6} M PQ and 10^{-7} to 10^{-5} M α-TH for 6 h. Group 9 leukocytes were incubated in medium containing 10^{-6} M PQ, 10^{-5} M α-TH, 1 μg mL^{-1} Mn(III)TMpyP and 0.1 U mL^{-1} CAT for 6 h. Control leukocytes were incubated in medium and not exposed either to Mn(III)TMpyP, CAT, α-TH or PQ. Data were analyzed by Chi-square test. *P*-values below 0.05 are considered significant. [a]Significantly greater than corresponding values for PQ only (group 1). [b]Significantly less than corresponding values for PQ plus α-TH (group 8).

Genotoxic Effects of BHT on Cultured *P. Nigromaculatus* Leukocytes

Normal (A) and aberrant karyotypes (B and C) of *P. nigromaculatus* is shown in Figure 9.

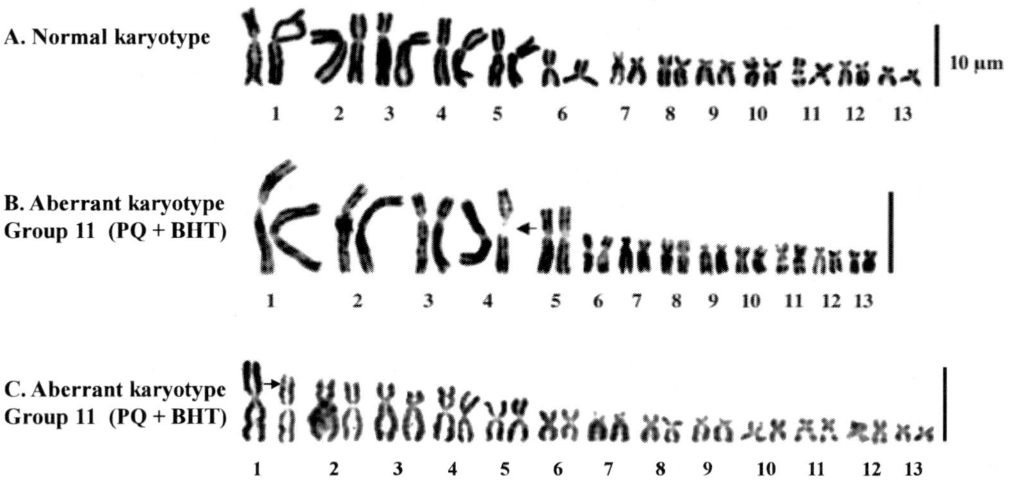

Figure 9. Normal (untreated control) and aberrant karyotype of *P. nigromaculatus* (10^{-6} M PQ + 10^{-4} M BHT treatment). Experimental conditions were the same as described in Figure 1. Normal karyotype of *P. nigromaculatus* (A) and aberrant karyotype of *P. nigromaculatus* (B and C). Arrows show break points in no. 1 and no. 4 chromosomes.

Figure 9A shows the same normal karyotype of *P. nigromaculatus* as the upper figure of Figure 1.

Figure 9B and C (group 11, 10^{-6} M PQ + 10^{-4} M BHT treatment, see legend of Figure 10 for further details) show two abnormal karyotypes marked by isochromatid break at proximal region of long arm of no.4 (Figure 9B) and the chromatid break at distal portion of long arm of no. 1 (Figure 9C).

No. 4 does not have second constriction and fragile site shown in Figure 9B, therefore empty space region observed in the proximal region of long arm of no .4 is abnormal. The length of left sister chromatid of aberrant no. 1 chromosome shown in Fig 9C is slightly longer than that of right sister chromatid because the left sister chromatid was elongated by the chromatid break. The karyotype is also abnormal.

Figure 10 shows the enhanced genotoxic effect of PQ plus BHT on cultured leukocytes of *P. nigromaculatus*. 10^{-6} M PQ only induced the increase in the frequency of leukocytes including structurally aberrant chromosomes by 23 %. NADPH-enhanced PQ cytogenetic toxicity was similar to the result shown in Figure 6, as expected. BHT only had no effect on genotoxic damage, but 10^{-6} M PQ plus 10^{-4} M BHT induced the increase in the frequency of the aberrant leukocytes more than PQ only and PQ plus NADPH, suggesting that BHT functions as electron donor to PQ.

In addition, PQ plus BHT-enhanced structural chromosomal damage was significantly suppressed by dual inhibitory action of the radical scavengers Mn(III)TMpyP and CAT. These results suggest that BHT donates electron to PQ, and thereby enhancing ROS generation involving superoxide and hydrogen peroxide generation, which induces excessive amount of structural chromosomal damage.

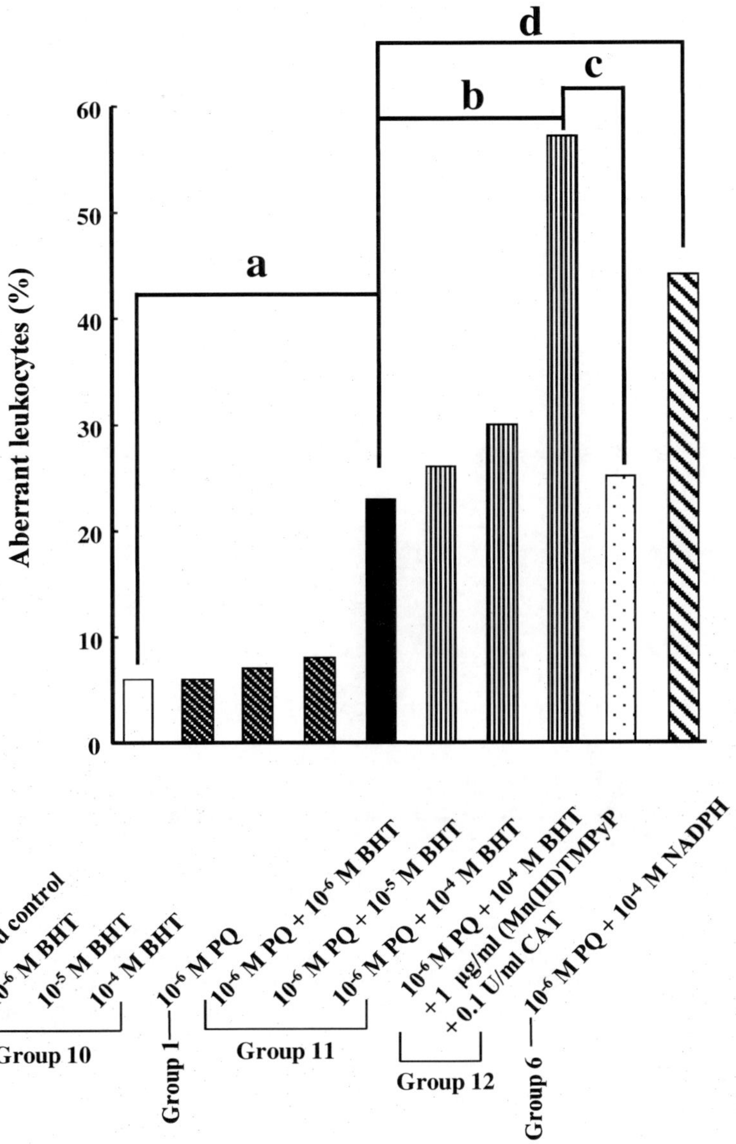

Figure 10. Genotoxic effect of PQ plus BHT on cultured *P. nigromaculatus* leukocytes. Experimental conditions were the same as described in Figure 1. Group-1 leukocytes were incubated in medium containing 10^{-6} M PQ for 6 h, and group-11 leukocytes were incubated in medium containing 10^{-6} M PQ and 10^{-6} to 10^{-4} M BHT for 6 h. Group-6 leukocytes (positive control) were incubated in medium containing 10^{-6} M PQ and 10^{-4} M NADPH for 6 h. Group-10 leukocytes were incubated in medium containing 10^{-6} to 10^{-4} M BHT for 6 h. Group-12 leukocytes were incubated in medium containing 10^{-6} M PQ, 10^{-4} M BHT, 1 μg mL^{-1} Mn(III)TMpyP and 0.1 U mL^{-1} CAT for 6 h. Untreated control leukocytes were incubated in medium and not exposed to BHT, Mn(III)TMpyP, CAT, NADPH or PQ. Data were analyzed by Chi-square test. *P*-values below 0.05 are considered significant. [a]Significantly greater than corresponding value for untreated leukocytes. [b]Significantly greater than corresponding value for group 1 (PQ only) leukocytes. [c]Significantly less than corresponding value for group 11 (10^{-6} M PQ + 10^{-4} M BHT) leukocytes. [d]Significantly greater than corresponding value for group 1 (PQ only) leukocytes.

CHEMICAL PQ^{+}• FORMATION BY BHT

α-TH antioxidant-function-disruption induced by PQ is shown in Figure 8, as described above. It was difficult to examine whether α-TH induces reduction of PQ^{2+} chemically to PQ^{+}• or not, due to insolubility of α-TH in water. PQ^{+}• formation-test was conducted using BHT that is more soluble in water than α-TH in order to confirm the hypothesis that BHT reduces PQ^{2+} chemically to PQ^{+}•. The level of PQ^{+}• chemically reduced by 10^{-4} M BHT was over 10^{-7} M (Figure 11). The result suggests that electron derived from BHT is chemically transferred to PQ^{2+}, resulting in forming PQ^{+}•.

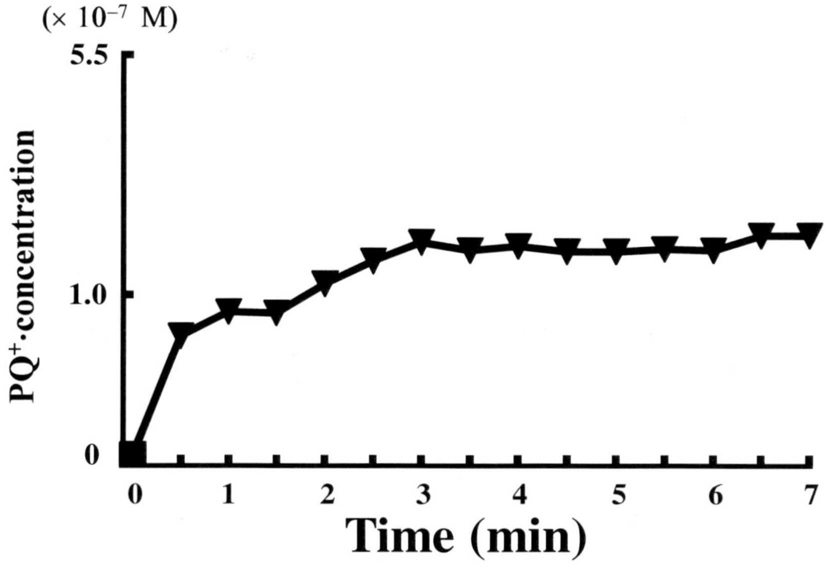

Figure 11. PQ^{+}• formation chemically induced by BHT. 1 μL of 10^{-1} M BHT was added to 1 mL of 0.9 M Tris-HCl buffer solution (pH 7.2) containing 10^{-2} M PQ at 25 °C, as previously described (Calderbank and Yuen 1965; Weidauer et. al. 2002). Tests were conducted under anaerobic conditions induced by replacing air with nitrogen gas. PQ^{+}• formation was spectrophotometrically monitored at 396 nm using UVIDEC-320H (Japan Spectroscopic Co. LTD). PQ^{+}• concentration was calculated from absorbance at 396 nm using alkali sodium dithionite-formed PQ^{+}• as the standard.

PRELIMINARY STUDY INVOLVED IN PQ TOXICITY ENHANCEMENT BY ENDOGENOUSLY PRODUCED NITRITE

PQ^{+}• formation by sodium nitrite ($NaNO_2$) is shown in Figure 12.

$NaNO_2$ was found to reduce PQ^{2+} chemically to PQ^{+}•. The PQ^{+}• concentration induced by 10^{-2} M $NaNO_2$ solution at pH 7.2 was 5.5×10^{-7} M and the levels at pH 8.2 and pH 9.2 was lower than 5.5×10^{-7} M, suggesting that $NaNO_2$ decreases PQ^{+}• formation in a pH-dependent manner (Figure 12). $NaNO_2$ in a near neutral solution exhibited a higher reducing power for PQ.

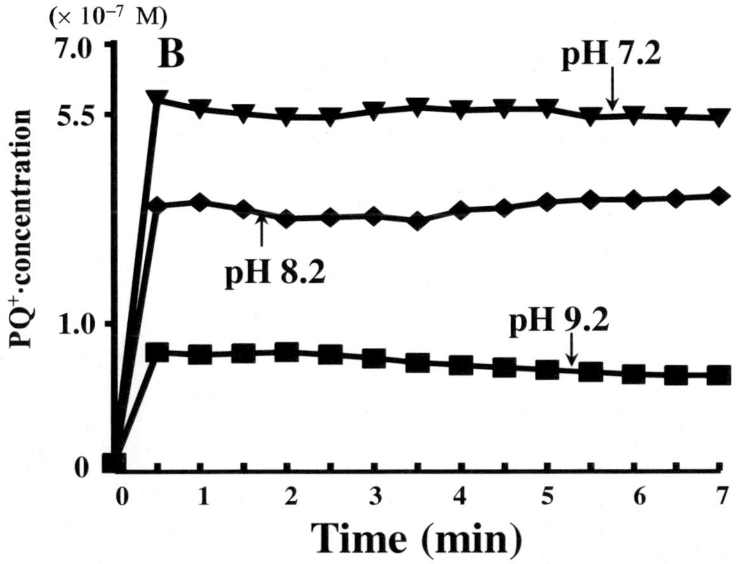

Figure 12. $PQ^{+}\bullet$ formation by $NaNO_2$. 5 μL of 2 M $NaNO_2$ was added to 1 ml of 0.9 M Tris-HCl buffer solutions (pH 7.2 ~ 9.2) containing 10^{-2} M PQ at 25 °C, as previously described (Calderbank and Yuen 1965; Weidauer et. al. 2002). All tests were conducted under anaerobic conditions induced by replacing air with nitrogen gas. $PQ^{+}\bullet$ formation was spectrophotometrically monitored at 396 nm using UVIDEC-320H (Japan Spectroscopic Co. LTD.). $PQ^{+}\bullet$ concentration was calculated from absorbance at 396 nm using alkali dithionite formed $PQ^{+}\bullet$ as the standard. Values given represent the mean value of three repetitions.

The concentration of $PQ^{+}\bullet$ induced by 10^{-4} M BHT at pH 7.2 reached 10^{-7} M level, and the approximate induction level of $PQ^{+}\bullet$ by 10^{-2} M NO_2^- at the same condition was five-fold higher than that of BHT.

Tanaka (1997, 2007) has reported that cytogenetic toxicity of nitrite is induced through ROS generation and moreover nitrite generated from the NO releaser enhances the PQ genetoxicity. As shown in Figure 12, generation of $PQ^{+}\bullet$ was enhanced by nitrite, suggesting that nitrite is the same electron donor as NADPH, α-TH and BHT to PQ.

10^{-6} M PQ plus 10^{-4} M BHT and 10^{-6} M PQ plus 10^{-5} M α-TH caused acute genotoxicity through ROS generation as shown in Figures 8 and 10. Figure 13 shows that PQ plus BHT enhances nitrite production in *R. rugosa* tadpole tails.

In order to investigate the possible involvement of nitrite in the mechanism of chromosomal damage, the nitrite levels by tails of PQ-treated tadpoles in the presence or absence of BHT were compared. Nitrite is one of the nitric oxide (NO) degradation products, and the evidence for NO synthase activation as well. The tails of group 1 (PQ only) tadpoles showed the increase in nitrite level.

No difference between control and group 10 (BHT only) was observed (Figure 13). In group 11 (PQ + BHT), nitrite level rose markedly during the treatment with 10^{-7} M PQ plus 10^{-5} M BHT. The approximate concentration was 2×10^{-2} M. Change in the pattern of nitrite levels is very similar to the change in the pattern of the frequency of leukocytes including structurally aberrant chromosomes (see Figure 10).

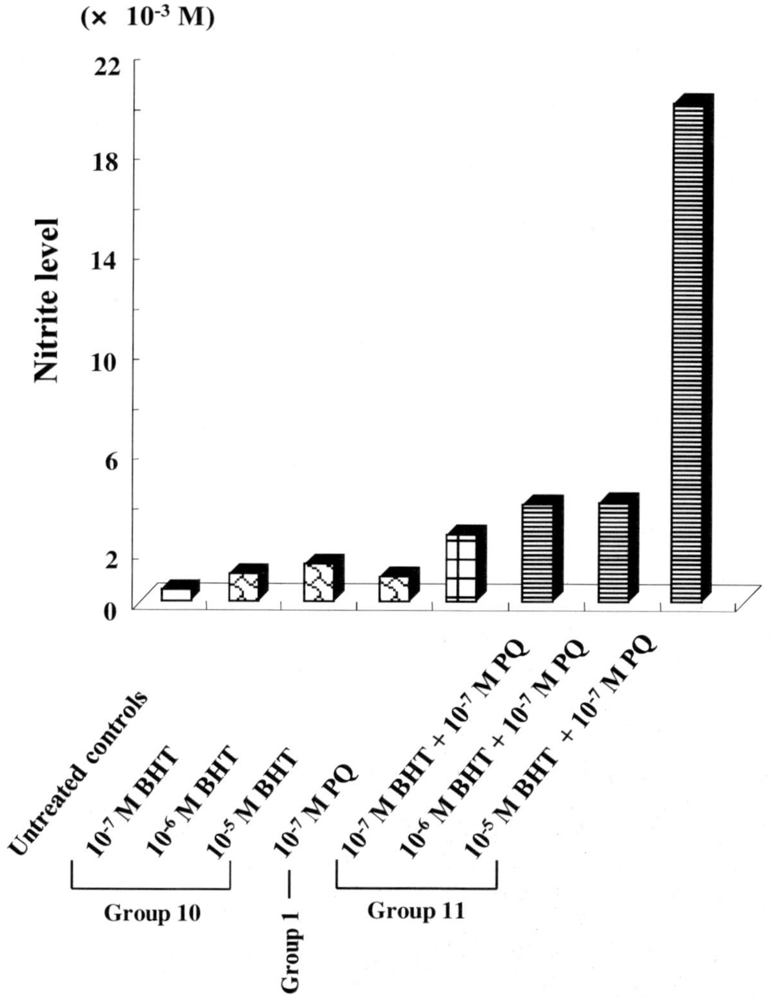

Figure 13. Effect of PQ plus BHT on production of nitrite in *R. rugosa* tadpole tails. Amounts of nitrite (NO_2^-) in *R. rugosa* tadpole tails were determined according to the method of nitrite-ferrous sulfate reaction (Vogel 1954). This study was conducted using the method having an advantage of measuring nitrite absorbed in tail tissues, because it was very difficult to measure nitrite absorbed into the tail tissues. *R. rugosa* tadpoles were fed on boiled spinach (Rugh 1934) and staged according to Taylor and Kolloros (1946). *R. rugosa* tadpole tails were homogenized in chilled distilled water. 25 μL of 1 M sodium carbonate was added to 80 μL of the lysate. After incubating at 95 °C for 10 min, 25 μL of silver sulfate saturated in 1 M sulfuric acid and 10 μL of 1 M sulfuric acid were added and centrifuged at 715 × *g* for 5 min. A reaction mixture was also prepared containing 0.25 M sulfuric acid and 1 M ferrous sulfate. Reactions were started at 20 °C by adding 30 μL of supernatant to 300 μL of reaction mixture. Amounts of NO_2^- in tails were spectrophotometrically measured at 600 nm. Concentration of NO_2^- was calculated from absorbance at 600 nm using sodium nitrite ($NaNO_2$) as the standard. Assays of *R. rugosa* tadpoles were conducted in saline solution containing 0.65 g L^{-1} sodium chloride, 0.014 g L^{-1} potassium chloride, 0.12 g L^{-1} calcium chloride and 0.02 g L^{-1} sodium bicarbonate. Stage X tadpoles were raised at a population density of three individuals per 100 ml in trays containing various solutions as follows: Group1 tadpoles were raised in saline solution containing 10^{-7} M PQ for 3 days. Group10 tadpoles were raised in saline solution containing 10^{-7} to 10^{-5} M BHT for 3 days. Group 11 tadpoles were raised in saline solution containing 10^{-7} to 10^{-5} M BHT and 10^{-7} M PQ for 3 days. Control tadpoles were raised in saline solution and not exposed either to BHT or PQ.

MECHANISM OF GENOTOXICITY ENHANCED BY PQ IN COMBINATION WITH ALPHA-TH OR BHT

Proposal mechanism for chromosomal damage enhancement by PQ in combination with phenolic antioxidants (take BHT, for example) is shown in Figure 14. PQ^{+}• formation is enhanced by reducing agents such as NADPH and NADH (Gage 1968; Dodge and Harris 1970; Bus et al. 1974), and then induces chromosomal aberrations (Nicotera et al. 1985; Tanaka and Amano 1989). These results suggest that reducing agents function as electron donor to PQ. This suggestion was also examined in the present study. Interestingly, PQ plus α-TH and PQ plus BHT induced increases in structural chromosomal damage acutely, regardless of the results which α-TH only and BHT only have no effect on induction of structural chromosomal damage. In addition, from the results that the acute chromosomal damage induced by PQ plus α-TH and PQ plus BHT was inhibited by combination of Mn(III)TMpyP and CAT, the chromosomal damage seems to be caused through ROS generation. These results suggest that phenolic antioxidants reduce PQ^{2+} chemically to PQ^{+}• and thereby generating ROS resulting in chromosomal damage. PQ has not only functions as ROS generator but also nitrite inducer, however.

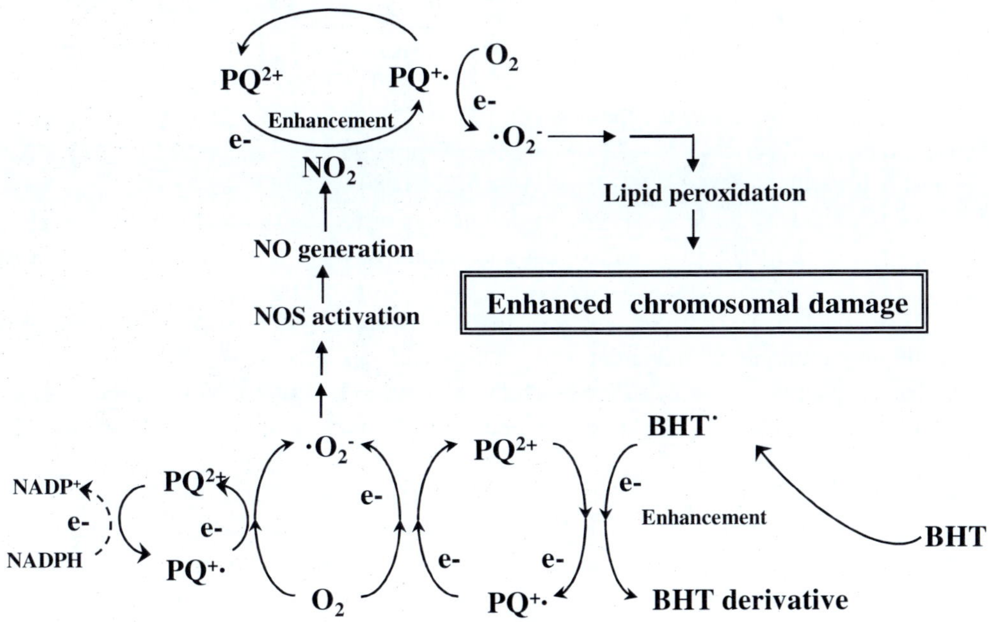

Figure 14. Proposal mechanism of chromosomal damage enhanced by phenolic antioxidants combined with PQ. e- = electron; NOS = NO synthase; → = Chemical reaction; ➔ Enzymatic reaction.

The results obtained from this preliminary study show the possibility that nitrite endogenously overproduced by PQ in combination with the phenolic antioxidants further enhances PQ^{+}• formation. Here, important chemical property of PQ toxicity is to take away electron generated from reducing agents, phenolic antioxidants and nitrite. Extra electron of free antioxidant radicals derived from the phenolic antioxidants enhances PQ^{2+}—PQ^{+}• formation—superoxide generation cycle, and thereby excessively accumulating nitrite, leading to inducing destructive increase in electron resulting in enhancement of ROS generation and chromosomal damage. Proposed mechanism (shown in Figure 14) is as follows: (1) $PQ^{+}\bullet$ formation enhanced by BHT• derived from BHT, (2) an increase in $\cdot O_2^{\,-}$ generated by electron transfer between molecular oxygen and $PQ^{+}\bullet$, (3) NOS activation, (4) nitrite overproduction, (5) excessive formation of $PQ^{+}\bullet$ by overproduced nitrite, (6) an accumulation of reactive oxygen species, such as superoxide, hydrogen peroxide and lipidic peroxides, (7) a marked increase in chromosomal damage.

CONCLUSION

Complexity of the mechanism involved in PQ toxicity-enhancement induced by free antioxidant radicals from α-TH and BHT arises from chemical property of PQ that tends to be activated by endogenously produced nitrite. Further research should be done in order to clarify the mechanism.

REFERENCES

Ahmad, I., Kumar, A., Shukla, S., Prasad Pandey, H., & Singh, C. (2008). The involvement of nitric oxide in maneb- and paraquat-induced oxidative stress in rat polymorphonuclear leukocytes. *Free Radic. Res.* 42, 849-862.

Anderson, K. D., & Scerri, G. V. (2003). A case of multiple skin cancers after occupational exposure to pesticides. *Br. J. Dermatol.* 149, 1088-1089.

Bauer, Dial, C. A., & Dial, N. A. (1995). Lethal effects of the consumption of field levels of paraquat-contaminated plants on frog tadpoles. *Bull. Environ. Contam. Toxicol.* 55, 870-877.

Berisha, H. I., Pakbaz, H., Absood, A., & Said, S. I. (1994a). Nitric oxide as a mediator of oxidant lung injury due to paraquat. *Proc. Natl. Acad. Sci. USA* 91, 7445-7449.

Berisha, H., Pakbaz, H., Absood, A., Foda, H. D., & Said, S. I. (1994b). Nitric oxide mediates oxidant tissue injury caused by paraquat and xanthine oxidase. *Ann. N. Y. Acad. Sci.* 723, 422-425.

Bomhard, E. M., Bremmer, J. N., & Herbold, B. A. (1992). Review of the mutagenicity/genotoxicity of butylated hydroxytoluene. *Mutat. Res.* 277, 187-200.

Bowry, V. W., Ingold, K. U., & Stocker, R. (1992). Vitamin E in human low-density lipoprotein. When and how this antioxidant becomes a pro-oxidant. *Biochem. J.* 288, 341-344.

Březina, P., Čepička, J., Davídek, J., Holas, J., Káš, J., Míková, K., Pipek, P., Pokorný, J., Rychtera, M., & Velíšek, J. (1990). Fats, oils and other lipids. In: Davídek. J., Velíšek, J., Pokorný, J., (Eds.), Developments in Food Science 21, Chemical changes during Food Processing. Elsevier, New York, pp 169-226.

Bus, J. S., Aust, S. D., & Gibson, J. E. (1974). Superoxide- and singlet oxygen-catalyzed lipid peroxidation as a possible mechanism for paraquat (methyl viologen) toxicity. *Biochem. Biophys. Res. Commun.* 58, 749-755.

Bus, J. S., Aust, S. D., & Gibson, J. E. (1976). Paraquat toxicity: proposed mechanism of action involving lipid peroxidation. *Environ. Health Perspect.* 16, 139-146.

Calderbank, A., & Yuen, S. H. (1965). An ion-exchange method for determining paraquat residues in food crops. *The Analyst* 90, 99-106.

Dial, N. A., & Bauer, C. A. (1984). Teratogenic and lethal effects of paraquat on developing frog embryos (*Rana pipiens*). *Bull. Environ. Contam. Toxicol.* 33, 592-597.

Dial, N. A., & Bauer Dial, C. A. (1987). Lethal effects of diquat and paraquat on developing frog embryos and 15-day-old tadpoles, *Rana pipiens*. *Bull. Environ. Contam. Toxicol.* 38, 1006-1011.

Di Lisa, F., Bobyleva-Guarriero, V., Jocelyn, P., Toninello, A., & Siliprandi., N. (1985). Stabilising action of carnitine on energy linked processes in rat liver mitochondria. *Biochem. Biophys. Res. Commun.* 131, 968-973.

Dodge, A. D., & Harris, N. (1970). The mode of action of paraquat and diquat. *Biochem. J.* 118, 43P-44P.

Frankel, E. N. (1991). Recent advances in lipid oxidation. *J. Sci. Food Agric.* 54, 495-511.

Gage, J. C. (1968). The action of paraquat and diquat on the respiration of liver cell fractions. *Biochem. J.* 109, 757-761.

Hanada, H. (2002). G and C banding show structural differences between the Z and W chromosomes in the frog *Buergeria buergeri*. *Hereditas* 136, 151-154.

Hanada, H., Kobuchi, H., Yamamoto, M., Kashiwagi, K., Katsu, K., Utsumi, T., Kashiwagi, A., Sasaki, J., Inoue, M., & Utsumi, K. (2013). Acetyl-L-carnitine suppresses thyroid hormone-induced and spontaneous anuran tadpole tail shortening. *Hereditas* 150, 1-9.

Jee, S. H., Kuo, H. W., Su, W. P. D., Chang, C. H., Sun, C. C., & Wang, J. D. (1995). Photodamage and skin cancer among paraquat workers. *Int. J. Dermatol.* 34, 466-469.

Kashiwagi, A., Kanno, T., Arita, K., Ishisaka, R., Utsumi, T., & Utsumi, K. (2001). Suppression of T_3- and fatty acid-induced membrane permeability transition by L-carnitine. *Comp. Biochem. Physiol.* B130, 411-418.

Kashiwagi, K., Shinkai, T., Kajii, E., & Kashiwagi, A. (2005). The effects of reactive oxygen species on amphibian aging. *Comp. Biochem. Physiol. C Toxicol. Pharmacol.* 140, 197-205.

Kontush, A., Finckh, B., Karten, B., Kohlschütter, A., & Beisiegel, U. (1996). Antioxidant and prooxidant activity of α-tocopherol in human plasma and low density lipoprotein. *J. Lipid Res.* 37, 1436-1448.

Kumar, B., Jha, M. N., Cole, W. C., Bedford, J. S., & Prasad, K. N. (2002). D-alpha-tocopheryl succinate (vitamin E) enhances radiation-induced chromosomal damage levels in human cancer cells, but reduces it in normal cells. *J. Am. Coll. Nutr.* 21, 339-343.

Mussi, M. A., & Calcaterra, N. B. (2010). Paraquat-induced oxidative stress response during amphibian early embryonic development. *Comp. Biochem. Physiol. C Toxicol. Pharmacol.* 151, 240-247.

Nicotera, T. M., Block, A. W., Gibas, Z., & Sandberg, A. A. (1985). Induction of superoxide dismutase, chromosomal aberrations and sister-chromatid exchanges by paraquat in Chinese hamster fibroblasts. *Mutat. Res.* 151, 263-268.

Nishimura, M., Okimura, Y., Fujita, H., Yano, H., Lee, J., Suzaki, E., Inoue, M., Utsumi, K., & Sasaki, J. (2008). Mechanism of 3-nitropropionic acid-induced membrane permeability transition of isolated mitochondria and its suppression by L-carnitine. *Cell Biochem. Funct.* 26, 881-891.

OECD SIDS, 2,6-di-tert-butyl-p-cresol (BHT), CAS N°: 128-37-0. (http://www.chem.unep.ch /irptc/sids/oecdsids/sidspub.html).

Osano, O., Oladimeji, A. A., Kraak, M. H. S., & Admiraal, W. (2002). Teratogenic effects of amitraz, 2,4-dimethylaniline, and paraquat on developing frog (*Xenopus*) embryos. *Arch. Environ. Contam. Toxicol.* 43, 42-49.

Paradies, G., Petrosillo, G., Gadaleta, M. N., & Ruggiero, F. M. (1999). The effect of aging and acetyl-L-carnitine on the pyruvate transport and oxidation in rat heart mitochondria. *FEBS Lett.* 454, 207-209.

Pasternack, R. F., Banth, A., Pasternack, J. M., & Johnson, C. S. (1981). Catalysis of the disproportionation of superoxide by metalloporphyrins. III. *J. Inorg. Biochem.* 15, 261-267.

Pokorny, J. (2008). Application of phenolic antioxidants in food products. *EJEAFChe* 7, 3320-3324.

Rugh, R. (1934). Induced ovulation and artificial fertilization in the frog. *Biol. Bull.* 66, 22–29.

Sofuni, T., & Ishidate, M. Jr. (1988). Induction of chromosomal aberrations in active oxygen-generating systems. I. Effects of paraquat in Chinese hamster cells in culture. *Mutat. Res.* 197, 127-132.

Śpiewak, R. (2001). Pesticides as a cause of occupational skin diseases in farmers. *Ann. Agric. Environ. Med.* 8, 1-5.

Tanaka, R. (1997). Induction of a sister-chromatid exchange by nitrogen oxides and its prevention by SOD. *J. Toxicol. Sci.* 22, 199-205.

Tanaka, R. (2007). Inhibitroy effects of xanthone on paraquat- and $NaNO_2$- induced genotoxicity in cultured cells. *J. Toxicol.* Sci. 32, 571-574.

Tanaka, R., & Amano, Y. (1989). Genotoxic effects of paraquat and diquat evaluated by sister-chromatid exchange, chromosomal aberration and cell-cycle rate. *Toxicol. In Vitro* 3, 53-57.

Taylor, A. C., & Kollros, J. J. (1946). Stages in the normal development of *Rana pipiens* larvae. Anat. Rec. 94, 7–24.

Upston, J. M., Terentis, A. C., & Stocker, R. (1999). Tocopherol-mediated peroxidation of lipoproteins: implications for vitamin E as a potential antiatherogenic supplement. FASEB J. 13, 977-994.

Valavanidis, A., Vlahogianni, T., Dassenakis, M., & Scoullos, M. (2006). Molecular biomarkers of oxidative stress in aquatic organisms in relation to toxic environmental pollutants. Ecotoxicol. Environ. Saf. 64, 178-189.

Vogel, A. I. (1954). A text-book of macro and semimicro qualitative inorganic analysis. Fourth edition. Longmans, Green and CO Ltd, London, pp. 338.

Wang, J. D., Li, W. E., Hu, F. C., & Hu, K. H. (1987). Occupational risk and the development of premalignant skin lesions among paraquat manufacturers. Br. J. Ind. Med. 44, 196-200.

Weidauer, E., Mörke, W., Foth, H., & Brömme, H. J. (2002). Does the anaerobic formation of hydroxyl radicals by paraquat monocation radicals and hydrogen peroxide require the presence of transition metals? Arch. Toxicol. 76, 89-95.

Wesseling, C., Hogstedt, C., Picado, A., & Johansson, L. (1997). Unintentional fatal paraquat poisonings among agricultural workers in Costa Rica: report of 15 cases. Am. J. Ind. Med. 32, 433-441.

Wesseling, C., van Wendel de Joode, B., Ruepert, C., León, C., Monge, P., Hermosillo, H., & Partanen, T. J. (2001). Paraquat in developing countries. Int. J. Occup. Environ. Health 7, 275-286.

Witter, A. E. (2005). The quantitative determination of butylated hydroxytoluene in chewing gum using GC-MS. J. Chem. Educ. 82, 1538-1541.

In: Handbook on Herbicides
Editors: Daiki Kobayashi and Eito Watanabe
ISBN: 978-1-62948-053-4

Chapter 10

GLYPHOSATE ADSORPTION IN A SOIL DEPTH PROFILE

***H. Tévez*[1] *and M. dos Santos Afonso*[2,*]**
[1]Facultad de Ciencias Forestales, Universidad Nacional de Santiago del Estero, Santiago del Estero, Argentina
[2]INQUIMAE, Facultad de Ciencias Exactas y Naturales-Universidad de Buenos Aires, Argentina

ABSTRACT

Glyphosate [N-(phosphonomethyl)glycine, PMG] is the most popular, non-selective, post-emergent and widely used herbicide in agriculture, especially in soybean crops. The Ap (0-18 cm), AB (18-50 cm) and BC (105-130 cm) horizons of a Typic Haplustoll soil, located at 62º06´W, 27º24´S, in Province of Santiago del Estero, Argentina, were characterized by chemical analysis, X-ray Diffraction (XRD), Point of Zero Charge (PZC) and surface area by water adsorption (S_w). The adsorption of glyphosate by soil samples of the three horizons was measured. Adsorption isotherms were fitted using the Langmuir and Freundlich model, The affinity constant (K_L) and maximum surface coverage (Γ_{max}) were determined and the results showed that Γ_{max} of the horizons increased with depth Ap<AB<BC and K_L followed the sequence BC< Ap < AB.

Keywords: Glyphosate; Adsorption isotherm; horizons profile

INTRODUCTION

The massive and unscrupulous use of pesticides led to the emergence of environmental problems, with direct impact on human health and the survival of many species. In line with the increased uses of pesticides, accidents and diseases related to them also increased

* Corresponding author: M. dos Santos Afonso. INQUIMAE, Facultad de Ciencias Exactas y Naturales-Universidad de Buenos Aires. Ciudad Universitaria, Pabellón II, 3er Piso, Ciudad Autónoma de Buenos Aires, C1428EHA, Argentina. E-mail: dosantos@qi.fcen.uba.ar.

significantly. Pesticides can be incorporated into the organism through the skin, by inhalation or intake in its various forms. The effects are of the most diverse nature depending on the pesticide, dosage, point of entrance and time of exposure. Effects such as vomiting, diarrhea, headaches, nervous disorders, drowsiness, convulsions, abortions or death can be immediate though there can also be chronic effects, cancer, liver necrosis, peripheral neuropathy or birth defects. These and other problems can be consequence of the accumulation of pesticides due to the slow biotransformation that some of them show and because of the synergistic effect resulting from the mixture of their active substances or adjuvants.

Pesticides permanence in natural systems produce biomagnification as a result of the accumulation in the human body, the endpoint of the food chain, by which they concentrate up to and sometimes beyond the amounts detected in the environment. Besides, this causes ecological imbalance since many beneficial species disappear and others harmful emerge.

THE PESTICIDES ROUTE

In relation to resource water pollution, pesticides have several entrance pathways. The direct routes of contamination correspond with fumigations and washing of the machinery and containers used in the operation as well as with their dumping in ponds, wells, and irrigation channels. In turn, the indirect route is that resulting from soil contamination where either by infiltration or runoff the undesirable substances eventually reach groundwater or surface waters reservoirs.

The polluted waters spread the pollutants on the flora and fauna causing the death of species, poisoning people in addition to the loss of water reservoirs. Some pesticides are resistant and do not breakdown easily so they can remain for long periods, even years, before disintegrating.

In an aqueous medium, pesticides may suffer chemical degradation by reactions like hydrolysis, and form low molecular weight products of equal or different degree of toxicity than the parent molecule. Hydrolysis can occur by acidic, basic or aqueous catalysis, where the reaction occur by the attack of the proton (H^+), hydroxyl ion (OH^-) or the water molecule onto the bond between two atoms of the pesticide molecule followed by the rupture of the chemical bond. Some pesticides that may suffer this kind of reactions belong to the sulfonylureas, carbamates, ureas and anilines families (Gatidou, et al., 2011; Nishiyama, et al., 2010; Tangri, et al., 2010).

Photodegradation, particularly by sunlight action, is another way by which pesticides are disposed (Wallace, et al., 2010). The ten most commonly used pesticides formulations in Argentina are eight different concentrations of glyphosate (N-phosphonomethyl glycine), one formulation derives from chlorpyrifos (an insecticide) and the last one is a mixture of pyraclostrobin and epoxiconazol that has fungicidal properties. Both, the glyphosate (Cantera, et al., 2008) and the chlorpyrifos (Trinelli, et al., 2006; Trinelli, 2011) are photodegraded through pseudo first-order mechanisms. The average half-life of some insecticides in aquatic media varies from hours to months, while the photolysis reactions in the soil are hard to determine due to the heterogeneity of the environment and the difficulties in the penetration of the radiation.

Moreover, the metals dissolved in the aqueous medium may also participate in photochemical degradation reaction such as the Fenton-like reactions (Kuo and Ho, 2010; Quivet et al. 2006).

The complexation reactions with metals ions present in natural water bodies are another type of reaction that should be considered in homogeneous aqueous media. This kind of reaction involves those pesticides that contain chemical groups as phosphate, carboxylate, amine, phenol, hydroxyl, and sulfurs since these groups are able to coordinate metals present in the water body like iron, copper, zinc, lead and aluminum among others (Barja, et al., 1998; Barja, et al., 2001; Ramstedt, et al., 2004; Paschevskaya, et al., 2006). As a consequence of that, once coordination compounds are formed, the total concentration of pesticide in aqueous solution may be either increased, because very soluble complex may occur, or decreased, because of the formation of sparingly soluble metallic salts that can settle on the sediments and become supply continued sources of pesticide in the water body.

In heterogeneous media (soils, sediments or particulate matter) three kinds of adsorption processes may exist in terms of the attraction between solute and adsorbent Mineral surfaces such as soils, sediments or particulate matter, may adsorb organic molecules weakly (physical adsorption) or strongly (chemisorption) depending on the strength of the adsorbate-adsorbent interaction. The characteristics of these processes are depicted in Table 1.

Table 1. Comparison of physical adsorption and chemisorption properties

Property \ Adsorption Type	Physisorption	Chemisorption
Force	Van der Waal	Chemical bond
Number of adsorbed layers	Multilayer	Monolayer
Adsorption heat	Low (10-40 kJ/mol)	High (> 40 kJ/mol)
Selectivity	Low	High
Temperature	Low	High
Surface complexes type	Outher-sphere	Inner-sphere

Study Area

The study area is located in the center-east of Santiago del Estero Province between 62°06´W and 61°52´W and, 27°24´S and 28°00´S (Figure 1).

The climate is semiarid mesothermal, with an average annual temperature of 19.6 °C and rainfall of between 600 and 750 mm per year concentrated in the spring-summer period.

The sampling area is a soil catena corresponding to that found in recent agriculture (9 years) and low for ancient agriculture (25-27 years).

The soil is derived from loessic sediments and it is located at the depression relief, and is classified as *Typic Pachic Haplustoll Haplustoll* with grasslands, *Elionurus muticus* (Lorenz, et al., 2000).

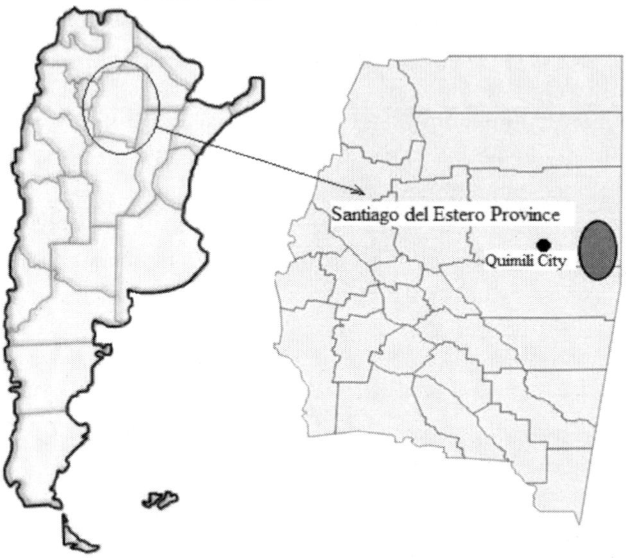

Figure 1. Argentina Map, Santiago del Estero Province and the ⬮: sampling area near Quimilí City.

In the early twentieth century, Santiago del Estero Province had 10 MHa of natural forest heritage, which was reduced to less than 2 MHa as a result of irrational forestry logging. The advance of the agricultural frontier in the arid areas of the Province of Santiago del Estero is produced over forest or range lands. When these lands are used for agricultural purposes biodiversity is reduced and, at the same time, the increased use of chemicals produced adverse effects on human health, water and soil pollution. The soybean sowing was developed at the expense of others crops, native forests and livestock (Pérez-Carrera, et al., 2008) (Figure 2).

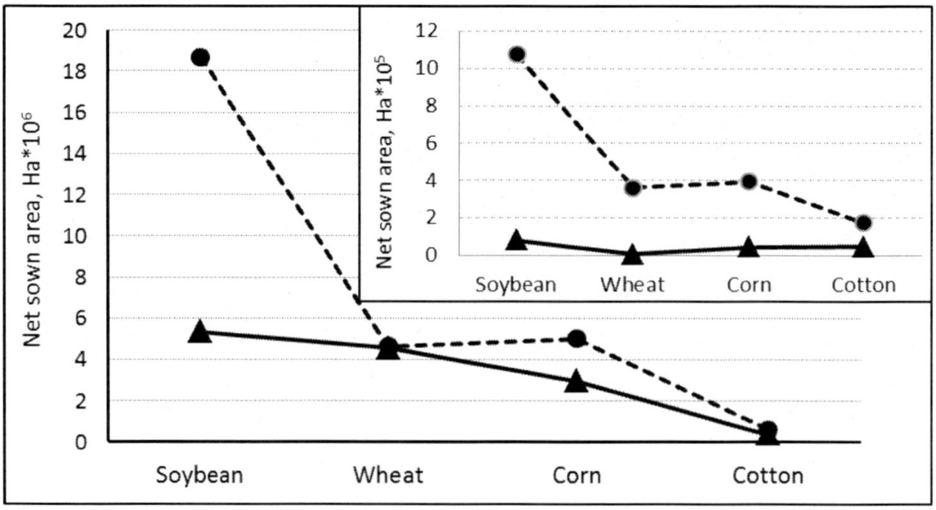

Figure 2. Area planted in season: —▲— 1992/93 and – – •– – 2011/12, in A: Argentina and B: Santiago del Estero Province (Sistema Integrado de Información Agropecuaria, 2013).

SAMPLE COLLECTION AND PREPARATION

A general survey of the study area for sampling procedure was made. Soil pits were open for identifying the horizons as well as the landscape and other external features. The characteristics of the soil profile, such as texture, structure, permeability, porosity, drainage, organic matter content, color, pH, calcium carbonate content, permeability, consistency, plasticity, shear resistance or shear strength, elasticity and deformation susceptibility due to volume changes were observed.

Soil used was classified as *Typic–Haplustoll* (Lorenz, et al., 2000) and sampled from North-west Argentine; Santiago del Estero Province (located at 62º06´W, 27º24´S). The main crop in this area is Soybean. The samples were taken up to 130 cm of depth (Figure 3), in horizons classified as Ap (0-18 cm), AB (18-50 cm) and BC (105-130 cm).

The fresh soil samples were air-dried and ground to pass a sieve of 2 mm before use.

Silt and clays fractions were separated by gravity (Carver, 1971). The clays particles sizes is less than 0.002 mm (2 μm) and they do not solely settle in water due to gravity.

All solutions and dispersions were prepared using reagent grade chemicals and Milli-Q water.

Figure 3. A view of a pit profile located at the sampling area.

Soil pH was measured in 0.01 M $CaCl_2$ solution at 1:2.5 ratio of soil dispersion using a combined glass electrode. Available phosphorus (P) is the inorganic P and extractable P at pH 8.5 (Page, et al., 1982; Olsen, et al., 1954) suitable for alkaline soils that are common in arid soils. The total surface area (Sw) was measured by H_2O adsorption (Torres Sánchez, et al., 1997) and the mineralogical composition and quantitative analysis of the soils were determined using XRD and the Rietveld method (Figure 4 and Table 2) (Rietveld, 1969).

The total iron oxides (F_{ed}) and amorphous iron oxides (F_{eo}) were established by dithionite (Holmgren, 1967) and oxalate method (Mc Keague, 1966) respectively.

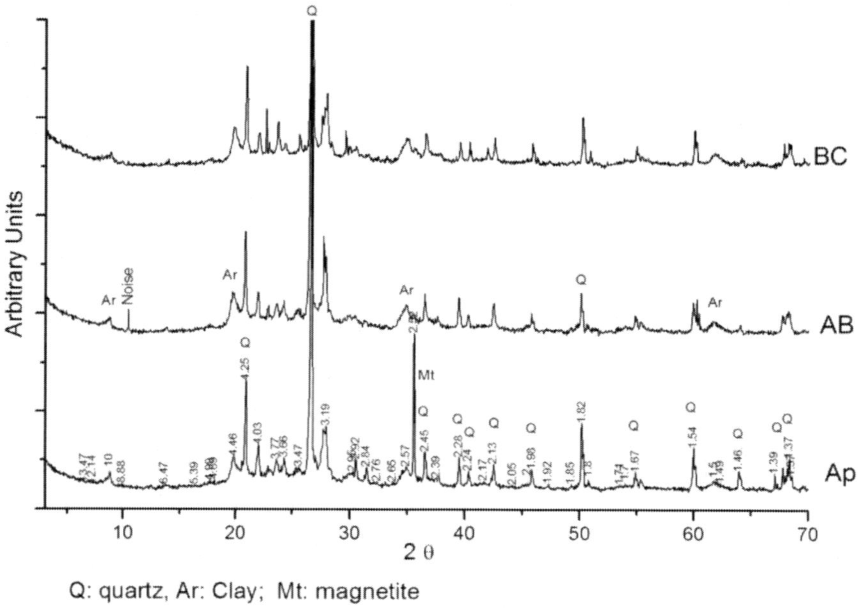

Figure 4. XRD diffraction patterns of Ap, AB and BC horizons.

Point of zero net proton charge (PZNPC) was performed by potentiometric titration of dispersions at different ionic strength (0.1, 0.01 and 0.001 M KCl) under N_2 atmosphere (Figure 5).

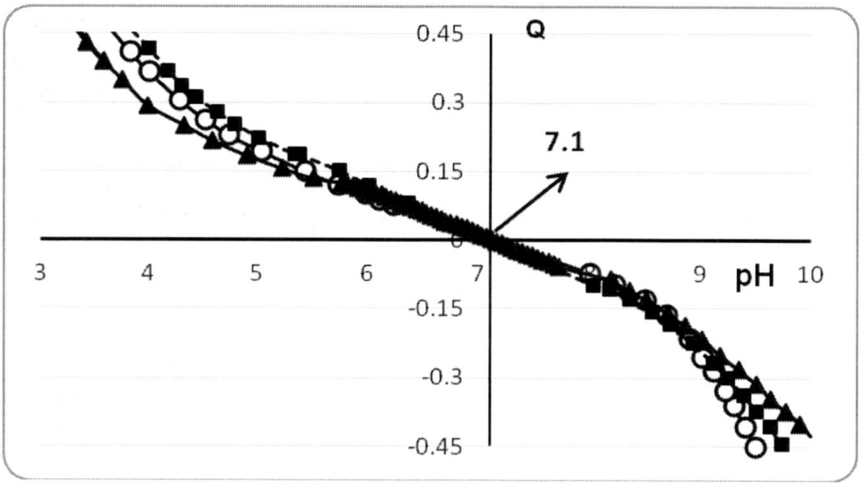

Figure 5. Potentiometric titration curves of the dispersions of the Ap horizon at three ionic strength: —▲— 0.001M, --■--, 0.01M and —○— 0.1M.

Organic matter (OM) content and soils chemical analysis were determined by the dichromate oxidation method (Schlichting, et al., 1995) and Lithium Metaborate/Lithium Tetraborate ($LiBO_2$ /$Li_2B_4O_7$) Fusion followed by Inductively Coupled Plasma - Atomic

Emission Spectroscopy (ICP-AES) analysis, respectively. The result is reported as oxide concentration percentage. The sample properties are shown in Table 2.

Table 2. Characteristics of agriculture soils profile from Santiago del Estero/Argentina

Horizon	Ap	AB	BC
pH ($CaCl_2$ 1:2.5)	5.90	5.75	6.02
OM (g C/Kg)	23.30	6.67	12.10
P (μg/g)	43.34	6.67	1.19
Sw (m^2/g)	188	259	242
Fe_o (mg/g)	0.239	0.158	0.095
Fe_d (mg/g)	1.66	1.91	0.99
PZNPC(pH)	7.1	7.4	8.1
Mineralogical Composition (%)			
Quartz	45.2	39.8	46.2
Sanidine Feldspar	9.6	9.6	9.3
Andesine Feldspar	24.7	23.5	19.9
Illite	18.6	25.9	24.7
Magnetite	1.3	1.2	1.2
Chemical Analysis (%)			
SiO_2	64.6	63.0	63.0
Al_2O_3	12.25	13.65	14.05
Fe_2O_3	3.51	4.10	4.42
CaO	1.47	1.34	1.44
MgO	1.14	1.40	1.62
Na_2O	1.70	1.50	1.56
K_2O	2.43	2.53	2.66

GLYPHOSATE

The herbicides are chemicals widely used in agriculture and their use increase with the increasing of crop areas. Then, this causes pollution problems in the environment. The glyphosate (N-phosphonomethylglicine) (PMG) is used to remove unwanted weeds such as annual grasses and perennial broadleaf weeds and woody species in agricultural, forestry and landscape. The PMG is no selective, post emergent and broad spectrum commercial herbicides worldwide used in soybean agriculture. At 25°C the solubility of glyphosate in water is 15.7 g/L at pH=7 and 11.6 g/L at pH=2.5 (Ahrens WH, 1994).

On the soil, the microbial is the main way of the PMG degradation while abiotic mineralization for the horizon A exceeds 1% reaching a maximum of 12% degradation for 60 days exposure (Jacobsen, et al., 2008).

PMG exhibits fast vertical mobility in soil, reaching high levels in deeper horizons where the degradation is slower (Veiga, et al., 2001).

The herbicide is a good chelating agent and can coordinate metal ion in solution especially at near-neutral pH levels where carboxylate and phosphonate chemicals groups are deprotonated forming strong complexes (Barja, et al., 2001). It can also be retained in the soil throughout adsorption onto aluminum and iron oxides (Nowack, et al., 1999; Barja, et al., 2005), clays ((Damonte, et al., 2007; Khoury, et al., 2010 and references therein cited) and organic matter (Piccolo, et al., 1996; Sposito, 1984).

Adsorption Experiment

Adsorption experiments were made using 0.100 g of soil samples dispersed in 11 mL of PMG (0, 70, 140, 220, 290, 360, 540, 700 and 1000 $mg.L^{-1}$). The pHs of samples were kept constant and adjusted with NaOH or HCl 0.10 M as required. The reaction vessels were left shaking overnight at 25 °C to reach the equilibrium.

The dispersions were filtered through a 0.45 μm pore membrane and the amount of PMG present in the supernatant was measured. The adsorbed glyphosate was calculated from the difference between the total added ligand and the supernatant concentration (C_e). PMG concentrations in the equilibrium were evaluated by ion chromatography (Zhua, et al., 1999) using a Dionex DX-100 instrument with a conductivity detector, a sample injection valve, and a 25 μL sample loop. Two plastic anion columns were coupled in series to serve both as a pre-column (Dionex AG-4) and an analytical chromatographic column (Dionex AS-4). The suppressor was regenerated with 50 mM H_2SO_4 with a flow rate of 12.5 $mL.min^{-1}$. A mixture of $NaOH/CO_3^{-2}$ 4 mM/9 mM was chosen as eluent with a flow rate of 1 $mL.min^{-1}$. The typical experimental error was lower than 5% for all the experimental results.

Results and Discussion

The relationship between the metal uptake and the sorbate equilibrium concentration at constant temperature is known as the adsorption isotherm. The adsorbent capacity of a certain material is related to the material balance adsorption: the sorbate that disappears from solution must be in the adsorbent. There are a considerable number of expressions that describe adsorption isotherms. Here, the Freundlich and Langmuir models were applied to the equilibrium data.

Freundlich Adsorption Isotherms Model

The Freundlich isotherm is an empirical equation that describes the adsorption on an heterogeneous surface with a non-uniform distribution of heat of adsorption that means that the surface on which the adsorbed molecules are interactive is energetically heterogeneous. This model does not predict any saturation of the adsorbent by the sorbate. Instead, infinite surface coverage is predicted, indicating multilayer adsorption on the surface. This isotherm fits, not all, but many soil adsorption systems and it is represented by equation (1)

$$\Gamma_e = K_F * C_{eq}^{1/n} \tag{1}$$

where, K_F is the Freundlich constant indicating adsorption capacity and represent a measure of the surface area of the adsorbent while 1/n is the index of the heterogeneity of the surface.

The PMG adsorption isotherms of soils dispersions equilibrated at different pH values are shown in Figure 6 where the solid lines are calculated using the Freundlich model and K_F and 1/n parameters values given in Table 3, 1/n values varies between 0.1 and 1 which indicates that this model could be used for interpreting the data.

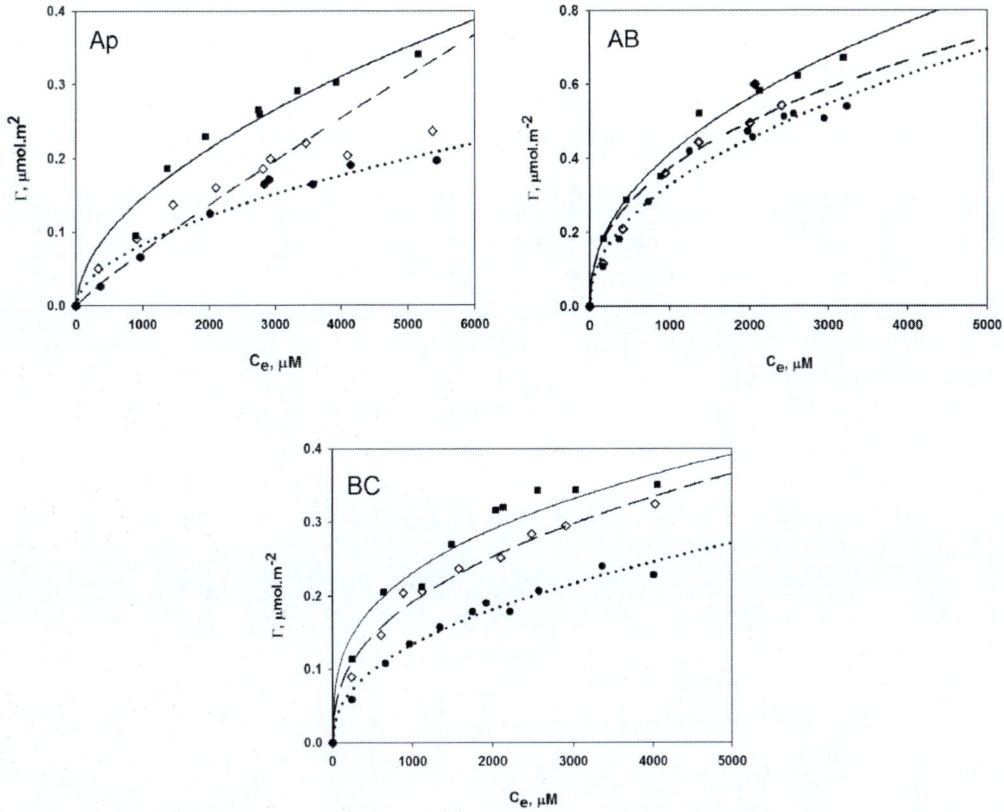

Figure 6. Adsorption isotherm of glyphosate on horizon Ap, AB and BC at pH 3 -■-, pH 6 -◇-, pH 8 -●-. Solid lines are calculated using Freundlich model with modeling parameters detailed in Table 3.

The correlation between experimental and calculated curves had a p-level of 0.001 and the determination coefficient is between 0.8554 and 0.9881 for different pHs and horizons. The value of the p-level represents a decreasing index of the reliability of a result. The higher p-level, the less we can believe that the observed relation between variables in the sample is a reliable indicator of the relation between the respective variables in the population. Specifically, the p-level represents the probability of error that is involved in accepting the observed results as valid. The p-level of 0.05 means a 5% probability that the relation between the experimental and calculated data is not a coincidence. In many areas of research, this parameter is customarily treated as a borderline acceptable error level.

Langmuir Adsorption Isotherm Model

The Langmuir isotherm is a well known model that indicates a decrease of the available surface sites as the adsorbent concentration increases. The Langmuir isotherm assumes monolayer adsorption and it is represented by the equation (2):

$$\Gamma = \Gamma_{max} \frac{K_L C_e}{1+ K_L C_e} \tag{2}$$

where Γ is the amount of glyphosate adsorbed (μmol.m^{-2}), Γ_{max} is the maximum amount of glyphosate adsorbed (μmol.m^{-2}), C_e is the equilibrium concentration of herbicide in the solution (mM), K_L is the Langmuir adsorption constant (mM^{-1}) which is related to the free energy of the reaction.

The PMG adsorption isotherms of soils dispersions equilibrated at different pH values are also shown in Figure 7 and solid lines are calculated using the Langmuir model and Γ_{max} and K_L, given in Table 3. The isotherm model parameters were obtained by a non-linear optimization using Solver-Excel tool. A first approximation to the parameters values was obtained from the plot of the inverse of the surface coverage as a function of the inverse of the equilibrium concentration.

The results of the adsorption and surface coverage calculations were normalized with S_w data and the various horizons were contrasted. The determination coefficients obtained were between 0.9100 and 0.9999 and were higher than those obtained using the Freundlich model. Then, Langmuir model is a better representation of adsorption process of glyphosate on the Santiago del Estero Province soil.

Table 3. Isotherm parameter for glyphosate adsorption on Santiago del Estero Province soils

	Freundlich Model (K_F*10^3 in μmol$^{1-1/n}$.m^{-2})						Langmuir Model (Γ_{max}, in μmol.m^{-2} and K_L in L.mmol^{-1})					
pH	3		6		8		3		6		8	
	1/n	K_F	1/n	K_F	1/n	K_F	Γ_{max}	K_L	Γ_{max}	K_L	Γ_{max}	K_L
Ap	0.54	3.4	0.90	0.1	0.54	2.0	0.40	0.87	0.31	0.63	0.24	0.61
AB	0.45	18.4	0.56	7.6	0.47	12.9	0.82	0.93	0.74	0.91	0.72	0.90
BC	0.33	23.9	0.35	18.4	0.40	12.0	1.49	0.42	1.23	0.39	1.10	0.28

The results show the dependence of the surface coverage with pH in the various horizons. The adsorption capacity increases from pH 8 to 3, the surface coverage decreases, $\Delta\Gamma_{max}$, for horizon Ap is around 40 % between pH 3 and pH 8 (Table 3). This difference is lower for horizons BC, 26 %, and AB, 12 %.

This variability may be linked to the organic matter content (Table 2) and denotes the substances affinity for attain charges and favoring the adsorption processes at a lower pH.

The highest adsorption capacity is obtained for the horizon BC followed by the horizon AB and the lowest for the horizon Ap . The difference between the horizons BC and AB is 55%, between horizons BC and Ap is 78% and between horizon AB and Ap is 51%.

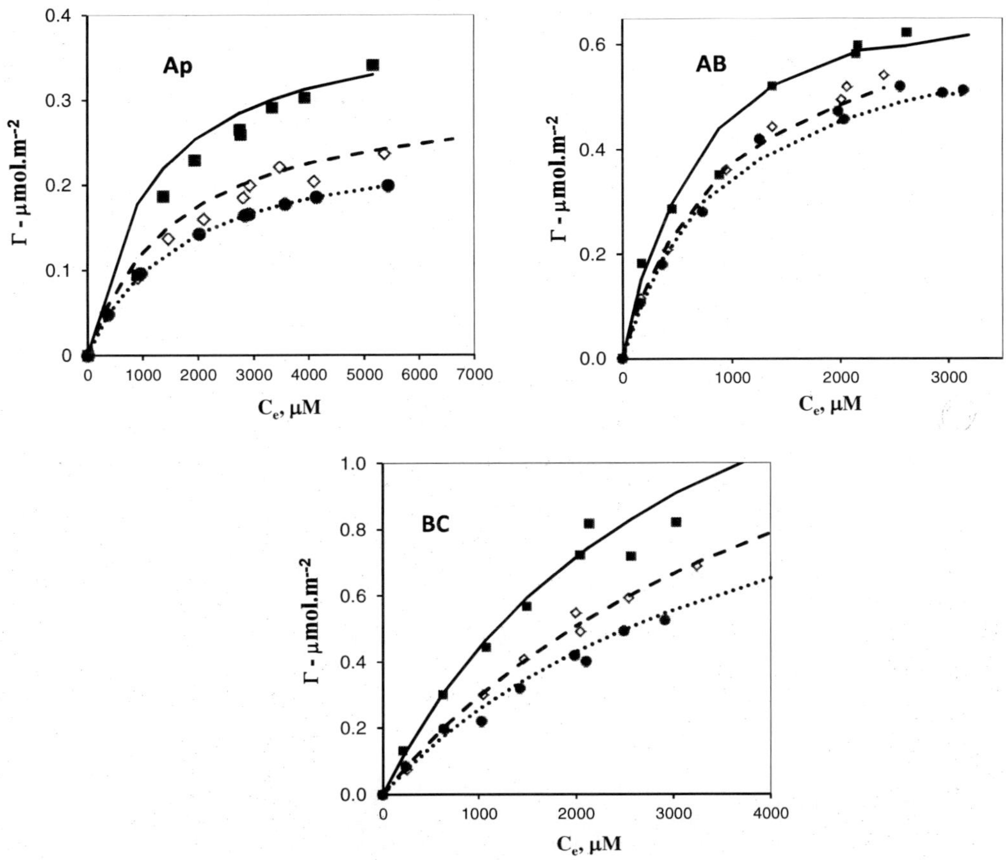

Figure 7. Adsorption isotherm of PMG on horizon Ap, AB and BC pH 3 -■-, pH 6 -◇-, pH 8 -●-. Solid lines are calculated using Langmuir model with constants and maximum surface coverage detailed in Table 3.

The predominant adsorption on deep horizons regard to the horizon Ap could be due to the competition of herbicide with phosphate groups for surface sites (Dion, et al., 2001). As opposed to the phosphate content that follows the order of Ap> AB> BC. Adsorption of glyphosate and phosphate in soil is similar to that which occurs on the clay minerals (Dion, et al., 2001; Gimsing, et al., 2005).

The highest adsorption constants correspond to horizon AB (Table 3). The changes in the adsorption affinity between horizon AB and Ap reaches $\Delta K_L = 38$ % while the horizon BC decreases 72% in respect to horizon AB.

The horizon Ap has an adsorption process in which the surface coverage reaches their maximum uptake at relatively low concentrations of PMG, showing a low capacity of adsorption behavior that could be due to the high phosphate content and low available surface area (Table 2) of this horizon with respect to the deeper horizons, AB and BC.

The horizon AB shows adsorption isotherms with a slope higher than the slope of the other horizons that is an indicative that PMG is more strongly bond to this horizon active sites.

Finally, the adsorption on the horizon BC does not reach the maximum coverage on the experimental conditions. The adsorption isotherms with a low initial slope describe an adsorption process with characteristic adsorption constants of low energy interaction (Figure 7).

Surface Complexation Model

The PMG adsorption on soils or clays was studied by several authors (Nomura, et al., 1977; McConnell, et al., 1985; Morillo, et al., 2000; Sheals, et al., 2002), in all the cases adsorption decreased with a pH increase following an anionic adsorption behavior.

The study of competitive adsorption between PMG and phosphate on iron oxides had shown that PMG could be exchange by phosphate (McBride, et al., 1989; Gimsing, et al., 2001), but the exchange on clays or soils are not so easy (Dion, et al., 2001).

The adsorption process in soils could be modeled in a first approximation as a linear combination of the minerals that composed them (Taubaso, et al., 2004).

In the heterogeneous media, pesticides interact with the suspended solid particles that can be organic macromolecules (organic matter), iron or aluminum (hydr)oxides, clays minerals or combinations of all of them. The adsorption processes are interpreted on the basis of several models, where one of the most popular is the surface complexation where the colloidal surface is protonated or deprotonated by the adsorption of OH^- or H^+, this phenomenon is represented by the following equation:

$$\equiv S\text{-}OH + H^+ \leftrightarrows \equiv S\text{-}OH_2^+$$

$$\equiv S\text{-}OH \leftrightarrows \equiv S\text{-}O^- + H^+$$

where $\equiv S\text{-}OH_2^+$ is a Lewis acid while the deprotonated hydroxyl of the surface are Lewis bases. The surface water molecule is labile and can be exchangeable by an anionic pesticide (ligand, L, or Lewis base) forming a chemical bond with the metallic center of the surface. Chemisorption occurs through a chemical reaction where the H^+ or the OH^- are exchanged by a metallic cation or an organic ligand, respectively, dealing to the formation of a surface complex represented by the following equations

$$\equiv S\text{-}OH_2^+ + Me^{+z} \leftrightarrows \equiv S\text{-}OMe^{+z-1} + 2H^+$$

$$\equiv S\text{-}OH + Me^{+z} \leftrightarrows \equiv S\text{-}OMe^{+z-1} + H^+$$

or

$$\equiv S\text{-}OH + L^- \leftrightarrows \equiv S\text{-}L + OH^-$$

$$\equiv S\text{-}OH_2^+ + L^- \leftrightarrows \equiv S\text{-}L + H_2O$$

Thus, iron or aluminum (hydr)oxides, clays minerals, carbonates or sulphates can adsorb pesticides. In particular, PMG can form predominating monodentate and bidentate phosphonate surface complexes while the carboxylate and amino group are not coordinated to the iron (hydr)oxides surface (Barja, et al., 2005; Gimsing, et al., 2005; Sheals, et al., 2003; Barja, et al., 2000). In particular, PMG forms complexes on the soil surface, in a similar way as do on surface of the minerals that comprise the soil (dos Santos Afonso, et al., 2004; Pessagno, et al., 2005; Pessagno, et al., 2006; Pessagno, et al., 2008; Tévez, et al., 2008). These processes are pH, herbicide concentration and temperature dependents.

The structure of the inner-sphere surface complexes were determined by different analytical methods: Infrared Spectroscopy by Fourier Transform (FTIR), X-ray photoelectron spectroscopy (XPS), X-ray Diffuse Scattering (XDS), X-ray Diffraction (XRD), Nuclear Magnetic Resonance Spectroscopy (NMR), etc. (Barja, et al., 2000; Sheals, et al., 2003; Dideriksen, et al., 2003; Barja, et al., 2005).

The acid-base constant for the deprotonation of the amine group of the surface complexes formed by PMG on the montmorillonite surface was determined by XPS obtaining a constant value with several orders of magnitude difference from the corresponding for the free PMG molecule (Khoury, et al., 2010), which shows the changes in the binding energies due to the coordination of the pesticide to the surface of a mineral.

$$(PO_3)^{-2}CH_2NH_2^+CH_2COO^- \leftrightharpoons (PO_3)^{-2}CH_2NHCH_2COO^- + H^+ \qquad pKa = 10.25$$

$$\equiv S\text{-}O(PO_2)^-CH_2NH_2^+CH_2COO^- \leftrightharpoons \equiv S\text{-}O(PO_2)^-CH_2NHCH_2COO^- + H^+ \quad pKa = 3.0$$

Thus, the surface coverage and the affinity constants for adsorption on iron oxides are highest than on clays minerals or soils but the experimental results suggest that the phosphonate is the liable chemical group for the surface coordination through inner-sphere surface complexes formation (dos Santos Afonso, et al., 2004; Pessagno, et al., 2005; Pessagno, et al., 2006; Pessagno, et al., 2008; Tévez, et al., 2008).

In expandable clays minerals, such as those classified as 2:1, PMG could penetrate into the interlayer (Damonte, et al., 2007; Khoury, et al., 2010). In consequence of PMG entrance there is a basal spacing increasing (Figure 8). This phenomenon is clearly observed on clay minerals (Damonte et al. 2007, Khoury, et al. 2010)

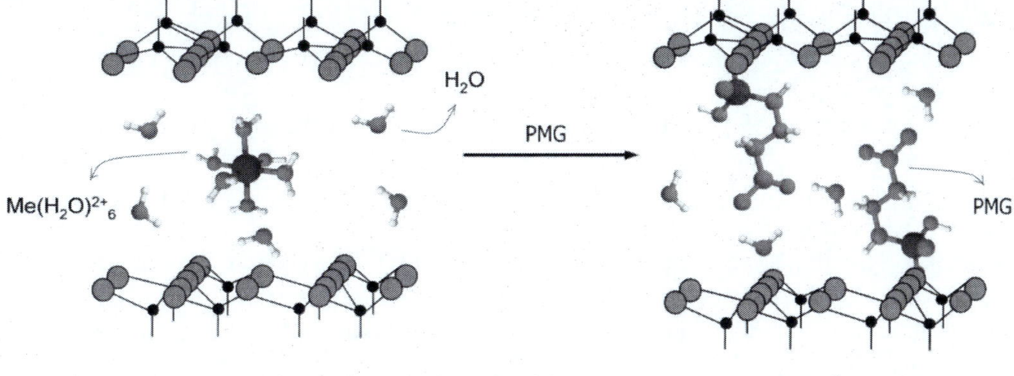

a b

Modified from McBride, 1994.

Figure 8. Adsorption reaction, A: hydrated divalent metal ion complex in the clay interlayer and B: glyphosate in the basal spacing of an expandable 2:1 clay mineral.

Conclusion

The major trend in the PMG adsorption on soil samples is given by the pH, that could be due to the influence of this parameter on the PMG molecule and on the surface charges of the soil particles. PMG adsorption increase as the acidity increases which corresponds to the adsorption of a ligand with a negative net charge. Regarding the relative adsorption capacity of the soil, the adsorption process has a different behavior profile, where the deeper profile (BC) has a higher capacity retention for this herbicide.

The AB and Ap horizons adsorption decrease could be influenced by the higher content of phosphorus. However, the strength of the interaction given by K_L (Langmuir Model Constant) is larger on the horizon AB and would be linked to the illite and iron oxide content that have a better distribution in the AB horizon in regard to Ap and BC horizons.

It can be noted that Langmuir is the model that best fitted the adsorption experimental results in these soils, although Freundlich model has a good adjustments for some pHs.

Acknowledgments

The authors acknowledge the Universidad de Buenos Aires, Secretaría de Ciencia y Técnica and MINCyT- ANPCyT- FONCyT for financial support. The authors are also grateful to Drs. Rosa M. Torres Sánchez, Susana Conconi and Jorge Maggi from Centro de Tecnología de Recursos Minerales y Cerámica (CETMIC) for Sw measurements and Rietveld analysis of soils.

References

Ahrens, W. H. (Editor) (1994). *Herbicide Handbook 7th Edition.* Weed Science Society of America. Illinois, US.

Barja, B. C., dos Santos Afonso, M. (2000). Surface complexes of glyphosate and AMPA onto goethite. *Div. Environ. Chem. Reprints Extend. Abstr.*, 40, 652–655.

Barja, B. C., Herszage, J., dos Santos Afonso, M. (2001). Iron (III)-phosphonate complexes. *Polyhedron*, 20, 1821–1830.

Barja, B. C., dos Santos Afonso, M. (1998). An ATR-FTIR study of glyphosate and its Fe(III) complex in aqueous solution. *Environ. Sci. Technol.*, 32, 3331–3335.

Barja, B. C., dos Santos Afonso, M. (2005). Aminomethylphosphonic acid and Glyphosate adsorption onto goethite: a comparative study. *Environ. Sci. Technol.*, 39, 585-592.

Cantera, C., Trinelli, A., dos Santos Afonso, M. (2008). Fotodegradación del Herbicida Glifosato. *V congreso iberoamericano de física y química ambiental,* 503-507. UNSAM (ed. M. Blesa, M. dos Santos Afonso and R. M. Torres Sánchez). Mar del Plata Argentina.

Carver, R. (1971). *Sedimentation Anal. Procedures in Sedimentary Petrology*, 4. Wiley and Interscience, New York,US.

Damonte, M., Torres Sánchez, R. M., dos Santos Afonso, M. (2007). Some aspects of the glyphosate adsorption on montmorillonite and its calcined form. *Appl. Clay Sci.*, 1-3, 36, 86-94.

Dideriksen, K., Stipp, L. S. (2003). The adsorption of glyphosate and phosphate to goethite: a molecular-scale atomic force microscopy study. *Geochim. Cosmochim. Acta*, 67, 3313–3327.

Dion, H. M., Harsh, J. B., Hill, Jr J. B. (2001). Competitive sorption between glyphosate and inorganic phosphate on clay minerals and organic matter soils. *J. Radional. Nucl. Chem.*, 49, 385-390.

Dos Santos Afonso, M., Barja, B. C., Pessagno, R. C., Tevez, H. R. (2004). Glyphosate adsorption on soils. *Division of Environmental Chemistry*, ACS, 43 (2), ISSN: 0093-3066. New York. US.

Gatidou, G., Iatrou, E. (2011). Investigation of photodegradation and hydrolysis of selected substituted ureas and organophospahte pesticides in water. *Environmental Science and Pollution Research.*, 18, 949-957.

Gimsing, A. L., Borggaard, O. K. (2011). Effect of KCl and CaCl2 as background electrolytes on the competitive adsorption of glyphosate and phosphate on goethite. *Clays Clay Miner.*, 49, 270–275.

Gimsing, A. L., dos Santos Afonso, M. (2005). Glyphosate (Book Section). *Biogeochemistry of Chelating Agents,* 910. ACS (ed. Van Briesen, J. M., Nowack, B.), ACS Symposium Series.

Holmgren, G. G. S. (1967). A rapid citrate-dithionite extractable iron procedure. *Soil Sci. Soc. Am. Proc.,* 31. - págs. 210-211.

Jacobsen, C. S., van der Keur, P., Ivensen, B. V., Rosenberg, P., Barlebo, H. C., Torp, S., Vosgerau, H., Juhler, R. K., Ernstsen, V., Rasmussen, J., Brinch, U. C., Jac, O. H. (2008). Variation of MCPA, metribuzine, methyltriazine-amine and glyphosate degradation, sorption, mineralization and leaching in different soil horizons. *Environmental Pollution*, 3, 156, 794-802.

Khoury, G. A., Gehris, T. C., Tribea, L., Torres Sánchez, R. M., dos Santos Afonso, M. (2010). Glyphosate adsorption on montmorillonite: An experimental and theoretical study of surface complexes. *Applied Clay Science*, 2, 50, 167–175.

Kuo, W., Wu, Ch. (2012). Treatment of Color Filter Wastewater by Fresnel Lens Enhanced Solar Photo-Fenton Process. *Advances in Materials Science and Engineering*, 1-7. Hindawi Publishing Corporation (ed. Huang Yu-Pei).

Liu, C. M., McLean, P. A., Sookdeo, C. C., Cannon, F. C. (1991). Degradation of the herbicide glyphosate by members of the family Rhizobiaceae. *Appl. Environ. Microbiol.*, 57, 1799–1804.

Lorenz, G., Bonelli, C. L., Roldán, S., Araya, C., Rondano, K. (2000). Soil quality changes due to land use in a Kastanozem-Phaeozem soilscape of semiarid Chaco. *Mitteilungen der Deutschen Bodenkundlichen Gesellschaft,* 92, 169-172.

McBride, M. B. (1994). *Environmental chemistry of soils*. Oxford Univ. Press, New York.

Mc Keague Day (1985). Dithionite and oxalate-extratable iron and aluminium in soils. *Can. J. Soil Sci.*, 47, 95-99.

McConnell, J. S., Hossner, L. R. (1985). pH-dependent adsorption isotherms of glyphosate. *J. Agric. Food Chem.*, 33.

Morillo, E., Undabeytia, T., Maqueda, C., Ramos, A. (2000). Glyphosate adsorption on soils of different characteristics.Influence of copper addition. *Chemosphere*. 40, 103–107.

Murphy and Riley (1962), Jackson (1958) (1996). Phosphorus. Methods of soil analysis. *Chemical Methods.* Part 3. In C.A. page (ed). ASA and SSSA. Madison, WI 53711, US.

Nishiyama, M., Suzuki, Y., Katagi, T. (2010). Hydrolysis and photolysis of insecticide metofluthrin in water. *Journal of Pesticide Science*, 4, 35, 447-455.

Nomura, N. S., Hilton, H. W. (1977). The adsorption and degradation of glyphosate in five Hawaiian sugarcane soils. *Weed Res.*, 17, 113–121.

Nowack, B., Stone, A. T. (1999). Adsorption of phosphonates onto the goethite-water interface. *J. Colloid Interface Sci.*, 214, 20-30.

Olsen, S. R., Cole, C. V., Watanabe, F. S., Dean, L. A. (1954). *Estimation of available phosphorous in soils by extrction with sodium bicarbonate* (Report). US Dep. of Agric.

Olson, R. V., Ellis, R. (1982). Iron. Methods of soil analysis (Book Section). *Methods of soil analysis*, 2. ASA and SSSA, ed. Page C.A. Madison, Wisconsin, US.

Page, A. L., Miller, R. H. and Keeney, D. R. (1982). *Chemical and microbilogical properties* (Book Section). Methods of soil analysis, Part 2. ASA, SSSA. Madison, Wisconsin, US.

Paschevskaya, N. V., Bolotin, S. N., Sklyar, A. A., Bukov, N. M., Panyushkin, V. T. (2006). Binary and ternary complexes of copper(II) with N-phosphonomethylglycine and valine. *Journal of Molecular Liquids*, 1-3, 126, 89–94.

Pérez-Carrera, A., Moscuzza, C. H. and Fernández-Cire, A. (2008). Efectos socioeconómicos y ambientales de la expansión agropecuaria. Estudio de caso: Santiago del Estero, Argentina. *Ecosistemas*, 1, 17, 5-15.

Pessagno, R., dos Santos Afonso, M., Torres Sánchez, R. M. (2005). N-(phosphonomethyl) glycine interactions with soil. *J. Argent Chem. Soc.*, 4-6, 93, 97–108.

Pessagno, R. C., Torres Sánchez, R. M., dos Santos Afonso, M. (2008). Glyphosate behavior at soil and mineral-water interfaces. *Environmental Pollution*, 153, 53-59.

Piccolo, A., Celano, G., Conte, P. (1996). Adsorption of glyphosate by humic substances. *Journal of Agricultural and Food Chemistry*, 44, 2442-2446.

Popov, K., Ronkkomaki, H., Lajunen, L. H. J. (2001). Critical evaluation of stability constants of phosphonic acids. *Pure Appl. Chem,* 73, 1641-1677. (IUPAC Technical Report).

Quivet, E., Faure, R., Georges, J., Paisse, J. O., Herbreteau, B., Lanteri, P. (2006). Photochemical degradation of imazamox in aqueous solution: Influence of metal ions and anionic. *J. Agr. Food Chem.*, 54, 3641-3645.

Ramstedt, M., Norgren, C., Sheals, J., Boström, D., Sjöberg, S., Persson, P. (2004). Thermodynamic and spectroscopic studies of cadmium(II)-N-(phosphonomethyl) glycine (PMG) complexes. *Inorg. Chim. Acta*, 357, 1185–1192.

Rietveld, H. M. A. (1969). Profile refinement method for nuclear and magnetic structures. *J. Appl. Crystal.*, 2, 65-71.

Schlichting, E., Blume, H., Stahr, K. (1995). Soils practical (Book Section). Wissenschafts-Verlag Blackwell, 81. *Pareys Studien texte*. Berlin, German.

Sheals, J., Granström, M., Sjöberg, S., Persson, P. (2003). Coadsorption of Cu(II) and glyphosate at the water–geothite (a-FeOOH) interface: molecular structures from FTIR and EXAFS measurements. *J. Colloid Interface Sci.,* 262, 38–47.

Sheals, J., Sjöberg, S., Persson, P. (2002). Adsorption of glyphosate on goethite: molecular charcterization of surface complexes. *Environ. Sci. Tech.*, 36, 3090-3095.

Sheals, J., Persson, P., Hedman, B. (2001). IR and EXAFS spectroscopic studies of glyphosate protonation and copper (II) complexes of glyphosate in aqueous solution. *Inorg. Chem.*, 40, 4302-4309.

Sistema Integrado de Información Agropecuaria, SIIA (2013). Ministerio de Agricultura, Ganadería y Pesca/Argentina [Online] // http://old.siia.gov.ar/. - Junio 18, 2013. - http:// old.siia.gov.ar/index.php/series-por-tema/agricultura.

Sposito, G. (1984). *Surface Chemistry of Soils*. Oxford Press. Clarendon, p. 234.

Sprankle, P., Meggit, W. F., Penner, D. (1975). Adsorption, obility and microbial degradation of gliphosate in soil. *Weed Sci.*, 23, 229-234.

Tangri, A., Mishra, P., Kumar, P. (2010). Kinetic study of hydrolysis of tri-phosphate ester of 2,5-diethoxy aniline an organ phosphorus pesticide in buffer medium. *International Journal of Applied Chemistry*, 1, 6, 139-150.

Taubaso, C., dos Santos Afonso, M., Torres Sánchez, R. M. (2004). Modelling soil surface charge density using mineral composition. *Geoderma*, 121, 123–133.

Tévez, H., dos Santos Afonso, M., Alvarez Escalada, F. C., Serrano, M., Ruiz Mostacero, N. (2008). Glifosato en fracciones de suelo de la provincia de Santiago del Estero. *V congreso Iberoamericano de física y química ambiental, 160, 223 (Conference).* UNSAM - Mar del Plata, Argentina.

Torres Sánchez, R., Falasca, S. (1997). Specific surface area and surface charges of some argentinian soils. *Z. Pflanz. Bodenk*, 160, 223.

Trinelli, A., dos Santos Afonso, M. (2006). Fotodegradación de pesticidas organofosforados. Clorpirifós: un caso en estudio (Book Section). Medioambiente en Iberoamérica. *Visión desde la Física y la Química en los albores del siglo XXI,* Vol. II, ed. Gallardo Lancho J.F. ISBN: 978-84-611-0352-2.

Trinelli, A. (2011) Estudios de Procesos de Fotodegradación y Adsorción de Plaguicidas Organofosforados para el Desarrollo de Tecnologías de Remediación. *Tesis Doctoral.* Buenos Aires: Universidad de Buenos Aires.

Veiga, F., Zapata, J. M., Marcos, M. L. F., Alvarez, E. (2001). Dynamics of glyphosate and aminomethylphosphonic acid in a forest soil in Galicia, north-west Spain. *Sci. Total Environ.*, 271, 135-144.

Wallace, D. F., Hand, L. H., Oliver, R. G. (2010). The role of indirect photolysis in limiting the persistence of crop protection products in surface waters. *Environmental Toxicology and Chemistry,* 3, 29, 575-581.

Zhua, Y., Zhanga, F-, Tong, Ch., Weiping, L. (1999). Determination of glyphosate by ion chromatography. *Journal of Chromatography*, 850, 297–301.

In: Handbook on Herbicides
Editors: Daiki Kobayashi and Eito Watanabe
ISBN: 978-1-62948-053-4

Chapter 11

ADSORPTION-DESORPTION PROCESSES AND MOBILITY OF (4-CHLORO-2-METHYLOPHENOXY) ACETIC ACID (MCPA) IN IRISH GRASSLAND SOILS

A. A. Piwowarczyk* and N. M. Holden
UCD Bioresources Research Centre / Biosystems Engineering,
UCD School of Biosystems Engineering , University College Dublin,
Belfield, Dublin, Ireland

ABSTRACT

The aim of this study was to quantify the adsorption-desorption isotherms of (4 – chloro-2-methylophenoxy) acetic acid (MCPA) in selected Irish grassland soils, and to assess its mobility potential to groundwater. The sorption isotherms were determined using a batch equilibrium method and were very well described by the linear and the Freundlich equation ($R^2 \geq 0.99$). The adsorption-desorption process tended towards linearity with the isotherm exponent (1/n) ranging from 0.90 to 0.97 for adsorption and from 0.92 to 1.03 for desorption. The adsorption coefficients calculated with the linear and the Freundlich equation were comparable. The Freundlich adsorption coefficient (K_f) ranged from 2.29 to 5.01 ($mg^{1-1/n}$ kg^{-1}) $(L)^{1/n}$ indicating low adsorption. Such behaviour was probably a result of electrostatic repulsion between negatively charged soil particles and herbicide anionic molecules formed in the soil/water system since the pH values of soils were greater than the dissociation constant of MCPA (pK_a = 3.07). Greater adsorption and consequently lower desorption was observed in the soil with the lowest pH, demonstrating some pH dependence, although this relationship was not significant (K_f, -0.732, P = 0.268). The organic carbon normalized coefficient (K_{oc},) ranged from 48.6 to 107.9 L kg^{-1}, suggesting that MCPA will be highly mobile in soils with similar characteristics to those studied, although this can be limited by environmental conditions in the field.

* Corresponding authors: agnieszka.piwowarczyk@ucd.ie, nick.holden@ucd.ie.

Keywords: Phenoxyalkanoic acid herbicides, MCPA, Freundlich isotherms, Adsorption-Desorption, Mobility, Grassland soils

INTRODUCTION

Pesticides that are commonly used in agriculture are receiving growing attention due to their migration and dispersion in the environment, and their possible adverse effects on non-target organisms. The capacity of the soil solid phase to retain such chemicals is an essential part of its function as a filter and reservoir [1]. An understanding of the nature of soil-pesticide interactions is a perquisite for any attempt to quantify and predict pesticide behaviour in soils [2], and consequently their final distribution in the wider environment [3]. One of the most common techniques used to assess pesticides mitigation in the environment is a laboratory batch equilibrium method [4], in which the adsorption-desorption isotherm data are obtained. The method is simple and reasonably rapid, but has disadvantages as described by Hance [4]. The sorption distribution coefficient K_d (L kg^{-1}) is the major parameter determined by the batch method. K_d describes the equilibrium partitioning of a solute (i.e. pesticide) between solid and liquid phases, and can be calculated from the linear isotherm as below:

$$K_d = C_s/C_e \qquad (1)$$

where C_s (mg kg^{-1}) is the equilibrium concentration of the solute in the solid phase, and C_e (mg L^{-1}) is the equilibrium concentration of the solute in the liquid phase. However, the sorption is often nonlinear [6], and in such cases the Freundlich isotherm approach is widely used:

$$x/m = K_f\, C_e^{1/n} \qquad (2)$$

where, x/m (mg kg^{-1}) is the quantity of the solute retained by a unit weight of the sorbent, C_e (mg L^{-1}) is the equilibrium concentration of the solute in the liquid phase, K_f and $1/n$ are empirical Freundlich constants representing the intercept and slope of the Freundlich isotherm respectively. The equation tends to a linear form only when $1/n = 1$ [7], but values of 0.7 to 1.0 have been reported [7,8].

(4-Chloro-2-methylphenoxy) acetic acid (MCPA) is an aryloxyalkanoic acid, systemic, polar herbicide with one polar carboxylic group and one lipophilic phenyl moiety group, and therefore is a suitable model substance to study adsorption and desorption processes in soils [9,10]. MCPA is a weak acid herbicide, therefore it ionizes in aqueous solutions to a form of anionic species, but as pH decreases, more molecular species are formed increasing its adsorption. Such pH dependence has been found for various acidic herbicides [11-14]. The anionic molecules are generally repelled by negatively charged fractions of organic matter and clays, minimising the adsorption. Such interaction may result in groundwater and/or surface water pollutions, as these ionisable pesticides are often found at concentrations exceeding the EU drinking water limit of 0.1 μg L^{-1} [15-17]. Nevertheless, MCPA in soils is subject to microbial degradation via 4-chloro-2-methylphenol to an ultimately harmless form [18,19].

In agriculture, MCPA is used post-emergence to control annual and perennial broad-leaved weeds in a range of crops and in grasslands [20], and is usually applied as salt or ester, although it readily hydrolyses to its acid form [21]. An average maximum application rate in Ireland for established grasslands, arable crops and forests is 1.75 kg ha^{-1} (Gordon Rennick, Department of Agriculture, Food and the Marine, personal communication). Since MCPA is extensively used on Irish grasslands to control noxious weeds (i.e. thistles, nettles or ragwort), which must be controlled by farmers to receive full payment of the EU financial supports [22], the objective of this research was (1) to quantify the adsorption-desorption isotherms of MCPA and (2) to assess its mobility and leachability potential in soils of typical grass management.

MATERIALS AND METHODS

Description of Soils and Analysis

Four soils from different grassland sites in Ireland were sampled: Soil 1 (52° 28´N, 008° 41´W; Co. Limerick), Soil 2 (52° 15´N, 006° 34´W; Co. Wexford), Soil 3 (52° 25´N, 006° 45´W; Co. Wexford) and Soil 4 (53° 51´N, 007° 54´W; Co. Carlow). The soil classification is presented in Table 1. Five sub-samples per site were randomly removed from a surface layer between 0 and 15 cm using a Dutch auger. Each sub-sample was air dried, lightly crushed and passed through a 2-mm sieve, and then bulked to create a composite sample for each site. The composite samples were stored in the dark at 20 ± 2 °C under low humidity. The soils were analysed for properties thought to be related to adsorptive capacity. Particle size distribution was determined by the pipette method [23], organic carbon content by dry combustion using a Skalar Primacs^{SLC} TOC analyzer (Skalar Analytical, Breda, The Netherlands), cation exchange capacity by the method of Metson [24], pH using slurries of 5 g of soil in 10 mL 0.01 M $CaCl_2$. The gravimetric water content was determined by oven drying for 24 h at 105 °C.

Chemicals

Analytical grade MCPA (99.5% purity) was obtained from Dr. Ehrenstorfer (Germany). MCPA has relatively high water solubility (0.825 g L^{-1} at 20 °C) and pK_a value of 3.07. Five standard working solutions were prepared in methanol and diluted in 10 mL 0.01 M $CaCl_2$ to initial MCPA concentrations of 1.01, 10.09, 20.22, 40.90 and 100.89 mg L^{-1}. In this study methanol represented only 0.01% of the final volume, 5% methanol addition had no effect on sorption as reported by Kookana et al. [25]. $CaCl_2$ (0.01 M) was used as an aqueous phase to keep the ionic strength similar to that of the natural soil solution before pesticide addition and for better phase separation. The standards in methanol were stored at -20 °C in capped glass flasks and the dilutions in 0.01 M $CaCl_2$ were freshly prepared for each stage of the analysis. All chemicals used were at the highest available grade.

MCPA Kinetic Experiment

The kinetic sorption study was performed in compliance with the standard batch equilibrium method [4] in Soils 3 and 4. Duplicate soil samples at two different soil-to-solution ratios (2 g and 5 g of soil) were pre-equilibrated overnight in 0.01 M $CaCl_2$ (10 mL) in 25 mL glass centrifuge tubes with Teflon-lined screw caps on an overhead shaker at 20 ± 2 °C. The kinetic study was run at a single MCPA concentration of 40.90 mg L^{-1}. At different time intervals (up to 24 hours) the tubes were removed from the shaker and centrifuged at 4500 rpm for 30 min. The supernatants were recovered and MCPA extracted. Desorption kinetics were studied afterwards. At established adsorption equilibrium, the supernatants were removed and replaced by an equal amount of herbicide free aqueous 0.01 M $CaCl_2$. The tubes were agitated to disperse the sediment pellets, shaken until an approximate desorption equilibrium time (up to 24 hours) and centrifuged as described earlier. Tubes with herbicide and $CaCl_2$ (0.01 M) solution but without soil served as control samples. Blank samples (soil, 0.01 M $CaCl_2$ and methanol) were handled identically to the experimental samples and the results showed no MCPA presence in the soils or interfering peaks. All supernatants were extracted immediately and analysed.

Adsorption-Desorption Isotherms

The adsorption isotherm study was performed in a similar manner to the kinetic study, except that the experimental samples were run at five initial MCPA concentrations described earlier, and analysed at the adsorption equilibrium time established by the kinetic experiment. The adsorption isotherm study was repeated twice giving four replicates for each soil × herbicide concentration combination. Desorption isotherms were determined immediately from the equilibrium points of the adsorption isotherm in duplicate as a single step desorption process at the desorption equilibrium time previously established by the desorption kinetic experiment. Control samples and blank samples were processed in the same manner.

Extraction and Quantification

The chemical analyses were undertaken in the Department of Agriculture, Food and the Marine Pesticide Control Laboratory (Celbridge, Co. Kildare, Ireland; ISO/IEC 17025). MCPA was extracted from supernatants manually using a vacuum manifold and reversed phase Strata X 60 mg cartridges (Phenomenex, UK). The supernatants were acidified with HCl to pH~2 prior to extraction. Isocratic solution (2 mL) of methanol and 0.025 M H_3PO_4 was applied for conditioning followed by the same volume of Milli-Q water (pH~2) at a flow rate of 6 mL min^{-1}. Samples were loaded at a flow rate of 2 mL min^{-1} and washed at the same rate with Milli-Q water (2 mL, pH~2). The cartridges were dried under vacuum for about 10 min after which, the herbicide residues were pre-concentrated by elution with 4 mL methanol (adsorption samples) or 2 mL methanol (desorption samples) at a flow rate of 1 mL min^{-1}. The elutions were homogenized using vortex and duplicate samples were injected to HPLC-DAD (Agilent Technologies). Aqueous 0.025 M H_3PO_4 and acetonitrile (ratio 55:45) were used as the mobile phases. The aqueous H_3PO_4 mobile phase was filtered beforehand under vacuum using HNWP 0.45 μm filter (Millipore, USA). The analyses were performed on a Synergi 4μm Hydro-RP, 150 x 4.6 mm column (Phenomenex, UK) at 22 °C and a flow rate of 1.5 mL min^{-1}. MCPA was analysed at the wavelength of maximum adsorption (230 nm). The lowest limit of detection achieved with this method was 100 μg L^{-1}. Several standard samples in 0.01 M $CaCl_2$ of the same herbicide concentration used to study adsorption-desorption isotherms were also extracted together with the experimental samples and injected to the HPLC-DAD to evaluate whether the extraction recovery was acceptable. The HPLC-DAD performance was examined by injecting the MCPA standards in methanol at different concentrations.

All the statistical analyses were performed using SPSS (PASW Statistics v. 18.0) at 95% level of confidence.

RESULTS AND DISCUSSION

Soils

Selected physical and chemical properties of the grassland soil samples used in the study are presented in Table 1. The samples had a good range of pH (pH = 4.6-6.0), OC (% OC = 39-4.7), CEC (CEC = 17.0-26.1), and texture as found in the surface layers of typical Irish grassland soils under livestock management [26].

Table 1. Soil classification and selected physico-chemical properties of the grassland soils used in the study

Soil	WRB/FAO	pH (0.01 M $CaCl_2$)	OC (%)	CEC (cmol kg^{-1})	sand (%)	silt (%)	clay (%)	Texture
1	Cutanic Luvisol	6.0	4.7	26.1	28	62	10	Silt loam
2	Luvic Stagnosol	5.6	4.3	19.9	35	46	19	Loam
3	Haplic Cambisol	4.6	4.1	19.9	38	52	10	Silt loam
4	Haplic Cambisol	5.1	3.9	17.0	68	19	13	Sandy loam

Adsorption and Desorption Kinetics

The adsorption kinetic experiment commenced at a 1:5 soil-to-solution ratio as shown in Figure 1. Approximately 22% adsorption occurred with Soil 4 and 44% with Soil 3. A new kinetic study at a lower 1:2 soil-to-solution ratio was therefore run and resulted in a greater adsorption of 40% and 69% in Soil 4 and Soil 3, respectively. The adsorption of MCPA was immediate and constant over the experimental period in both cases. The mean standard deviations calculated from % MCPA adsorbed ranged from 1.0 to 3.0% only, and therefore it can be assumed that the adsorption equilibrium was reached within 24 hours, with 2 hours being sufficient.

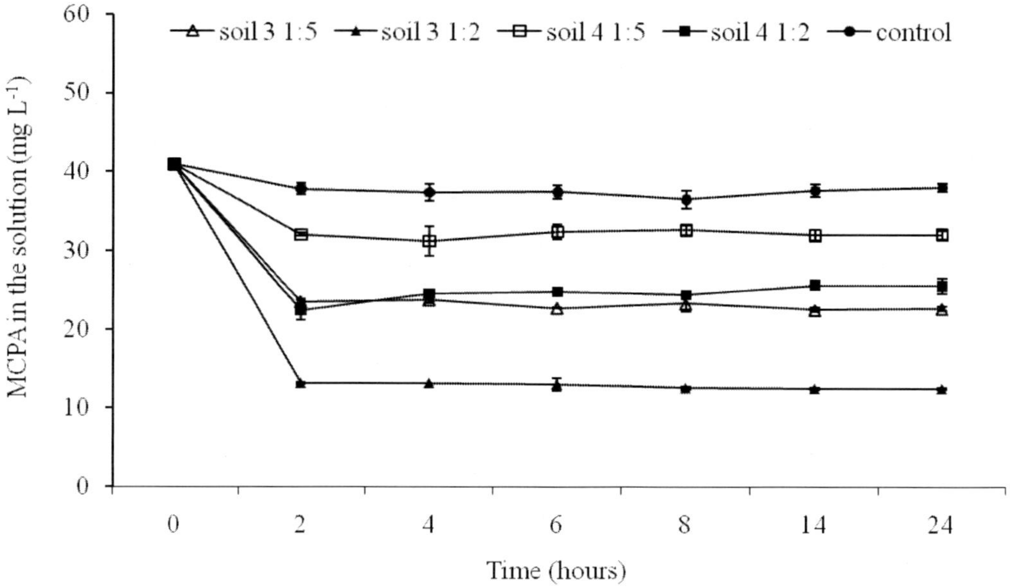

Figure 1. Adsorption kinetics of MCPA in two grassland soils at 1:2 and 1:5 soil-to-solution ratio together with control samples. (*Error bars* represent standard deviation of duplicates for soil samples and of 10 replicates for control samples. Some *error bars* are smaller than the *symbols*).

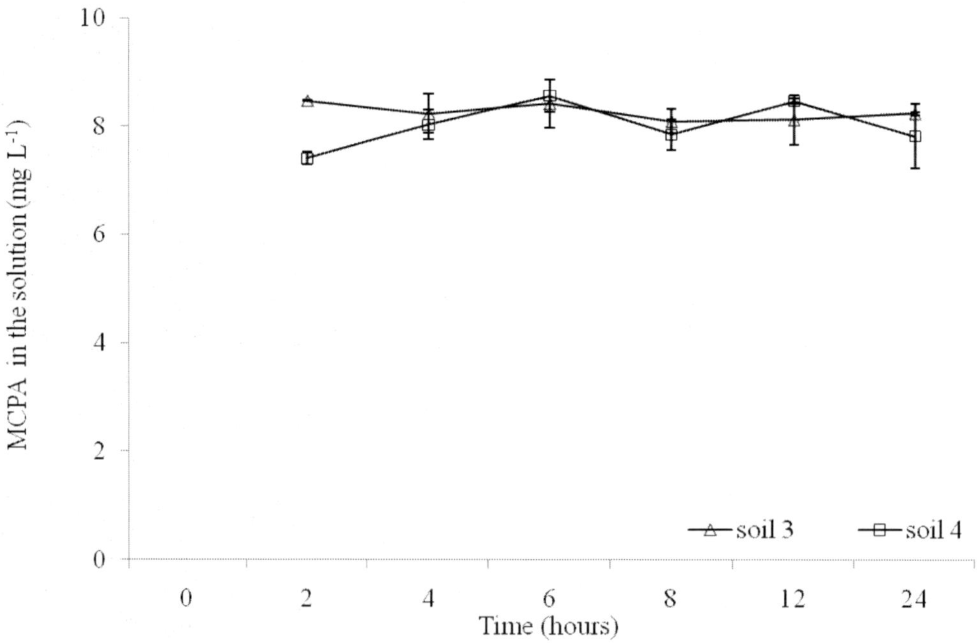

Figure 2. Desorption kinetics of MCPA in two grassland soils. (*Error bars* represent standard deviation of duplicates. Some *error bars* are smaller than the *symbols*).

Following the outcome of the adsorption kinetic experiment, desorption kinetics were performed at 1:2 soil-to-solution ratio. Results showed that the desorption process was less consistent over time as shown in Figure 2, and especially in Soil 4. Approximately 29% and 43% of adsorbed MCPA was desorbed from Soil 3 and Soil 4, respectively. The mean value of MCPA desorbed into solution was 8.27 mg L^{-1} (SD = 0.15 mg L^{-1}) from Soil 3 and 8.03 mg L^{-1} (SD = 0.43 mg L^{-1}) from Soil 4. The data were interpreted to indicate that desorption equilibrium was reached within 24 hours with 2 hours being sufficient.

The stability test examined by processing the control samples (Figure 1) showed no losses of MCPA due to absorption onto the surface of the test tubes, degradation or volatilization. The initial MCPA concentration used was 40.90 mg L^{-1}. The mean concentration measured over the 24 hour experimental period was 37.5 mg L^{-1} (SD = 0.5 mg L^{-1}). This difference indicated a 92% recovery, which was associated with the solid phase extraction as found from the standard samples of known MCPA concentration (results not shown).

Sorption Isotherms

The Freundlich and the linear isotherm fitted the adsorption (Table 2) and desorption (Table 3) data very well ($R^2 \geq 0.99$). The exponent (1/n) of the Freundlich isotherm ranged from 0.90 to 0.97 for adsorption and 0.92 to 1.03 for desorption. The adsorption process was not exactly linear but tended towards linearity.

A linear adsorption suggested that there is probably a constant partitioning of MCPA between the solution and the soil sorption sites, while a non-linear adsorption suggested that the process of MCPA adsorption may be limited at higher herbicide concentrations [27]. The graphical adsorption-desorption isotherms of MCPA are presented in Figure 3. In the literature, both linear and non-linear isotherms of phenoxyalkanoic acid herbicides have been reported [28-30].

From Table 2 it can be seen that the Freundlich adsorption coefficient (K_f) ranged from 2.29 to 5.01 (mg $^{1-1/n}$ kg^{-1}) (L) $^{1/n}$ and the adsorption distribution coefficient from the linear isotherm (K_d) ranged from 2.07 to 4.39 L kg^{-1}, suggesting low MCPA adsorption in the soils. This low adsorption probably resulted from electrostatic repulsion between negatively charged soil particles and anionic molecules of MCPA, since the pH of the soils was greater than the dissociation constant of the herbicide ($pK_a = 3.07$) [31].

Table 2. MCPA adsorption coefficients for the linear and the Freundlich isotherm. Number in brackets represent standard deviation (n = 20)

Soil	Linear isotherm		Freundlich isotherm		
	K_d (L kg^{-1})	R^2	K_f ($mg^{1-1/n}$ kg^{-1}) $(L)^{1/n}$	1/n	R^2
1	2.45 (0.23)	1.000	2.61	0.97	1.000
2	2.07 (0.25)	0.999	2.31	0.95	0.999
3	4.39 (0.90)	0.999	5.01	0.90	1.000
4	1.96 (0.42)	1.000	2.29	0.91	0.998

The highest adsorption was observed in Soil 3 and this could be related to its relatively low pH, although the relationship between adsorption and pH was not significant (K_f, -0.732, $P = 0.268$; K_d, -0.667, $P = 0.323$).

Nevertheless, strong correlations have been reported by others [32-34], showing an increase of adsorption with decreasing soil pH. For example, Haberhauer et al. [29] observed that the strongest adsorption of phenoxyalkanoic acid herbicides occurred in the soil with the lowest pH (pH = 4.6), whilst the soils had a very similar percentage organic carbon. Relationships between the organic carbon and adsorption of phenoxyalkanoic herbicides have also been reported [35,36].

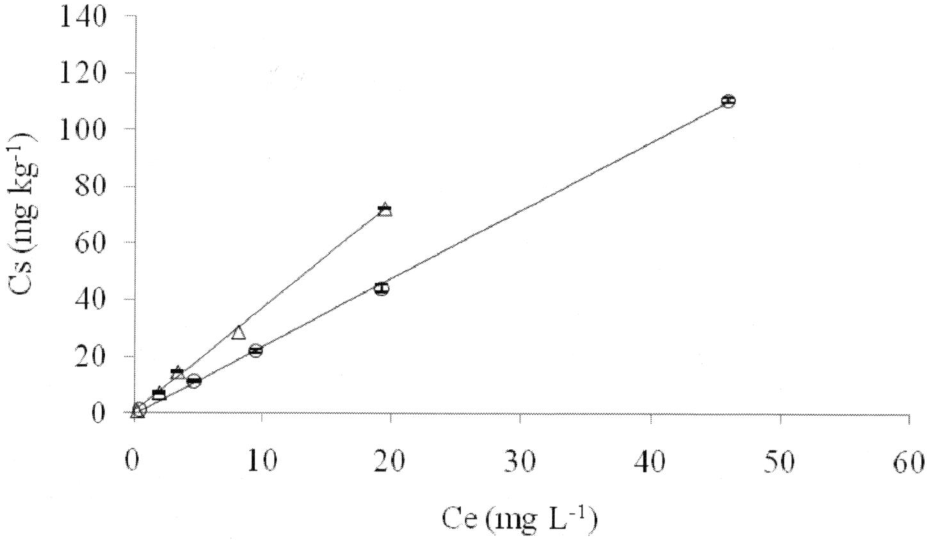

A

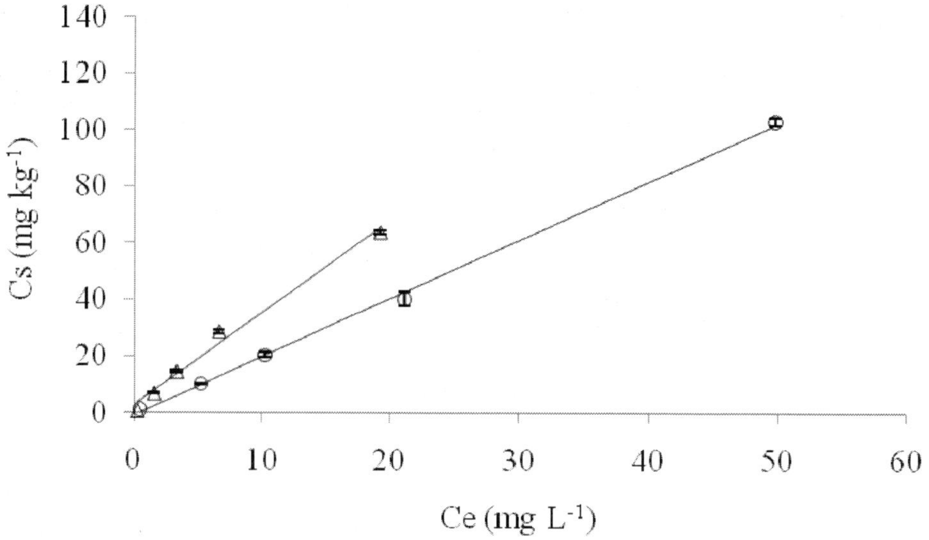

B

Figure 3. (Continued).

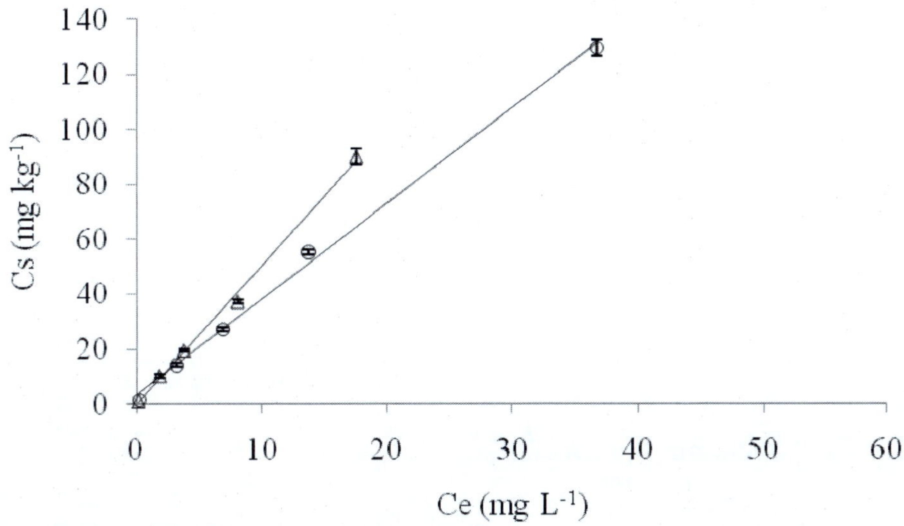

C

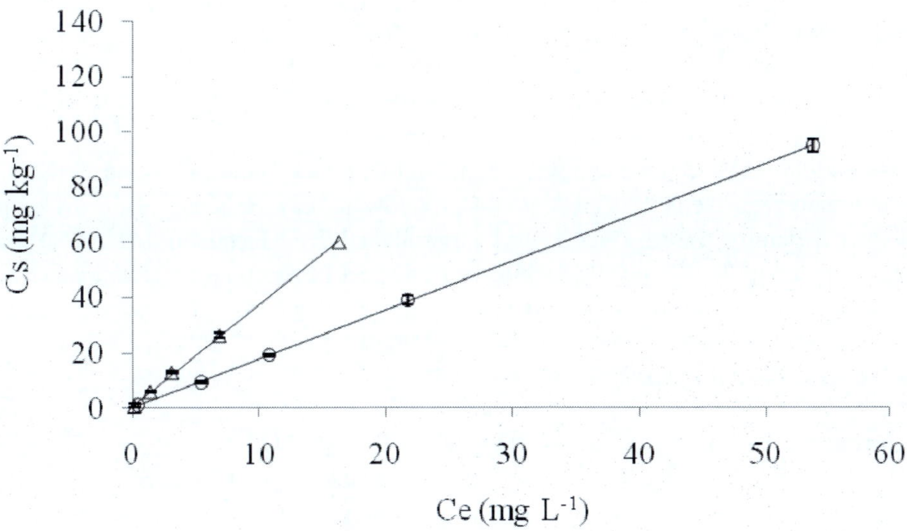

D

Figure 3. Linear adsorption (○) and desorption (Δ) isotherms of MCPA on: (A) Soil 1, (B) Soil 2, (C) Soil 3 and (D) Soil 4. Points are the observed values and the lines are the fit to the linear regression model. (*Error bars* represent standard deviation of four replicates for adsorption and of duplicates for desorption. Some *error bars* are smaller than the *symbols*).

Only between 27% and 35% of MCPA was desorbed from the soils during one washing cycle (Table 3). Desorption coefficients from the Freundlich and the linear isotherm (Table 3) were greater than the corresponding values for adsorption (Table 2), meaning than the adsorption process was not fully reversed. Since this study used only one desorption step, it would be inappropriate suggesting hysteretic behaviour, although the hysteresis has been previously reported for MCPA and many other pesticides in various soils [37-39].

Table 3. Desorption isotherms of MCPA in the soils studied. Number in brackets represent standard deviation (n = 10)

Soil	Linear isotherm		Freundlich isotherm			% des
	K_d (L kg^{-1})	R^2	K_f ($mg^{1-1/n}$ kg^{-1}) $(L)^{1/n}$	1/n	R^2	
1	3.72 (0.48)	0.999	3.68	1.02	0.999	34.98 (2.36)
2	4.24 (0.85)	0.994	3.77	1.03	0.995	34.47 (3.21)
3	5.55 (1.00)	0.998	5.88	0.92	0.999	27.38 (1.21)
4	4.09 (0.77)	0.999	4.35	0.93	0.996	33.06 (1.72)

Mobility and Leachability of MCPA

The distribution coefficients from the linear isotherm (K_d) were used to calculate the organic carbon adsorption constants (K_{oc}, L kg^{-1}) as follows:

$$K_{oc} = (K_d/\%OC) \times 100 \tag{3}$$

The K_{oc} values were used to predict the mobility of MCPA in each soil with respect to leaching as proposed by McCall et al. [40]. According to this classification, MCPA is very highly mobile (class 5) in Soils 2 and Soil 4, and mobile (class 4) in Soils 1 and Soil 3.

Further analysis of MCPA occurrence in groundwater can be made using the empirical Groundwater Ubiquity Score (GUS) index introduced by Gustafson [41]. The method classifies the potential for chemical leaching with respect to combined effects of degradation ($t_{½}$, half-life in soil) and sorption processes (K_{oc}, L kg^{-1}):

$$GUS = \log_{10}(t_{½}) \times [(4 - \log_{10}(K_{oc})] \tag{4}$$

According to this method, organic compounds with GUS > 2.8 are considered as potential leachers (L), those with GUS < 1.8 are considered non-leachers (NL), and those with GUS between 1.8 and 2.8 have transient properties (T). MCPA has relatively low persistence with a reported half-life in the field of between 7 days and one month, depending on the environmental factors such as soil moisture, temperature and organic matter content [42-44]. The GUS index for MCPA (Table 4) was calculated using the K_{oc} values obtained in this study and reported half-life values (between 7 and 28 days), since no degradation study of MCPA was performed. The results showed that MCPA will be considered a non-leacher in Soil 3 with a field degradation half-life up to 7 days, but can become a transient if it persists longer. In the remaining soils, MCPA will have transient properties for half-life values between 7 and 14 days, but from 21 days onwards it can become a potential leacher. The GUS index was used by e.g. Bailey et al. [45] to assess pesticides to be included in monitoring programmes for private water supplies in eastern England and a similar approach could be taken in Ireland and elsewhere with data such as those presented in this paper.

Table 4. Calculated organic carbon adsorption constant (K_{oc}) and GUS index taking into account reported soil degradation half-life values of 7, 14, 21 and 28 days. Letters in brackets represent GUS index as follows: L - leacher, T - transient, and NL - non-leacher

Soil	K_{oc} (L kg^{-1})	7 days	14 days	21 days	28 days
1	51.80	1.93 (T)	2.62 (T)	3.02 (L)	3.31 (L)
2	48.59	1.96 (T)	2.65 (T)	3.06 (L)	3.35 (L)
3	107.86	1.66 (NL)	2.03 (T)	2.60 (T)	2.85 (T, L)
4	49.94	1.95 (T)	2.64 (T)	3.04 (L)	3.33 (L)

CONCLUSION

The objective of this research was to quantify the adsorption-desorption isotherms of MCPA in Irish grassland soils and to assess MCPA mobility potential to groundwater under typical farming conditions.

The sorption was generally not dependent on the concentration of MCPA in the solution or there were still available sorption sites at the highest measured concentration, although processes other than partitioning between the solution and the sorption sites could take place. Measured adsorption of MCPA was much lower than that of chlorothalonil previously reported for Irish soils [46].

The adsorption-desorption process was mainly governed by the soil pH, as a greater adsorption and consequently lower desorption was observed in the soil with the lowest pH, although the organic carbon content and cation exchange capacity could also be responsible. The majority of Irish grassland soils have pH values greater than the dissociation constant of MCPA [47], and therefore the compund will be generally repelled by negatively charged soil particles (organic matter and clays mainly), suggesting a threat for groundwater pollution. However, the adsorption-desorption data are not the only indicator of pesticide mobility to groundwater.

A simple estimation using the GUS index suggested that in most Irish grassland soils MCPA will have transient properties being neither a leacher nor a non-leacher. Local environmental factors i.e. soil moisture, temperature and organic matter content will probably have to be taken into consideration when assessing local leaching risk.

ACKNOWLEDGMENTS

The authors wish to acknowledge the Department of Agriculture, Food and the Marine Pesticide Control Laboratory (Celbridge, Co. Kildare, Ireland) for access to facilitate for carrying out the chemical analysis, and in particular Dennis Carr and Noel Cosgrove for their technical support as well as other PCL technical and office staff. This work was funded by the Irish Government through Research Stimulus Fund contract RSF 07 544.

REFERENCES

[1] Yaron, B.; Calvet, R.; Prost, R. Soil Pollution: Processes and dynamics. Germany: Springer; 1996.

[2] de Jonge, H., & de Jonge, L.W. (1999). Influence of pH and solution composition on the sorption of glyphosate and prochloraz to a sandy loam soil. *Chemosphere, 39,* 753-163.

[3] Dyson, JS; Beulke, S; Brown, CD; Lane, MCG. Adsorption and degradation of the weak acid Mesotrione in soil and environmental implications. *Journal of Environmental Quality,* 2002, 3, 613-618.

[4] OECD. (2000). *Test guideline 106: adsorption-desorption in soils.* OECD guidelines for testing of chemicals. Paris: Organisation for Economic Co-operation and Development.

[5] Hance, RJ. Environmental Chemistry of Herbicides. In: Grover R, editor. *Adsorption and Bioavailability.* Florida: CRC Press; 1988; Vol. 1; 2-19.

[6] Richter, O; Diekkrüger, B; Nörtersheuser, P. Environmental fate modelling of pesticides: From the laboratory to the field scale. Weinheim: VCH; 1996.

[7] Hartley, GS. Physical behavior in the soil. In: Audus LJ, editor. *Herbicides physiology, biochemistry, ecology.* London, UK: Academic Press; 1976; Vol. 2.

[8] O'Connor, D.J., & Connolly, J.P. (1980). The effect of concentration of adsorbing solids on the partition coefficient. *Water Research, 14,* 1517-1523.

[9] Bolan, N.S., & Baskaran, S. (1996). Biodegradation of 2,4-D herbicide as affected by its adsorption-desorption behaviour and microbial activity of soils. *Australian Journal of Soil Research, 34*, 1041-1053.

[10] Celis, R; Hermosín, MC; Cox, L; Cornejo, J. Sorption of 2,4 Dichlorophenoxyacetic acid by model particles simulating naturally occurring soil colloids. *Environmental Science and Technology,* 1999, 33, 1200-1206.

[11] Nearpass, DC. (1976). Adsorption of picloram by humic acids and humin. *Soil Science, 121*, 272-277.

[12] Goetz, AJ; Walker, RH; Wehtje, G; Hajek, BF. Sorption and mobility of chlorimuron in Alabama soils. *Weed Science,* 1989, 37, 428-433.

[13] Walker, A; Cotterill, EG; Welch, SJ. Adsorption and degradation of chlorsulfuron and metsulfuron-methyl in soils from different depths. *Weed Research*, 1989, 29, 281-287.

[14] Grey, TL; Wehtje, GR; Hajek; BF; Walker RH. Sorption and mobility of bentazon in Coastal Plain soils. *Weed Science,* 1996, 44, 166-170.

[15] EEA. *Groundwater quality and quantity in Europe*. Copenhagen: EEA; 1999; 1-123.

[16] Environmental Agency. *Pesticides 2002*. The annual report of the Environment Agency pesticide monitoring programme. Wallingford: Environmental Agency; 2003.

[17] EHS. *Groundwater Monitoring Review 2004.* Regional Groundwater Monitoring Network. Northern Ireland: Environment and Heritage Service; 2005; 1-25.

[18] Tostnensson, NTL; Stark, J; Göranson, B. The effect of repeated applications of 2,4-D and MCPA on their breakdown in soil. *Weed Research,* 1975, 38, 159-164.

[19] Crespín, MA; Gallego, M; Valcárcel, M; González, JL. Study of the degradation of the herbicides 2,4-D and MCPA at different depths in contaminated agricultural soil. *Environmental Science and Technology*, 2001, 35, 4265-4270.

[20] Roberts, TR. Part 1: Herbicides and Plant Growth Regulators. 1st edition. *Metabolic Pathways of Agrochemicals*. Cambridge, UK: The Royal Society of Chemistry; 1988, 1-870.

[21] Tadeo, JL; Sanchez-Brunete, C; García-Valcarcer, AI; Martínez, L; Pérez RA. Determination of cereal herbicide residues in environmental samples by gas chromatography. *Journal of Chromatography A*, 1996, 754, 347-365.

[22] Teagasc. *Grassland for horses*. A handbook on best grazing/forage management practices and techniques. Carlow, Ireland: Teagasc; 2010; 1-46.

[23] Gee, G.W., & Or, D. (2002). Particle-size analysis. In Dick WA editor. *Methods of soil analysis, Physical Methods part 4* (201-228). Madison, WI: Soil Science Society of America, Inc.

[24] Metson, A.J. Methods of Chemical Analysis for Soil Survey Samples. New Zealand: DSIR; Soil Bureau Bulletin; 1956, No. 12; 193-204.

[25] Kookana, RS.; Baskaran, S; Naidu, R. Pesticide fate and behavior in Australian soils in relation to contamination and management of soil and water: a review. *Australian Journal of Soil Research,* 1988, 36, 715-764.

[26] Brennan, FP; O'Flaherty, V; Kramers, G; Grant, J; Richards, KG. Long-Term Persistence and Leaching of Escherichia coli in Temperate Maritime Soils. *Applied and Environmental Microbiology*, 2010, 76, 1449-1455.

[27] Giles, CH; MacEwan, TH; Nakhwa, SN; Smith, D. Studies in sorption. Part XI. A system of classification of solution adsorption isotherms and its use in diagnosis of adsorption mechanisms and in measurement of specific areas of solids. *Journal of Chemical Society*, 1960, 111, 3973-3993.

[28] Bolan, N.S., & Baskaran, S. (1996). Biodegradation of 2,4-D herbicide as affected by its adsorption-desorption behaviour and microbial activity of soils. *Australian Journal of Soil Research, 34,* 1041-1053.

[29] Haberhauer, G; Pfeiffer, L; Gerzabek, MN. Influence of molecular structure on sorption of phenoxyalkanoic herbicides on soil and its particle size fractions. *Journal of Agricultural and Food Chemistry,* 2000, 48, 3722-3727.

[30] Hiller, E; Krascsenits, Z; Čerňanský, S. Sorption of acetochlor, atrazine, 2,4-D, chlorotoluron, MCPA, and trifluralin in six soils from Slovakia. *Bulletin of Environmental Contamination and Toxicology*, 2008, 80, 412-416.

[31] Nicholls, PH, & Evans, AA. 1991. Sorption of ionisable organic compounds by field soils. Part 1: Acids. *Pesticide Science,* 33, 319-330.

[32] Madsen, L; Lindhardt, B; Rosenberg, P; Clausen, L; Fabricius I. Pesticide sorption by low organic carbon sediments: A screening for seven herbicides. *Journal of Environmental Quality*, 2000, 29, 1488-2000.

[33] Dubus, IG; Barriuso, E; Calvet, R. Sorption of weak organic acids in soils: clofencet, 2,4-D and salicylic acid. *Chemosphere,* 2001, 45, 767-774.

[34] Alister, CA; Araya, MA; Kogan, M. Adsorption and desorption variability of four herbicides used in paddy rice production. *Journal of Environmental Science and Health Part B*, 2011, 46, 62-68.

[35] Jensen, PH; Hansen, HChrB; Rasmussen, J; Jacobsen, OS. Sorption-controlled degradation kinetics of MCPA in soil. *Environmental Science and Technology*, 2004, 38, 6662-6668.

[36] Boivin, A; Cherrier, R; Schiavon, M. A comparison of five pesticides adsorption and desorption processes in thirteen contrasting field soils. *Chemosphere*, 2005, 61, 668-676.

[37] Kan, AT; Fu, G; Tomson, MB. Adsorption/desorption hysteresis in organic pollutants and soil/sediment interaction. Environmental Science and Technology, 1994, 28, 859-867.

[38] Morrica, P; Barbato, F; Giordano, A; Seccia, S; Ungaro, F. Adsorption and desorption of imazosulfuron by soil. *Journal of Agricultural and Food Chemistry*, 2000, 48, 6132-6137.

[39] Hiller, E; Tatraková, V; Šimonovičová, A; Bartaľ, M. Sorption, desorption, and degradation of (4-chloro-2-methylphenoxy) acetic acid in representative soils of the Danubian Lowland, Slovakia. *Chemosphere*, 2012, 87, 437-444.

[40] McCall, PJ; Swann, RL; Laskowski, DA; Unger, SM; Vrona, SA; Dishburger, HJ. Estimation of chemical mobility in soil from liquid chromatographic retention times. Bulletin of Environmental Contamination and Toxicology, 1980, **24**, 190-195.

[41] Gustafson, DI. (1989). Groundwater Ubiquity Score: A simple method for assessing pesticide leachability. *Environmental Toxicology and Chemistry, 8*, 339-357.

[42] Tomlin, CDS. The Pesticide manual. 14th edition. Hampshire: British Crop Protection Council; Hampshire; 2006.

[43] Wauchope, RD; Buttler, TM; Hornsby, AG; Augustijn-Beckers, PWM; Burt, JP. The SCS/ARS/CES pesticide properties database for environmental decision-making. *Reviews of Environmental Contamination and Toxicology,* 1992, 123, 1-155.

[44] Vencill, WK. WSSA Herbicide handbook. 8th edition. USA: Lawrence; Weed Science Society of America; 2002.

[45] Bailey, S.W.; Allsopp, J.; Martin, B. Pesticides selection for monitoring private water supplies. In: Walker A, editor. *Pesticide movement to water*. BCPC Monograph 63, 1995; 363 - 368.

In: Handbook on Herbicides
Editors: Daiki Kobayashi and Eito Watanabe

ISBN: 978-1-62948-053-4

Chapter 12

PESTICIDES AND CANCER: STUDIES ON THE INTERACTION OF PHENOXY ACID HERBICIDES WITH DNA

Sofia Benfeito[1,2], Jorge Garrido[1,2], M. J. Sottomayor[2], Fernanda Borges[2] and E. Manuela Garrido[1,2]

[1]Departamento de Engenharia Química, Instituto Superior de Engenharia do Porto (ISEP), Instituto Politécnico do Porto, Porto, Portugal
[2]CIQ, Departamento de Química e Bioquímica, Faculdade de Ciências, Universidade do Porto, Porto, Portugal

ABSTRACT

Herbicides are important for the control of weed growth but increasing use causes environmental problems and undesirable side effects in crops. Phenoxy acid herbicides with auxin-like activity have been used against grass and broad leaf weeds in many crops, such as rice, winter wheat and soybean. Carcinogenicity in humans and embryotoxicity in animals have been described as the main hazards of phenoxy acid herbicides.

Thermal denaturation of double stranded DNA has been used as a measure of the effect of harmful chemical substances on the structure and stability of the DNA molecule. Hence, the interaction between several phenoxy acid herbicides (phenoxyacetic acid, 4-chlorophenoxyacetic acid, MCPA, mecoprop, 2,4-D and dichlorprop) and Calf thymus DNA was assessed by UV–Vis absorption spectroscopy. UV spectra and melting curves have been recorded for solutions at constant DNA concentration and using different concentrations of each of the herbicides under study. The transition temperature values and the thermodynamic parameters of DNA thermal denaturation have been determined. The studies performed so far help to understand herbicide–DNA interactions gathering suitable information for the (re)design of new molecules with lower potential risk to humans.

1. INTRODUCTION

Human exposure to environmental and occupational chemicals has increased considerably in the past decades. Pesticides, a broad group of biologically active chemicals used for pest management, are among the most widely used chemicals in the world, and also among the most dangerous to human health and environment. In the European Union (EU) more than 200,000 tonnes of pesticides (active ingredients) are used annually [1]. The adverse health effects of pesticide use have long been established, with links to neurologic and endocrine (hormone) system disorders, birth defects, cancer, and other diseases [2-4]. Carcinogenic pesticides may increase the risk of cancer through a variety of mechanisms, including genotoxicity, tumor promotion, hormonal action, and immunotoxicity [5]. Pesticides can be non-genotoxic (i.e. epigenetic) and capable of initiating a number of the steps in the onset of cancer without actually causing genetic damage or genotoxic, triggering the onset of cancer causing damage to DNA directly [6]. Farmers and industrial workers employed in the manufacture of pesticides are considered a high-risk group to developing cancer following pesticide exposure. Epidemiological data on cancer risk are conflicting. In most studies, the overall cancer incidence and mortality was lower in farmers and pesticide applicators than in the general population [5]. However, it has often been reported excessive incidence and mortality in several specific types of cancer, especially for soft tissue sarcoma, malignant lymphomas, leukemia and cancer of the skin, prostate, testis, lung, and brain [3, 5].

The phenoxy herbicides are frequently used in agricultural and household sectors worldwide. This chemical family encompasses MCPA, involved in several episodes of intentional self-poisoning with pesticide in rural areas, and 2,4-D, one the most commonly used pesticide active ingredients in the U.S. agricultural market sector [7, 8]. Additionally, 2,4-D and MCPA are both in the top five of the most commonly used pesticide active ingredients in the home and garden sector in the US and herbicide active ingredients in the EU [1, 8]. Based on epidemiologic studies, phenoxy acid herbicides have been classified by International Agency for Research on Cancer (IARC) as possibly carcinogenic to humans (category 2B) [9]. Actually, exposure to phenoxy herbicides has been linked with an increased risk of cancer in humans, namely the appearance of soft-tissue sarcoma, non-Hodgkin's lymphoma, and other cancers [10-15]. However, the experimental data, particularly from case-control studies, is inconsistent [3, 11].

Genotoxic potential is considered to be a primary risk factor for long-term effects and its evaluation has become a crucial parameter in the safety evaluation of harmful chemicals, such as environmental pollutants and pesticides. Chemicals that interact with DNA can cause direct damage by covalent modifications, such as adducts or strand breaks, or can perturb DNA and chromatin function by non-covalent binding [16-19]. Thus, the preliminary evaluation of DNA damage is considered to be of high significance for the study of the putative interactions of pesticides-DNA and for the screening of their mutagenic properties. A plethora of methods are currently available for the study of DNA interactions, namely gel electrophoresis, NMR, electrochemistry, fluorescence and UV/Vis spectroscopy [18-24]. Among them, UV–Visible absorption spectroscopy is perhaps the simplest and most commonly employed instrumental technique to study the changes induced in DNA by small molecules.

In this work, the interaction of several phenoxy acid herbicides (phenoxyacetic acid, 4-chlorophenoxyacetic acid, MCPA, mecoprop, 2,4-D and dichlorprop) with calf thymus DNA has been investigated by UV–visible absorption spectroscopy (Figure 1). Thermodynamic parameters of DNA thermal denaturation, in solutions containing the herbicides, have been evaluated to interpret the mode of interaction. The study described along this chapter is expected to provide structure–toxicity relationship data and thus assist in the design of less toxic herbicides.

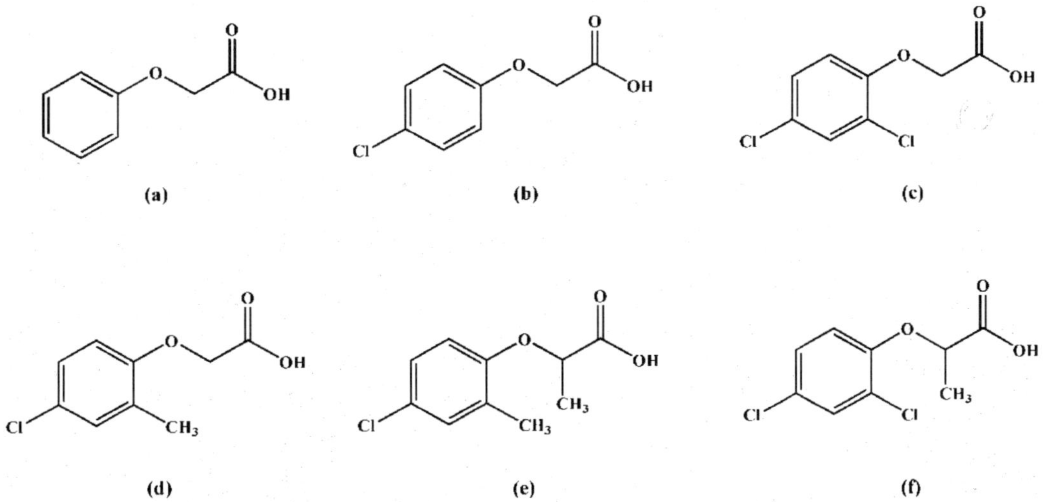

Figure 1. Chemical structures of the phenoxy acid herbicides: (a) Phenoxyacetic acid, (b) 4-CPA, (c) 2,4-D, (d) MCPA, (e) mecoprop and (f) dichlorprop.

2. Experimental

2.1. Chemicals and Reagents

The phenoxy acid herbicides, as well as NaCl, Na_2HPO_4 and NaH_2PO_4 were supplied by Sigma-Aldrich Química (Sintra, Portugal) and used without further purification. Calf thymus DNA, as Type I calf thymus DNA sodium salt (protein content < 3%), was also purchased from Sigma-Aldrich. Ultrapure (Type 1) water (Millipore, Milli Q Gradient) was used throughout the experiments.

A stock solution (1.0×10^{-3} mol dm^{-3}) of DNA was prepared by dissolving an appropriate amount of DNA in 0.05 mol dm^{-3} phosphate buffer solution (pH 7.2; ionic strength adjusted to 0.1 mol dm^{-3} with NaCl). Stock solutions (2.0×10^{-4} mol dm^{-3}) of herbicides were prepared by dissolution of a suitable quantity in ultrapure water.

The UV absorbance measurements were performed in pH 7.2 phosphate buffer solutions with ionic strength 0.01 mol dm^{-3}. Working solutions of DNA (7.0×10^{-5} mol dm^{-3}), herbicides (5.0×10^{-5} mol dm^{-3} or 1.0×10^{-4} mol dm^{-3}) and DNA/herbicides were prepared by simple dilution of the appropriate amounts of the stock solutions in phosphate buffer and ultrapure water.

The concentration of DNA solutions, expressed in moles of base pairs, was determined at 20 °C by UV spectroscopy at 260 nm, using a molar absorption coefficient $\varepsilon_{260} = 13200$ mol^{-1} dm^{3} cm^{-1} [25]. A ratio of absorbance at 260 nm to that at 280 nm, (A_{260}/A_{280}) greater than 1.8 indicated that DNA was sufficiently pure and free from protein [26].

Control experiments under the same conditions and initial concentrations of herbicides were carried out in parallel for comparison. All the solutions were stored in the refrigerator at 4 °C.

2.2. UV Spectroscopy Experiments

The absorption spectra, as well as the UV melting curves, were recorded on a hermetic quartz cell with a 1 cm path length, using an Agilent 8453 UV-vis spectroscopy system equipped with a thermostatic cell holder. The temperature of the samples was controlled using a Julabo F25/HP thermostatic bath. The UV absorption spectra of herbicides and DNA/herbicides solutions were acquired in the wavelength range of 200-400 nm at 20 °C. For DNA melting studies, the temperature of the cell was changed from 20 to 95 °C, with a heating rate of 1 °C min^{-1}.

The fraction of melted base pairs, θ, at each temperature, has been calculated from the curves of absorbance *vs.* temperature, using equation 1 [27]:

$$?? = \frac{A\,(T) - A_{\mathrm{L}}\,(T)}{A_{\mathrm{U}}\,(T) - A_{\mathrm{L}}\,(T)} \quad (1)$$

A is the experimental absorbance, A_{L} and A_{U} are the lower and upper absorbance baselines, respectively, obtained by extrapolating the linear regions of the experimental curve before and after the transition.

The melting temperature, T_{m}, defined as the temperature at which half of the amount of DNA is denatured [28], was determined by interpolation in the curves $\theta = \mathrm{f}\,(T)$ for $\theta = 0.5$.

The hyperchromicity at 260 nm, H_{260}, was calculated using equation 2 [29]:

$$H_{260} = \frac{A_{\mathrm{U}}\,(T_m) - A_{\mathrm{L}}\,(T_m)}{A_{\mathrm{L}}\,(T_m)} \quad (2)$$

2.3. Thermodynamic Parameters of DNA Thermal Denaturation

The thermodynamic parameters of DNA thermal denaturation were obtained from the denaturation curves, by two different methods based on the van't Hoff equation [30-32]. Both methods assume that denaturation is a two-state transition and the values of enthalpy and entropy are not dependent of temperature.

The first method is based on the dependence of the equilibrium constant, K, of DNA denaturation on temperature, T. The value of K at each temperature can be expressed as a function of the broken base pairs fraction, θ. The van´t Hoff denaturation enthalpy and entropy values were obtained by linear regression of - ln *K* *vs.* 1/*T* and fitting the data to values of θ from 0.25 to 0.75, according to equation 3:

$$-\ln K = \frac{\Delta H^{0}_{vH}}{RT} - \frac{\Delta S^{0}_{vH}}{R} \quad (3)$$

The Gibbs energy of DNA denaturation was calculated using equation 4:

$$\Delta G^{0}(T) = \Delta H^{0} - T\,\Delta S^{0} \quad (4)$$

In the second method, the van´t Hoff denaturation enthalpy and entropy values were calculated using the peak height maxima obtained from the derivative melting curves, $(d\theta/dT)_{max}$, using equations 5 and 6 (*c* is the concentration of DNA) [30, 31].

$$\Delta H^{0}_{vH} = 6\,R\,T_{m}^{2}\left(\frac{d\theta}{dT}\right)_{T_m} \quad (5)$$

$$\Delta S^{0}_{vH} = \frac{\Delta H^{0}_{vH}}{T_m} - R\,\ln\left(\frac{c}{2}\right) \quad (6)$$

The Gibbs energy of DNA denaturation was calculated as described earlier (Equation 4).

Values of ΔH^{o}_{vH}, ΔS^{o}_{vH} and $\Delta G^{o}(T)$ obtained from both methods differed less than ±1%. Thus, the values given correspond to the average calculated from the values obtained using both methods.

3. Results and Discussion

3.1. UV absorption Spectra

The interaction of small molecules with DNA involves two modes of binding, covalent and non-covalent. The drug–DNA covalent mode of binding is irreversible and invariably causes the complete inhibition of DNA processes and subsequent cell death. Non-covalent DNA interacting agents, DNA-groove binders and DNA intercalators, are generally considered less cytotoxic than agents producing covalent DNA adducts and other DNA damage. While most DNA-interacting agents selectively bind to DNA either by groove binding or intercalation, some compounds can exhibit both binding modes. Understanding the forces involved in the binding of small molecules, namely pesticides, to DNA is of prime importance.

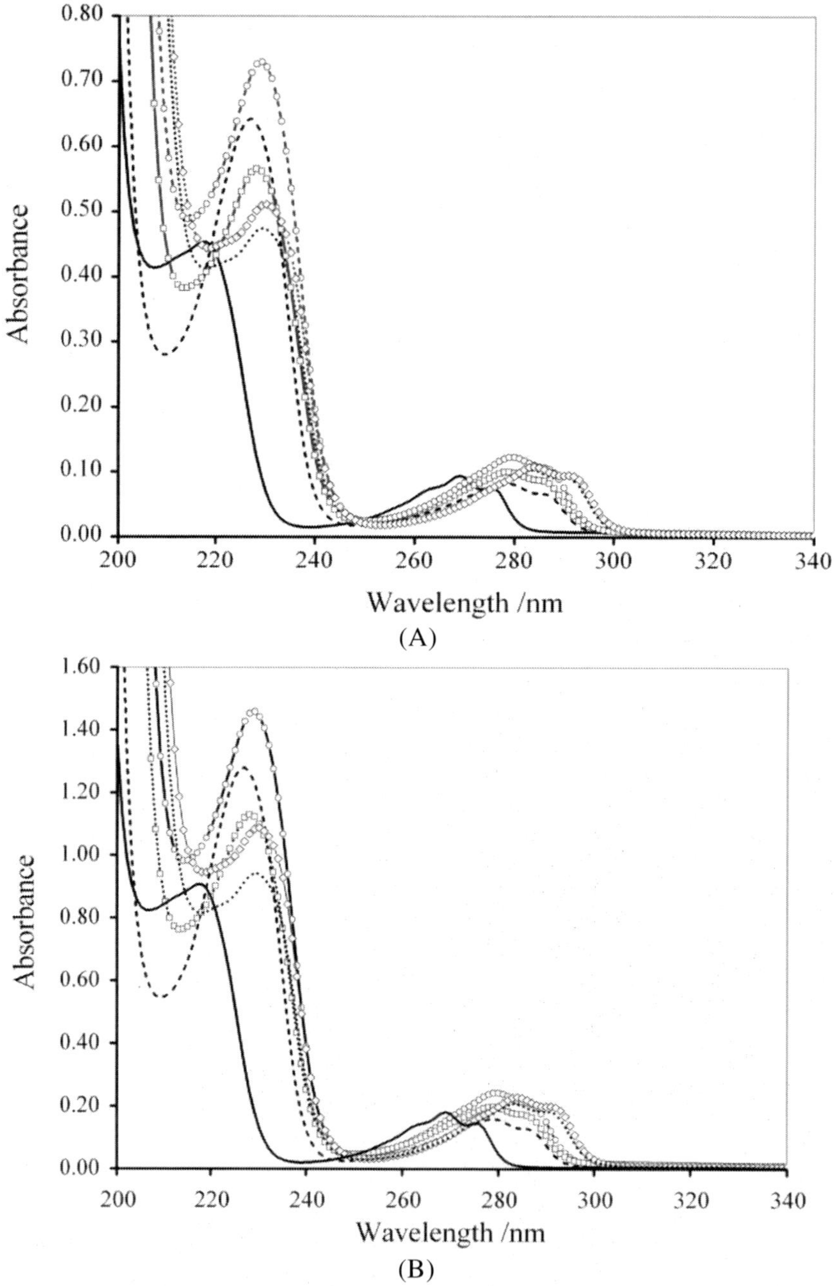

Figure 2. UV-Vis absorption spectra of (A) 5.0×10^{-5} mol dm^{-3} and (B) 1.0×10^{-4} mol dm^{-3} of phenoxy acid herbicides standard solutions in pH 7.2 phosphate buffer electrolyte. **(—) Phenoxyacetic acid,** (---) 4-CPA, (•••) 2,4-D, (-○-) mecoprop, (-□-) MCPA and (-◊-) dichlorprop.

The study of drug–DNA interactions could be carried out by UV–Visible absorption spectroscopy by monitoring the changes in the absorption properties of the drug or the DNA molecules. An easy way to determine whether there is any interaction between the DNA and the drug is therefore to examine the shifting of the position of the absorption bands when the drug is free in solution to when the drug is bound to DNA [33-35]. The magnitude of this shifting could be interpreted as an indication of the strength of the DNA-drug interaction [36, 37].

The UV absorption spectra of the phenoxy acid herbicides, at two different concentrations, in the absence of DNA are presented in figure 2. As can be seen, each phenoxy acid herbicide presents three peaks: the two weaker bands, in the 270-300 nm range, correspond to $\pi \rightarrow \pi^*$ bands characteristic of aromatic compounds and $n \rightarrow \pi^*$ bands typical of carbonyl compound; the band corresponding to the maximum absorption, occurring below 240 nm, is also related with the existence of an aromatic system [38-40]. With the exception of phenoxyacetic acid, all other herbicides studied have substituents in the aromatic ring (namely the presence of a Cl group in *p*-position) that causes the observed bathochromic shift in the UV absorption bands [41].

The UV absorption spectra obtained for DNA in the absence or presence of herbicides (DNA/herbicides solutions) are presented in figure 3. The DNA spectrum displays a strong absorption band with a maximum at about 260 nm, derived from $\pi \rightarrow \pi^*$ transitions occurring in the nucleotide bases (purines and pyrimidines). Spectra registered for DNA/herbicides solutions exhibit the absorption bands common to both DNA and each of the phenoxy acid herbicides under study.

To evaluate the effect of herbicides on the UV spectrum of DNA, the spectra of DNA/herbicides solutions were compared with the sum of the individual spectra of DNA and herbicide at the same concentration (Figure 4).

The UV spectra of DNA/herbicides solutions present only slight variations in the intensity of the bands relative to the sum of the spectra of DNA and each of the herbicides (Figure 4). In all cases a slight bathochromic shift (2-3 nm) is observed for the band at ~ 260 nm of DNA which might indicate that the herbicides associate with the DNA base pairs [42]. Changes in absorbance and wavelength shifts of this characteristic band reflect the corresponding structural changes of the DNA, including for instance changes of stacking pattern, disruption of the hydrogen bonds between complementary strands, covalent binding to the DNA bases, intercalation of aromatic rings of molecules between adjacent base pairs, etc [43-45]. Nevertheless, as no significant bathochromic effect has been observed, one can assume that the binding between these herbicides and DNA base pairs should not be strong.

3.2. DNA Melting Analysis

DNA consists of two polynucleotide chains twisted around each other to form a double helix. The double-helical structure of DNA is remarkably stable due to the hydrogen bonds and also to base stacking interactions. When a DNA solution is heated, the non-covalent forces that hold the two strands together weaken and finally break. When this happens, the two strands come apart in a process known as DNA denaturation, and the temperature at which the DNA strands are half-denatured is called the melting temperature, T_m [46]. Due to the disruption of the double helical structure, bases become unstacked and thus the UV absorption increases [29]. Therefore, UV–vis spectroscopy is a valuable technique to determine melting temperatures and to study the interaction of small molecules with DNA

[47, 48]. Neutralization of the negative charges of the phosphate groups through external binding as well as the stacking interactions of intercalation contribute to the increase of DNA melting temperature.

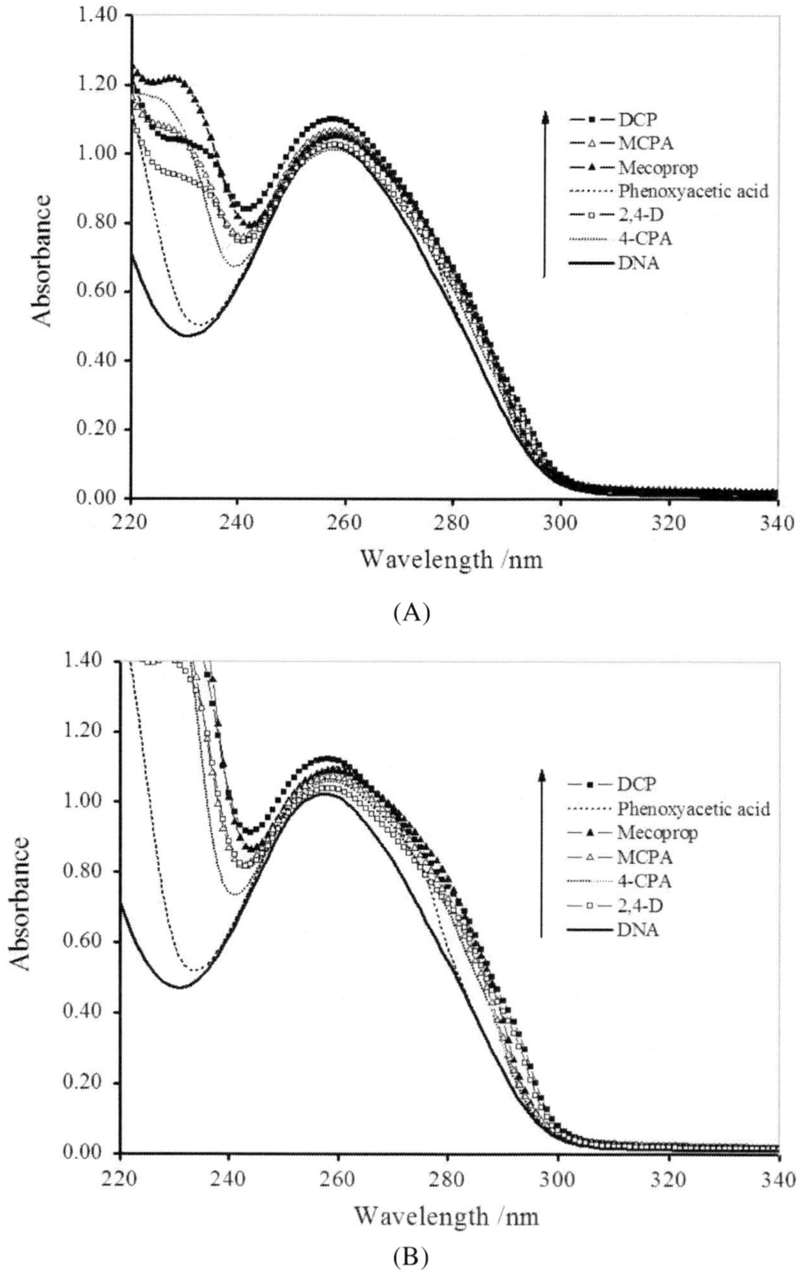

Figure 3. UV-Vis absorption spectra of 7.67×10^{-5} mol dm^{-3} of DNA in the absence and presence of (A) 5.0×10^{-5} mol dm^{-3} and (B) 1.0×10^{-4} mol dm^{-3} of phenoxy acid herbicides in pH 7.2 phosphate buffer electrolyte.

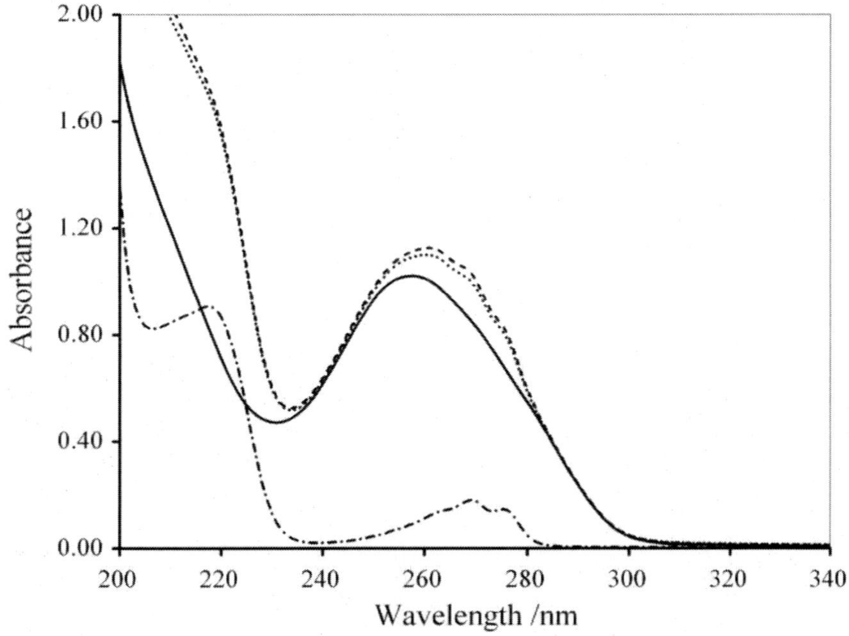

(A)

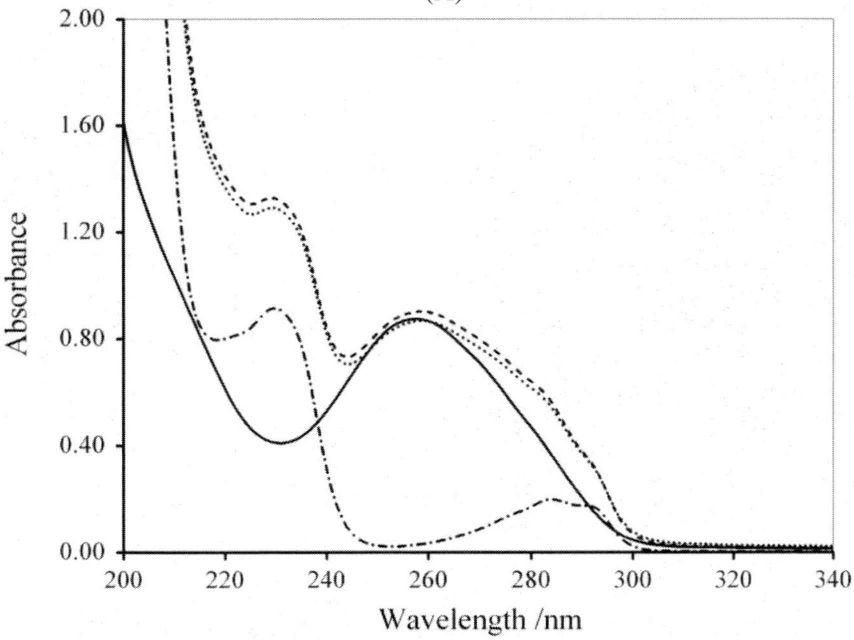

(B)

Figure 4. UV-Vis absorption spectra of 7.67×10^{-5} mol dm^{-3} of DNA in the absence and presence of (A) 1.0×10^{-4} mol dm^{-3} of phenoxyacetic acid and (B) 1.0×10^{-4} mol dm^{-3} of dichlorprop herbicides in pH 7.2 phosphate buffer electrolyte. (—) DNA, (-•-) herbicide, (---) sum of free herbicide and DNA and (•••) DNA/herbicide solution.

The DNA melting curves at 260 nm for DNA and DNA/mecoprop solutions are shown in figure 5. The data obtained for the other herbicides have been found to be similar.

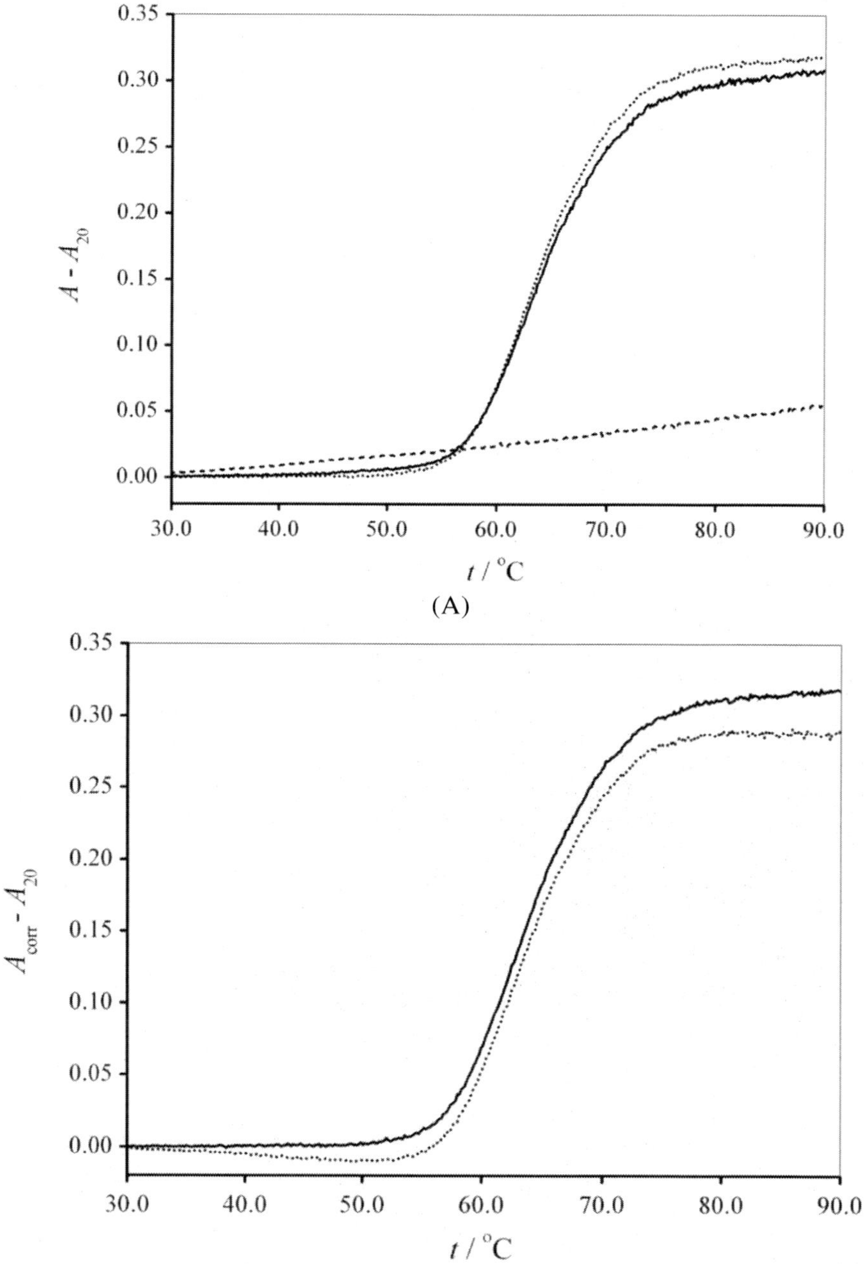

(B)

Figure 5. (A) Absorbance and (B) corrected absorbance at 260 nm vs. temperature for (—) 7.67×10^{-5} mol dm^{-3} of DNA, (---) 5.0×10^{-5} mol dm^{-3} mecoprop and (•••) DNA/mecoprop solution in pH 7.2 phosphate buffer electrolyte.

As absorbance obtained for the herbicides at 260 nm depends on temperature (Figure 5A), all DNA/herbicide solutions denaturation curves were corrected to this effect by subtracting from the melting curves of DNA/herbicide solutions the values of absorbance corresponding to the solutions containing only herbicides. The resulting curves (A_{corr} in Figs. 5B and 6) reflect only the change in absorbance due to DNA denaturation resulting from temperature increase.

The data allow concluding that, except for phenoxyacetic acid, all the phenoxy acid herbicides under study reduce the hyperchromism along the DNA denaturation process (Figure 6). Moreover, the hyperchromism is influenced by the herbicide concentration: an increase of concentration leads to a decrease of the DNA absorbance upon denaturation (Figs. 5B and 6). Above 80 °C (temperature at which it is assumed that the strands of DNA have been totally separated), there is a reduction in the hyperchromism as a result of the presence of phenoxy acid herbicides (Figure 6). As mentioned before the herbicide phenoxyacetic acid does not influence the hyperchromism observed upon DNA denaturation (the absorbance obtained for DNA/phenoxyacetic acid solution is equal to that attained for DNA alone). This result may be related to the absence of substituents on the aromatic ring in the molecular structure of phenoxyacetic acid. Apparently the presence of chloro and methyl substituents on the aromatic ring of the other herbicides under study may be responsible for the observed interactions in DNA/herbicide solutions.

The values of hyperchromism, calculated at T_m, are presented in Table 1.

The reduction of the hyperchromicity observed in the DNA melting curves indicates that an interaction of the studied herbicides with DNA occurs. However, it is expected that the interactions which occur are weak. This can also be confirmed observing the curves of molar fraction of DNA denaturated *vs.* temperature (Figure 7), which clearly show that the denaturation temperature values are not significantly affected by the herbicides (Table 1).

The thermodynamics of drug–DNA association reactions is essential for a complete understanding of the process. In fact, thermodynamics can provide quantitative data that is important for elucidating the driving forces for binding interactions [49-52].

Spectroscopy was used to assess the thermal stability of the DNA in comparison with the DNA/herbicides solutions. The van't Hoff analysis assumes a two state model. Therefore, it describes a DNA molecule without intermediate states that might influence the thermodynamics. A two-state model predicts a molecule that remains completely annealed (double-strand) at lower temperatures, becomes completely unannealed (single-strand) at high temperatures.

From the calculated thermodynamic parameters it is possible to infer that the presence of the herbicides causes a slight increase in entropy and enthalpy (Table 1). In fact, although the denaturation temperature values are not significantly different, the denaturation enthalpy values for DNA/herbicides solutions are generally higher than that observed for DNA (Table 1). These results show that a stabilization of the DNA double helix occurs in the presence of herbicides and no influence on the relative stability of base pairs is observed.

The data allow concluding that for this type of herbicides no intercalation occurs with DNA. So, it is very unlikely that these herbicides interact with DNA bases by intercalation.

On the other hand it is well-known that particular small compounds bind to the groove of DNA. Hydrophobic and/or hydrogen bonding are usually important driving forces of this binding process providing DNA stabilization. Unlike to intercalating compounds, groove-binding drugs induce little or no structural rearrangement of the DNA helix.

Table 1. Thermodynamic parameters of DNA thermal denaturation, obtained by UV spectroscopy, in solutions with different herbicides[a]

Herbicides	$c_{herbicides}$ / mol dm^{-3}	pH	t_m / °C	H_{260} (t_m)	H_{260} (85 °C)	ΔH^o_{vH} / kJ mol^{-1}	ΔS^o_{vH} / kJ K^{-1} mol^{-1}	ΔG^o_{vH} / kJ mol^{-1}
–	0	7.28	63.4 ± 0.5	0.31	0.32	421 ± 5	1.17 ± 0.01	27.2 ± 0.7
Phenoxyacetic	5.09×10^{-5}	7.17	63.4 ± 0.4	0.29	0.30	421 ± 4	1.17 ± 0.01	27.2 ± 0.5
acid	1.02×10^{-4}	7.06	63.4 ± 0.5	0.30	0.32	438 ± 5	1.22 ± 0.01	27.4 ± 0.7
4-CPA	5.09×10^{-5}	7.18	63.2 ± 0.4	0.27	0.25	416 ± 3	1.16 ± 0.01	25.8 ± 0.5
	1.02×10^{-4}	7.10	63.2 ± 0.5	0.28	0.26	447 ± 5	1.25 ± 0.01	26.6 ± 0.7
2,4-D	5.09×10^{-5}	7.15	63.0 ± 0.3	0.30	0.32	435 ± 3	1.21 ± 0.01	28.2 ± 0.4
	1.02×10^{-4}	7.12	63.2 ± 0.6	0.27	0.29	430 ± 6	1.20 ± 0.02	26.3 ± 0.8
MCPA	5.11×10^{-5}	7.13	63.4 ± 0.6	0.26	0.26	442 ± 6	1.23 ± 0.02	28.1 ± 0.8
	1.02×10^{-4}	7.04	63.9 ± 0.4	0.26	0.25	421 ± 3	1.17 ± 0.01	26.7 ± 0.5
Mecoprop	5.12×10^{-5}	7.15	63.7 ± 0.6	0.31	0.29	428 ± 5	1.19 ± 0.02	27.1 ± 0.8
	1.02×10^{-4}	7.03	63.7 ± 0.6	0.28	0.26	452 ± 6	1.26 ± 0.02	27.6 ± 0.8
DCP	5.11×10^{-5}	7.19	64.1 ± 0.5	0.28	0.29	467 ± 5	1.30 ± 0.01	28.6 ± 0.7
	1.02×10^{-4}	7.09	64.7 ± 0.5	0.27	0.27	450 ± 5	1.25 ± 0.01	27.7 ± 0.7

[a] $c_{DNA} = 7.67 \times 10^{-5}$ mol dm^{-3}, phosphate buffer solution, ionic strength = 0.01 mol dm^{-3}.

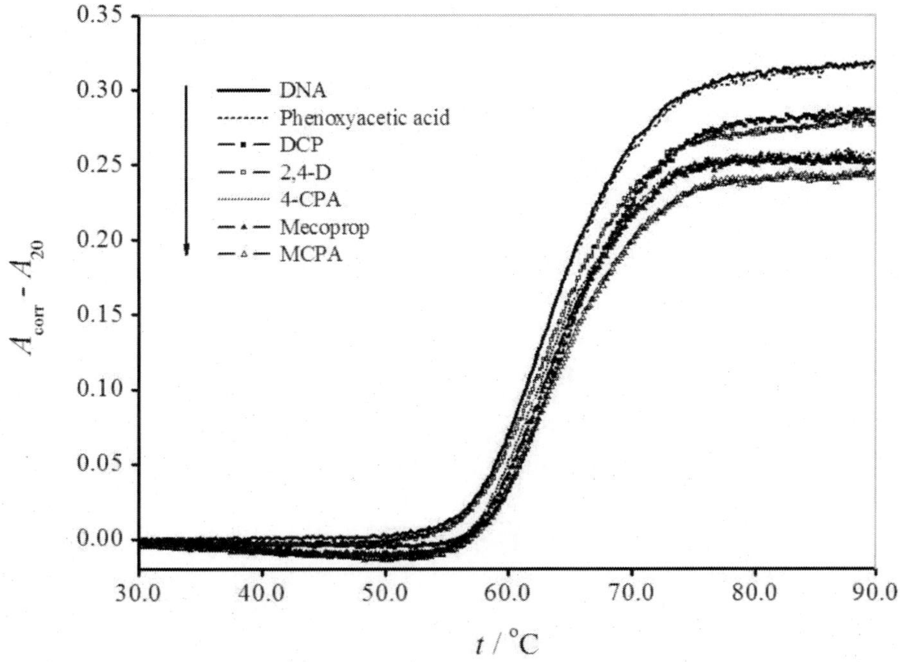

Figure 6. Corrected absorbance at 260 nm vs. temperature for 7.67×10^{-5} mol dm^{-3} of DNA in the absence and presence of 1.0×10^{-4} mol dm^{-3} of phenoxy acid herbicides in pH 7.2 phosphate buffer electrolyte.

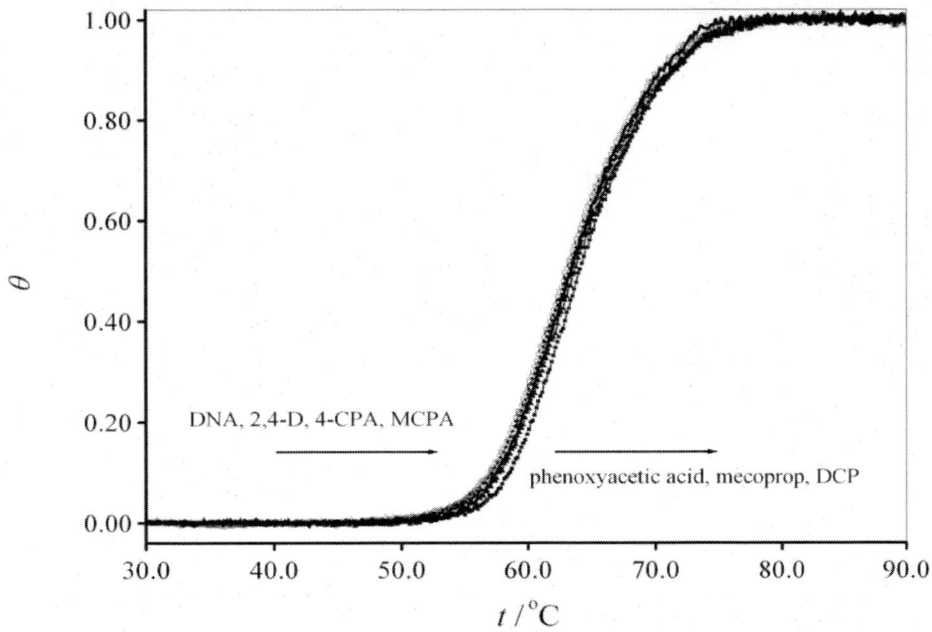

Figure 7. Molar fraction of DNA denatured vs. temperature for 7.67×10^{-5} mol dm^{-3} of DNA in the absence and presence of 5.0×10^{-5} mol dm^{-3} of phenoxy acid herbicides in pH 7.2 phosphate buffer electrolyte.

The data herein obtained is in agreement with a reported study of the interaction of 2,4-D with DNA. In fact, from the results of the analysis of fluorescence spectra, viscosity measurements and alternative current voltammetry, it was concluded that 2,4-D is a groove binder of DNA [53].

CONCLUSION

Pesticides constitute a heterogeneous category of chemicals specifically designed for the control of pests, weeds or plant diseases. Some pesticides have been considered potential chemical mutagens, due to their capacity to induce mutations, chromosomal alterations or DNA damage.

DNA is one of the most critical cellular targets for hazardous chemicals and wastes. Accordingly, understanding the DNA-chemical (drug) interactions has become an active research area at the interface between chemistry, molecular biology and medicine. Thus, in this study the interaction of phenoxy acid herbicides with DNA was assessed at a physiological pH of 7.2 using UV-vis spectroscopy. The data acquired allow concluding that a slight effect of the phenoxy acid herbicides on the stability of the double helix take place.

From the experimental data obtained so far it can be concluded that no intercalation processes are present in herbicide-DNA binding. External binding to the major or minor DNA groove could be responsible for the observed disturbance on DNA conformation.

The results obtained also indicate a significant influence of the substituents present in the aromatic moiety of the phenoxy herbicides on the extent of the interaction, although it is not possible to establish a global order for the stabilizing effect.

The studies performed so far help to understand herbicide–DNA interactions gathering important information for the (re)design of new molecules with lower human potential risk.

ACKNOWLEDGMENTS

Thanks are due to Fundação para a Ciência e Tecnologia FCT/MCTES for financial support to CIQ-UP (PEst-C/QUI/UI0081/2011) and project PTDC/AGR-AAM/105044/2008, National Funds PIDDAC also co-financed by the European Community Fund FEDER through COMPETE–Programa Operacional Factores de Competitividade (POFC).

REFERENCES

[1] Eurostat. *Office for official publications of the European Communities*: Luxembourg, 2007.

[2] Alavanja, M.C.R.; Hoppin, J.A.; Kamel, F., Health effects of chronic pesticide exposure: Cancer and neurotoxicity. *Annu. Rev. Public Health,* 2004, 25, 155–197.

[3] Miligi, L.; Costantini, A.S.; Veraldi, A.; Benvenuti, A.; Will; Vineis, P., Cancer and Pesticides. *Annals of the New York Academy of Sciences,* 2006, 1076 (1), 366–377.

[4] Perry, M.J., Effects of environmental and occupational pesticide exposure on human sperm: A systematic review. *Human Reproduction Update,* 2008, 14 (3), 233–242.

[5] Dich, J.; Zahm, S.H.; Hanberg, A.; Adami, H.O., Pesticides and cancer. *Cancer Causes and Control,* 1997, 8 (3), 420–443.

[6] Ratitsky, V.N.; Koblyakov, V.A.; Turusov, V.S., Nongenotoxic (epigenetic) carcinogens: Pesticides as an example. A critical review. *Teratogenesis, Carcinogenesis, and Mutagenesis,* 2000, 20 (4), 229–240.

[7] Roberts, D.M.; Seneviratne, R.; Mohammed, F.; Patel, R.; Senarathna, L.; Hittarage, A.; Buckley, N.A.; Dawson, A.H.; Eddleston, M., Intentional self-poisoning with the chlorophenoxy herbicide 4-chloro-2-methylphenoxyacetic acid (MCPA). *Annals of Emergency Medicine,* 2005, 46 (3), 275–284.

[8] Kiely, T.; Donaldson, D.; Grube, A., Pesticides industry sales and usage – 2000 and 2001 market estimates. *U.S. Environmental Protection Agency,* Washington DC, USA, 2004.

[9] International Agency for Research on Cancer (IARC). *Chlorophenoxy herbicides.* 1987, 41 (Suppl. 7) 156–160.

[10] Zahm, S.H.; Blair, A., Pesticides and non-Hodgkin's lymphoma. *Cancer Research,* 1992, 52 (19 SUPPL.), 5485s–5488s.

[11] Kogevinas, M.; Becher, H.; Benn, T.; Bertazzi, P.A.; Boffetta, P.; Bueno-de-Mesquita, H.B.; Coggon, D.; Colin, D.; Flesch-Janys, D.; Fingerhut, M.; Green, L.; Kauppinen, T.; Littorin, M.; Lynge, E.; Mathews, J.D.; Neuberger, M.; Pearce, N.; Saracci, R., Cancer mortality in workers exposed to phenoxy herbicides, chlorophenols, and dioxins: An expanded and updated International Cohort Study. *American Journal of Epidemiology,* 1997, 145 (12), 1061–1075.

[12] McDuffie, H.H.; Pahwa, P.; McLaughlin, J.R.; Spinelli, J.J.; Fincham, S.; Dosman, J.A.; Robson, D.; Skinnider, L.F.; Choi, N.W., Non-Hodgkin's lymphoma and specific pesticide exposures inmen: Cross-Canada study of pesticides and health. *Cancer Epidemiology Biomarkers and Prevention,* 2001, 10 (11), 1155–1163.

[13] Fleming, L.E.; Gómez-Marín, O.; Zheng, D.; Ma, F.; Lee, D., National health interview survey mortality among US farmers and pesticide applicators. *American Journal of Industrial Medicine,* 2003, 43 (2), 227–233.

[14] Knopper, L.D.; Lean, D.R.S., Carcinogenic and genotoxic potential of turf pesticides commonly used on golf courses. Journal of Toxicology and Environmental Health - *Part B: Critical Reviews,* 2004, 7 (4), 267–279.

[15] Eriksson, M.; Hardell, L.; Carlberg, M.; Åkerman, M., Pesticide exposure as risk factor for non-Hodgkin lymphoma including histopathological subgroup analysis. *International Journal of Cancer,* 2008, 123 (7), 1657–1663.

[16] Chaires, J.B., Drug-DNA interactions. *Current Opinion in Structural Biology,* 1998, 8 (3), 314–320.

[17] Strekowski, L.; Wilson, B., Noncovalent interactions with DNA: An overview. *Mutation Research - Fundamental and Molecular Mechanisms of Mutagenesis,* 2007, 623 (1-2), 3–13.

[18] Sirajuddin, M.; Ali, S.; Badshah, A., Drug-DNA interactions and their study by UV-Visible, fluorescence spectroscopies and cyclic voltametry. *Journal of Photochemistry and Photobiology B: Biology,* 2013, 124, 1–19.

[19] Ahmadi, F. In Pesticides - *The Impacts of Pesticides Exposure.* Stoytcheva, M., Ed.; InTech: Rijeka, Croatia, 2011, pp. 446

[20] Araya, F.; Huchet, G.; McGroarty, I.; Skellern, G.G.; Waigh, R.D., Capillary electrophoresis for studying drug-DNA interactions. *Methods,* 2007, 42 (2), 141–149.

[21] Nakamoto, K.; Tsuboi, M.; Strahan, G.D. In Drug–DNA Interactions; John Wiley & Sons, Inc., 2008, pp i-xvi.

[22] Ahmadi, F.; Jafari, B., Voltammetry and spectroscopy study of in vitro interaction of fenitrothion with DNA. *Electroanalysis,* 2011, 23 (3), 675–682.

[23] Campagne, S.; Gervais, V.; Milon, A., Nuclear magnetic resonance analysis of protein-DNA interactions. *Journal of the Royal Society Interface,* 2011, 8 (61), 1065–1078.

[24] Kashanian, S.; Shariati, Z.; Roshanfekr, H.; Ghobadi, S., DNA binding studies of 3, 5, 6-trichloro-2-pyridinol pesticide metabolite. *DNA and Cell Biology,* 2012, 31 (7), 1341–1348.

[25] Reichmann, M.E.; Rice, S.A.; Thomas, C.A.; Doty, P., A Further examination of the molecular weight and size of desoxypentose nucleic acid. *Journal of the American Chemical Society,* 1954, 76 (11), 3047–3053.

[26] Marmur, J., A procedure for the isolation of deoxyribonucleic acid from micro-organisms. *Journal of Molecular Biology,* 1961, 3 (2), 208-218.

[27] Wartell, R.M.; Benight, A.S., Thermal denaturation of DNA molecules: A comparison of theory with experiment. *Physics Reports,* 1985, 126 (2), 67–107.

[28] Owczarzy, R., Melting temperatures of nucleic acids: Discrepancies in analysis. *Biophysical Chemistry,* 2005, 117 (3), 207–215.

[29] Bloomfield, V.A.; Crothers, D.; Ignacio Tinoco, J., *Nucleic acids: structures, properties, and functions.* University science books, 2000.

[30] Marky, L.A.; Breslauer, K.J., Calculating thermodynamic data for transitions of any molecularity from equilibrium melting curves. *Biopolymers - Peptide Science Section,* 1987, 26 (9), 1601–1620.

[31] Moreira, B.G.; You, Y.; Behlke, M.A.; Owczarzy, R., Effects of fluorescent dyes, quenchers, and dangling ends on DNA duplex stability. *Biochemical and Biophysical Research Communications,* 2005, 327 (2), 473–484.

[32] Vitorino, J.; Sottomayor, M.J., DNA interaction with flavone and hydroxyflavones. *Journal of Molecular Structure,* 2010, 975 (1-3), 292–297.

[33] Nakamoto, K.; Tsuboi, M.; Strahan, G.D. In Drug–DNA Interactions; John Wiley & Sons, Inc., 2008, pp. 1–118.

[34] González-Ruiz, V.; Olives, A.I.; Martín, M.A.; Ribelles, P.; Ramos, M.T.; J. Carlos Menéndez, J.C., In *Biomedical Engineering, Trends, Research and Technologies.* Komorowska, M.A.; Olsztynska-Janus, S., Ed.; InTech: Madrid, 2011, pp. 644.

[35] Zhang, G.; Hu, X.; Zhao, N.; Li, W.; He, L., Studies on the interaction of aminocarb with calf thymus DNA by spectroscopic methods. *Pesticide Biochemistry and Physiology,* 2010, 98 (2), 206–212.

[36] Zhang, L.Z.; Tang, G.-Q., The binding properties of photosensitizer methylene blue to herring sperm DNA: a spectroscopic study. *Journal of Photochemistry and Photobiology B: Biology,* 2004, 74 (2–3), 119–125.

[37] Zhang, G.; Hu, X.; Pan, J., Spectroscopic studies of the interaction between pirimicarb and calf thymus DNA. *Spectrochimica Acta - Part A: Molecular and Biomolecular Spectroscopy,* 2011, 78 (2), 687–694.

[38] Gauglitz, G.; Vo-Dinh, T., *Handbook of Spectroscopy*. Wiley, 2006.

[39] Feigenbrugel, V.; Le Calvé, S.; Mirabel, P., Molar absorptivities of 2,4-D, cymoxanil, fenpropidin, isoproturon and pyrimethanil in aqueous solution in the near-UV. *Spectrochimica Acta - Part A: Molecular and Biomolecular Spectroscopy*, 2006, 63 (1), 103–110.

[40] Breuer, D. In *The MAK-Collection for Occupational Health and Safety;* Wiley-VCH Verlag GmbH & Co. KGaA, 2002.

[41] Pavia, D.L., *Introduction to spectroscopy*. Brooks/Cole, Cengage Learning, 2009.

[42] Sun, Y.; Ji, F.; Liu, R.; Lin, J.; Xu, Q.; Gao, C., Interaction mechanism of 2-aminobenzothiazole with herring sperm DNA. *Journal of Luminescence,* 2012, 132 (2), 507–512.

[43] Dolatabadi, J.E.N.; Kashanian, S., A review on DNA interaction with synthetic phenolic food additives. *Food Research International*, 2010, 43 (5), 1223–1230.

[44] Armitage, B. In *DNA Binders and Related Subjects*. Waring, M.; Chaires, J., Eds.; Springer Berlin Heidelberg, 2005; Vol. 253, pp. 55–76.

[45] Suh, D.; Chaires, J.B., Criteria for the mode of binding of DNA binding agents. *Bioorganic and Medicinal Chemistry*, 1995, 3 (6), 723–728.

[46] Kalfas, C.A.; Loukakis, G.K.; Georgakilas, A.G.; Sideris, E.G.; Anagnostopoulou-Konsta, A., Flexibility and thermal denaturation (melting) of irradiated DNA. *Journal of Biological Systems,* 1996, 4 (3), 405–423.

[47] Kashanian, S.; Askari, S.; Ahmadi, F.; Omidfar, K.; Ghobadi, S.; Tarighat, F.A., In vitro study of DNA interaction with clodinafop-propargyl herbicide. *DNA and Cell Biology*, 2008, 27 (10), 581–586.

[48] Kabir, A.; Hossain, M.; Kumar, G.S., Thermodynamics of the DNA binding of biogenic polyamines: Calorimetric and spectroscopic investigations. *Journal of Chemical Thermodynamics,* 2013, 57, 445–453.

[49] Lane, A.N.; Jenkins, T.C., Thermodynamics of nucleic acids and their interactions with ligands. *Quarterly Reviews of Biophysics,* 2000, 33 (3), 255–306.

[50] Haq, I.; Ladbury, J., Drug–DNA recognition: energetics and implications for design. *Journal of Molecular Recognition,* 2000, 13 (4), 188–197.

[51] Haq, I., Thermodynamics of drug–DNA interactions. *Archives of Biochemistry and Biophysics,* 2002, 403 (1), 1–15.

[52] Borea, P.A.; Dalpiaz, A.; Varani, K.; Gilli, P.; Gilli, G., Can thermodynamic measurements of receptor binding yield information on drug affinity and efficacy? *Biochemical Pharmacology,* 2000, 60 (11), 1549–1556.

[53] Ahmadi, F.; Bakhshandeh, F., In vitro study of damaging effects of 2,4-dichlorophenoxyacetic acid on DNA structure by spectroscopic and voltammetric techniques, *DNA and Cell Biology*, 2009, 28 (10), 527–533.

In: Handbook on Herbicides
Editors: Daiki Kobayashi and Eito Watanabe

ISBN: 978-1-62948-053-4

Chapter 13

SYNTHESIS, STRUCTURE AND HERBICIDAL ACTIVITY OF PYRAZOLE DERIVATIVES

Yuhan Zhou*[*], *Xiaoliang Dong and Jingping Qu
State Key Laboratory of Fine Chemicals, School of Pharmaceutical Science and Technology, Dalian University of Technology, P. R. China

ABSTRACT

Herbicides are widely used in crop production all over the world. Among them, protoporphyrinogen oxidase (Protox) inhibitors are one of the most important classes of herbicides, which inhibit Protox in the chlorophyll biosynthetic pathway, resulting in light-induced membrane lipid peroxidation. Targeting the porphyrin pathway, these herbicides show high activity and low toxicity, and thus have become a hot-point of novel pesticides research.

To search for novel Protox inhibitors, several pyrazole derivatives were synthesized from 1-(4-chloro-2-fluoro-5-methoxyphenyl)ethanone, *via* a condensation, ring closure reaction, methylation, chlorination, demethylation and alkylation. The single crystal of 4-chloro-3-[4-chloro-2-fluoro-5-(2-methyl)allyloxyphenyl]-1-methyl-5-trifluoromethyl-1*H*-pyrazol was prepared, and its structure was further determined by X-ray analysis. The preliminary bioassay showed some promising results to tested gramineous weeds and latifoliate weeds. The study on the structure-activity relationship suggests that chlorine substituent in 4-position of pyrazole is important for high activity. When hydrogen is located in 4-position of pyrazole, there is no herbicidal activity or low activity, but when chlorine is on pyrazole, most of them show good herbicidal efficiency.

Keywords: Pyrazole derivatives, herbicidal activity, protox-inhibitor, synthesis

[*] Corresponding author: Yuhan Zhou (Email: zhouyh@dl.cn).

INTRODUCTION

Protoporphyrinogen oxidase (Protox) is a key enzyme in the chlorophyll/heme biosynthetic pathway. This enzyme catalyzes the oxidation of protoporphyrinogen IX to protoporphyrin IX. It is the last common step in the production of heme and chlorophyll. While the production of chlorophyll, a light-harvesting pigment, is an essential process for all green photosynthetic organisms, heme is an essential cofactor in cytochromes, hemoglobin, oxygenases, peroxidases and catalases. When Protox is inhibited by the herbicides, protoporphyrin IX accumulates and causes light-dependent membrane damage. This characteristic makes Protox an excellent target for herbicide development. Targeting the porphyrin pathway, the herbicides inhibiting Protox show high activity and low toxicity, and become a hot-point of novel pesticides research [1-10].

Diphenyl ether (DPE) based herbicides are the first and widely used family of Protox inhibitors, and Nitrofen (Figure 1) is the leading compound of this type. During the period covering the years 1970-1980, a great deal of work had been done on several areas of Protox herbicides, though at that time the exact mode of action was not known. Later on, during the 1980s further work was done at the FMC Corporation in the area of Protox herbicides, which resulted in the discovery of several new classes of Protox inhibitors, including two new aryl triazolinone herbicides, Sulfentrazone [11] and Carfentrazone-ethyl [12]. These two herbicides were commercialized in the late 1990s. Among these Protox inhibiting herbicides, substituted phenyl 5-membered heterocycles were developed practically and the compounds of this class are under development. Substituted phenyl heterocycle compounds are thought to be potent Protox-inhibitors, because they are similar to one half of the protoporphyrinogen IX, which is the target of Protox. Many researchers have studied on these compounds, and a large number of compounds having high bioactivity were reported [13-20]. Earlier work at our laboratory involved some isoxazole and benzoxazole derivatives and tricyclic compounds with high herbicidal activity [21-24]. Herein, we describe the synthesis and herbicidal activity of some phenyl pyrazole derivatives.

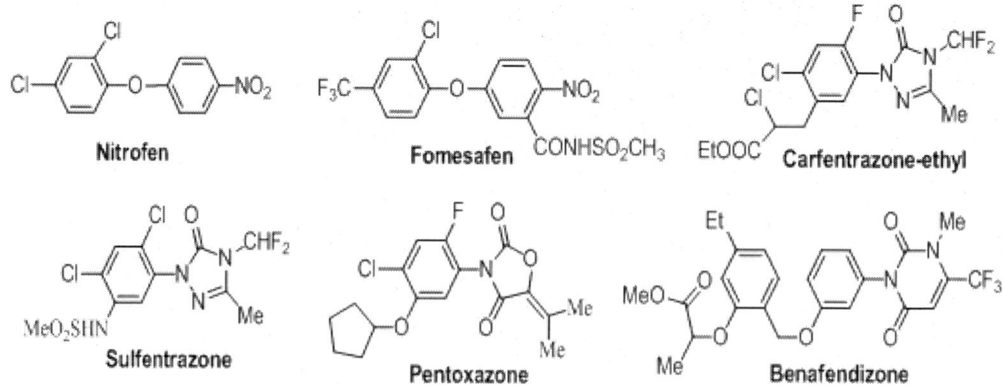

Figure 1. Samples of commercial Protox inhibitors.

Results and Discussion

The title compounds were obtained following the sequences described in scheme 1. At first, 1-(4-chloro-2-fluoro-5-methoxyphenyl)ethanone (**1**) condensed with ethyl trifluoroacetate in sodium methoxide/methanol solvent to form 1-(4-chloro-2-fluoro-5-methoxyphenyl)-4',4',4'-trifluoro-1',3'-butandione (**2**). Then, compound **2** cyclized with aqueous hydrazine and gave a substituted phenyl pyrazole derivative (**3**), from which compound **4** was obtained through an alkylation reaction. The chlorine atom was introduced to the 4-position of the pyrazole cycle by the *N*-chlorosuccimide (NCS), and the methoxyl group in the 5-position of the phenyl can be change to other alkyl groups through a dealkylation followed by an alkylation with different alkylating reagents.

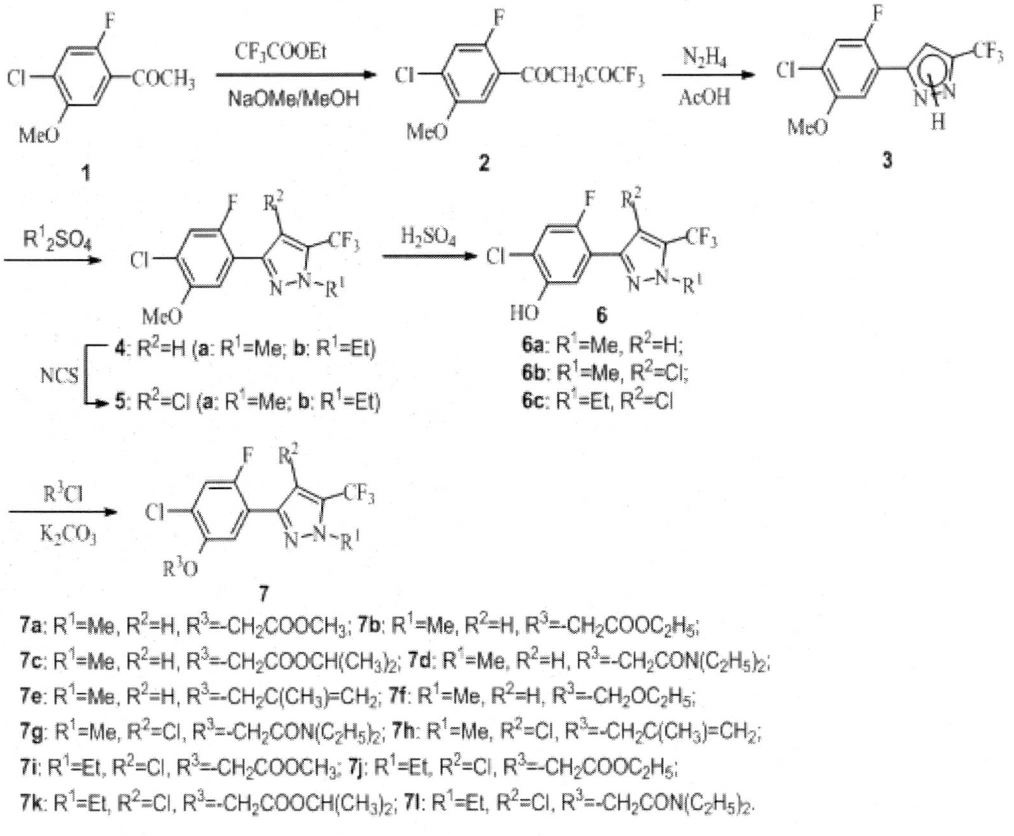

7a: R^1=Me, R^2=H, R^3=-CH_2COOCH_3; **7b**: R^1=Me, R^2=H, R^3=-$CH_2COOC_2H_5$;
7c: R^1=Me, R^2=H, R^3=-$CH_2COOCH(CH_3)_2$; **7d**: R^1=Me, R^2=H, R^3=-$CH_2CON(C_2H_5)_2$;
7e: R^1=Me, R^2=H, R^3=-$CH_2C(CH_3)$=CH_2; **7f**: R^1=Me, R^2=H, R^3=-$CH_2OC_2H_5$;
7g: R^1=Me, R^2=Cl, R^3=-$CH_2CON(C_2H_5)_2$; **7h**: R^1=Me, R^2=Cl, R^3=-$CH_2C(CH_3)$=CH_2;
7i: R^1=Et, R^2=Cl, R^3=-CH_2COOCH_3; **7j**: R^1=Et, R^2=Cl, R^3=-$CH_2COOC_2H_5$;
7k: R^1=Et, R^2=Cl, R^3=-$CH_2COOCH(CH_3)_2$; **7l**: R^1=Et, R^2=Cl, R^3=-$CH_2CON(C_2H_5)_2$.

Scheme 1. Synthesis of pyrazole derivatives.

The herbicidal activities of the title compounds (**7**) were assayed. Their postemergence and pre-emergence herbicidal activity to different weeds at a dose of 15 g/hm^2, 150 g/hm^2 and 1500 g/hm^2, contrasting with Fomesafen and Pentoxazone, which were widely used as high-activity herbicides, were shown in Table 1.

Table 1. Herbicidal activity of title compounds[1]

Comp.	Dose/g (a.i.)/hm^2	*Echinochloa crusgalli*		*Setaria viridis*		*Artemisia mongolica*		*Acalypha australis*	
		Post[2]	Pre-[3]	Post	Pre-	Post	Pre-	Post	Pre-
7a	15	2	0	0	0	0	0	1	2
	150	4	0	1	1	1	1	2	4
	1500	5	0	3	2	4	2	4	6
7b	15	3	0	0	0	1	0	2	3
	150	5	0	1	1	2	0	3	6
	1500	6	1	2	1	4	0	4	7
7c	15	4	0	0	0	1	0	1	3
	150	5	0	0	1	1	1	2	5
	1500	6	1	1	3	3	1	4	6
7d	15	2	0	0	0	1	0	3	6
	150	5	0	1	0	3	0	3	6
	1500	6	0	1	1	5	0	6	7
7e	15	5	0	1	0	5	0	1	5
	150	6	0	1	1	6	0	2	8
	1500	7	1	3	1	6	0	5	8
7f	15	2	0	1	1	3	0	3	6
	150	4	1	6	5	3	1	4	8
	1500	7	2	8	6	7	1	6	8
7g	15	5	2	7	3	6	4	10	8
	150	6	4	8	10	8	5	10	10
	1500	9	8	9	10	8	7	10	10
7h	15	6	4	7	10	7	6	10	10
	150	9	6	9	10	8	10	10	10
	1500	10	10	9	10	8	10	10	10
7i	15	7	4	4	1	8	4	10	3
	150	8	5	7	1	9	4	10	10
	1500	8	8	8	1	9	9	10	10
7j	15	6	5	5	1	9	2	10	4
	150	9	5	5	2	9	4	10	10
	1500	9	7	9	3	9	10	10	10
7k	15	6	6	6	1	8	5	10	4
	150	7	6	6	1	9	6	10	10
	1500	8	6	9	2	9	7	10	10
7l	15	6	3	7	1	8	6	10	10
	150	8	5	9	5	9	7	10	10
	1500	10	7	9	8	9	7	10	10
Fomesafen	15	4	0	4	1	5	9	6	10
	150	5	0	4	2	7	10	6	10
	1500	9	7	9	10	8	10	10	10
Pentoxazone	15	8	0	5	0	6	0	5	6
	150	8	0	6	9	7	0	9	8
	1500	9	6	10	10	7	6	9	10

[1] 0 = no acitivity, 10 = complete death of weeds; [2] Postemergence herbicidal activity; [3] Pre-emergence herbicidal activity.

From Table 1 we can conclude that some of them have good herbicidal activity. The methyl group or the ethyl group substitution in the 1-position of the pyrazole cycle has little difference in the activity (**7g~7l**). **7g~7l** have good activity to various weeds in both postemergence and pre-emergence test. Especially, **7g~7l** show excellent postemergence herbicidal activity to *Acalypha australis* at a dose of 15 g/hm^2, which are much higher than Fomesafen and Pentoxazone. In general, the postemergence herbicidal activity of **7g~7l** is a little higher than their pre-emergence herbicidal activity. The chlorine atom in the 4-position of the pyrazole cycle is necessary for high activity. When no substituent in 4-position of pyrazole, there were no herbicidal activity or low activity (**7a-7f**). However, when chlorine is in pyrazole, most of them showed good herbicidal efficiency, that indicated halogen substituent in the pyrazole moiety was very important for high biological activity. Similar results have been reported by Meazza [25]. Their study on pyrroles showed that the halopyrrole nucleus was important for biological activity; if there was only hydrogen in pyrrole the compounds were virtually inactive.

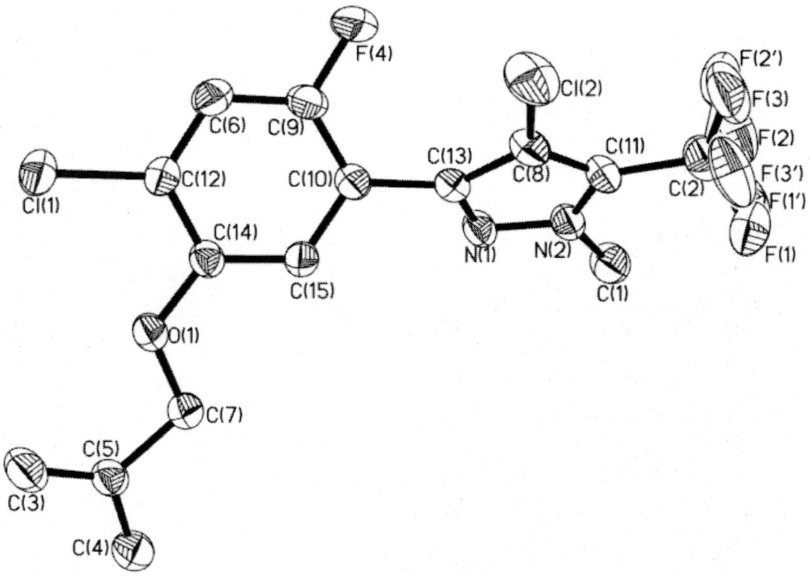

Figure 2. ORTEP (ellipsoids at 30% probability) diagram of 7h (Selected bond distances (Å): C(10)-C(13) 1.482(4), C(9)-F(4) 1.350(3), C(12)-Cl(1) 1.730(3), C(14)-O(1) 1.354(3), C(7)-O(1) 1.413(4), C(5)-C(7) 1.495(4), C(8)-Cl(2) 1.712(3), C(2)-C(11) 1.476(5), C(1)-N(2) 1.466(4); angles (°): C(8)-C(13)-C(10) 129.2(3), C(9)-C(10)-C(13) 122.1(3), O(1)-C(14)-C(12) 115.5(2), O(1)-C(7)-C(5) 109.3(2), C(9)-C(10)-C(13)-C(8) 52.35.

It was noted that the ortho-position (to the phenyl ring) substituent in the heterocyclic moiety was necessary for high activity. The introduction of halogen can force the heterocyclic moiety out of planarity with the attached phenyl ring and to match closely the angle of the methylene bridge between two pyrrole rings of the protoporphyrinogen structure [26]. To further explore the structure-activity relationship of the phenyl pyrazole derivatives, a single crystal of compound **7h** was prepared, and its structure was determined. The structure, selected bond distances (Å), angles (°) and torsion angles (°) of **7h** were shown in Fig 2. The dihedral angle of the phenyl ring and the pyrazole ring (shown as C(9)-C(10)-C(13)-C(8)) is

52.35 °, which is similar with the angle of the methylene bridge between two pyrrole rings of the protoporphyrinogen structure (60 °) [26].

CONCLUSION

Substituted phenyl pyrazole derivatives can be prepared from 1-(4-chloro-2-fluoro-5-methoxyphenyl)ethanone. Preliminary bioassay shows that most of them exhibit good herbicidal activity toward various weeds, and the compounds with a chlorine atom in 4-position of their pyrazole ring have much better activity than those compounds with a hydrogen atom in that position. The methyl group or the ethyl groups in 1-position of the pyrazole cycle has little effect on the activity.

EXPERIMENTAL

General

^{1}H NMR spectra were measured on a Varian VA400MHz spectrometer with TMS as an internal standard. ^{13}C NMR spectra were obtained with broadband proton decoupling. CIMS were performed on a HP1100 High Performance Liquid Chromatography/Mass Selective Detector. Melting points were determined using a YanacoMP-500 apparatus and were uncorrected. IR spectra were run on Nicolet 20DBX FT-IR. Elemental analysis was measured on a Vario EL III (Elementar, German) elemental analysis instrumentation.

1-(4-chloro-2-fluoro-5-methoxyphenyl)-4',4',4'-trifluoro-1',3'-butandione (2)

To a solution of 4-chloro-2-fluoro-5-methoxyacetophenone (**1**) (13 g, 0.064 mol) in methanol (60 mL) was added 25% sodium methoxide in methanol (26 mL), and then ethyl trifluoroacetate (13.5 g, 0.095 mol) was added. After refluxed for 1 h, the reaction mixture was cooled and poured into ice water, then adjusted to acidic using aqueous hydrochloric acid, filtered, washed with water, dried to give a yellow solid, 18.6 g (97%). mp 122.0~123.0 °C ; ^{1}H NMR (400 MHz, $CDCl_3$), δ: 3.96 (s, 3H, CH_3), 6.73 (s, 1H, =CH-), 7.27 (d, 1H, J = 10.8 Hz, Ph-3-H), 7.50 (d, 1H, J = 6.4 Hz, Ph-6-H); CIMS (API-ES, negative), *m/z*: 297 ($[M-H]^-$).

3(5)-(4-chloro-2-fluoro-5-methoxyphenyl)-5(3)-trifluoromethylpyrazole (3)

To a solution of **2** (2 g, 6.7 mmol) in acetic acid (20 mL), 50% aqueous hydrazine (1 mL, 10 mmol) was added. The solution was heated to 110 °C and maintained at that temperature for 1 h. The reaction mixture was then cooled to room temperature and poured into water, resulting in a white solid precipitate. The precipitated product was isolated by filtering, and then dried to afford a white solid, 2 g (100%). mp 194.0~198.0 °C; ^{1}H NMR (400 MHz, $CDCl_3$), δ: 3.94 (s, 3H, OCH_3), 6.90 (s, 1H, Pyr-H), 7.22 (d, 1H, J = 10.0 Hz, Ph-3-H), 7.41 (d, 1H, J = 6.0 Hz, Ph-6-H).

General Procedure for the Preparation of 4

To a solution of **3** (12 g, 0.04 mol) in toluene (120 mL), was added dialkylsulfate (0.05 mol). Then the solution was heated to reflux for 2~6 h. After the reaction was completed (monitor by TLC), the solution was cooled, washed with 2 mol/L aqueous sodium hydroxide (20 mL) and water (20 mL). While toluene was stripped from the solution, the product was precipitated, recrystallized from alcohol to give a white solid.

3-(4-chloro-2-fluoro-5-methoxyphenyl)-1-methyl-5-trifluoromethyl-1H-pyrazole (4a)

This compound was obtained as a white needles, yield 91%, mp 115.5~116.0 °C ; ^{1}H NMR (400 MHz, $CDCl_3$), δ: 3.94 (s, 3H, OCH_3), 4.04 (s, 3H, Pyr-CH_3), 7.02 (d, 1H, J = 3.0 Hz, PyrH), 7.17 (d, 1H, J = 8.8 Hz, Ph-3-H), 7.55 (d, 1H, J = 5.8 Hz, Ph-6-H) ; ^{13}C NMR (100 MHz, $CDCl_3$), δ: 38.4 (s, Pyr-CH_3), 56.9 (s, OCH_3), 108.1 (d, J = 12.3 Hz, Pyr-4'-C), 112.2 (s, Ph-6-C), 118.2 (d, J = 27.4 Hz, Ph-3-C), 118.9 (d, J = 13.0 Hz, Ph-1-C), 120.0 (q, J = 268.5 Hz, CF_3), 122.7 (d, J = 10.7 Hz, Ph-4-C), 133.2 (q, J = 39.7 Hz, Pyr-5'-C), 144.2 (s, Pyr-3'-C), 151.8 (s, Ph-5-C), 153.6 (d, J = 245.7 Hz, Ph-2-C); CIMS (API-ES, positive), *m/z*: 309 ($[M+H]^+$), 331 ($[M+Na]^+$); (negative) *m/z*: 307 ($[M-H]^-$).

3-(4-chloro-2-fluoro-5-methoxyphenyl)-1-ethyl-5-trifluoromethyl-1H-pyrazole (4b)

This compound was obtained as a white needles, yield 89%, mp 74.0~75.0 °C; CIMS (API-ES, positive), *m/z*: 323 ($[M+H]^+$), 345 ($[M+Na]^+$).

General Procedure for the Preparation of 5

To a solution of **4** (0.051 mol) in *N,N*-dimethylformamide (125 mL), was added *N*-chlorosuccinimide (7 g, 0.053 mol). The solution was heated to 80 °C for 4 h, allowed to cool and poured into ice water. The aqueous mixture was extracted twice with ethyl acetate (30 mL), the combined organic extracts washed with water, dried with magnesium sulfate, concentrated and recrystallized from alcohol to give a white solid as product.

4-chloro-3-(4-chloro-2-fluoro-5-methoxyphenyl)-1-methyl-5-trifluoromethyl-1H-pyrazole (5a)

This compound was obtained as a white needles, yield 83%, mp 72.0~72.5 °C; ^{1}H NMR (400 MHz, $CDCl_3$), δ: 3.91 (s, 3H, OCH_3), 4.07 (s, 3H, PyrCH_3), 7.03 (d, 1H, J = 6.4 Hz, Ph-6-H), 7.26 (d, 1H, J = 9.2 Hz, Ph-3-H); CIMS (API-ES, positive), *m/z*: 343 ($[M+H]^+$).

4-chloro-3-(4-chloro-2-fluoro-5-methoxyphenyl)-1-ethyl-5-trifluoromethyl-1H-pyrazole (5b)

This compound was obtained as brown oil, and used directly without any purification. CIMS (API-ES, positive), *m/z*: 357 ($[M+H]^+$).

General Procedure for the Preparation of 6

A solution of **4** or **5** (0.039 mol) in sulfate acid (120 mL) was heated to 100 °C for 3 h, allowed to cool and poured into ice water. The aqueous mixture was extracted twice with ethyl acetate (30 mL), the combined organic extracts washed with 1 mol/L aqueous sodium hydroxide (20 mL) and water, dried with magnesium sulfate, concentrated to afford a hoar solid.

3-(4-chloro-2-fluoro-5-hydroxyphenyl)-1-methyl-5-trifluoromethyl-1H-pyrazole (6a)

This compound was obtained as white needles (recrystallized from alcohol), yield 93%, ^{1}H NMR (400 MHz, $CDCl_3$), δ: 4.02 (s, 3H, NCH_3), 6.98 (d, 1H, J = 3.2 Hz, PyrH), 7.11 (d, 1H, J = 10.0 Hz, Ph-3-H), 7.58 (d, 1H, J = 6.8 Hz, Ph-6-H), 8.1 (br, 1H, OH).

4-chloro-3-(4-chloro-2-fluoro-5-hydroxyphenyl)-1-methyl-5-trifluoromethyl-1H-pyrazole (6b)

This compound was obtained as white needles (recrystallized from alcohol), yield 93%.

4-chloro-3-(4-chloro-2-fluoro-5-hydroxyphenyl)-1-ethyl-5-trifluoromethyl-1H-pyrazole (6c)

This compound was obtained as white needles (recrystallized from alcohol), the united yield of chlorination and demethylation was 88%.

General Procedure for the Preparation of 7

To a slurry of **6** (5.1 mmol), anhydrous potassium carbonate (1.5 g, 10.8 mmol) in acetone (15 mL), was added alkylating reagent (10 mmol). The mixture was refluxed for 3~10 h. After the reaction was completed (monitor by TLC), the mixture was allowed to cool and poured into water, filtered, washed, dried to give a white solid.

Methyl 2'-[2-chloro-4-fluoro-5-(1-methyl-5-trifluoromethyl-1H-pyrazole-3-yl) phenoxy]acetate (7a)

This compound was obtained as white solid (alcohol), yield 90%; mp 172.5~173.5 °C; IR (KBr) ν (cm^{-1}): 1769 (C=O); ^{1}H NMR (400 MHz, $CDCl_3$), δ: 3.82 (s, 3H, OCH_3), 4.03 (s, 3H, NCH_3), 4.75 (s, 2H, OCH_2COO), 7.01 (d, 1H, J = 2.4 Hz, Pyr-H), 7.21 (d, 1H, J = 10.4 Hz), Ph-3-H), 7.54 (d, 1H, J = 6.4 Hz, Ph-6-H); MS (API-ES, positive) *m/z*: 367 ($[M+H]^+$), 405 ($[M+K]^+$).

Anal

Calcd. for $C_{14}H_{11}ClF_4N_2O_3$: C, 45.86; H, 3.02; N, 7.64. Found: C, 45.68; H, 2.96; N, 7.49.

Ethyl 2'-[2-chloro-4-fluoro-5-(1-methyl-5-trifluoromethyl-1H-pyrazole-3-yl)phenoxy] acetate (7b)

This compound was obtained as white solid (alcohol), yield 89%; mp 128.5~129.5 ºC; IR (KBr) ν (cm^{-1}): 1765 (C=O); ^{1}H NMR (400 MHz, $CDCl_3$), δ: 1.31 (t, 3H, *J* = 7.2 Hz, $COOCH_2CH_3$), 4.03 (s, 3H, NCH_3), 4.31 (q, 2H, *J* = 7.2 Hz, $COOCH_2CH_3$), 4.74 (s, 2H, OCH_2COO), 7.01 (d, 1H, *J* = 3.2 Hz, Pyr-H), 7.20 (d, 1H, *J* = 10.4 Hz, Ph-3-H), 7.54 (d, 1H, *J* = 6.4 Hz, Ph-6-H); MS (APCI, positive) *m/z*: 381 ($[M+H]^+$).

Anal

Calcd. for $C_{15}H_{13}ClF_4N_2O_3$: C, 47.32; H, 3.44; N, 7.36. Found: C, 47.26; H, 3.37; N, 7.27.

Isopropyl 2'-[2-chloro-4-fluoro-5-(1-methyl-5-trifluoromethyl-1H-pyrazole-3-yl) phenoxy]acetate (7c)

This compound was obtained as white solid (alcohol), yield 93%; mp 130.0~130.5 ºC; IR (KBr) ν (cm^{-1}): 1759 (C=O); ^{1}H NMR (400 MHz, $CDCl_3$), δ: 1.29 (d, 6H, *J* = 6.4 Hz, $CH(CH_3)_2$), 4.03 (s, 3H, NCH_3), 4.72 (s, 2H, OCH_2COO), 5.17 (sept, 1H, *J* = 6.4 Hz, $CH(CH_3)_2$), 7.02 (d, 1H, *J* = 3.6 Hz, Pyr-H), 7.21 (d, 1H, *J* = 10.4 Hz, Ph-3-H), 7.52 (d, 1H, *J* = 6.8 Hz, Ph-6-H); MS (APCI, positive) *m/z*: 395 ($[M+H]^+$); 417 ($[M+Na]^+$).

Anal

Calcd. for $C_{16}H_{15}ClF_4N_2O_3$: C, 48.68; H, 3.83; N, 7.10. Found: C, 48.89; H, 3.89; N, 7.03.

N,N-diethyl 2'-[2-chloro-4-fluoro-5-(1-methyl-5-trifluoromethyl-1H-pyrazole-3-yl) phenoxy]acetamide (7d)

This compound was obtained as white solid (alcohol), yield 91%; mp 86.0~86.5 ºC; IR (KBr) ν (cm^{-1}): 1652 (C=O); ^{1}H NMR (400 MHz, $CDCl_3$), δ: 1.15 (t, 3H, *J* = 6.8 Hz, NCH_2CH_3), 1.26 (t, 3H, *J* = 6.8 Hz, NCH_2CH_3), 3.2~3.6 (m, 4H, $CONCH_2$), 4.03 (s, 3H, NCH_3), 4.80 (s, 2H, OCH_2CO), 7.01 (d, 1H, *J* = 2.8 Hz, Pyr-H), 7.20 (d, 1H, *J* = 10.0 Hz, Ph-3-H), 7.61 (d, 1H, *J* = 6.4 Hz, Ph-6-H); MS (APCI, positive) *m/z*: 408 ($[M+H]^+$).

Anal

Calcd. for $C_{17}H_{18}ClF_4N_3O_2$: C, 50.07; H, 4.45; N, 10.30. Found: C, 50.18; H, 4.55; N, 10.36.

3-[4-chloro-2-fluoro-5-(2'-methyl)allyloxyphenyl]-1-methyl-5-trifluoromethyl-1H-pyrazole (7e)

This compound was obtained as white solid (alcohol), yield 86%; mp 61.0~61.5 ºC; IR (KBr) ν (cm^{-1}): 1658 (C=C); ^{1}H NMR (400 MHz, $CDCl_3$), δ: 1.87 (s, 3H, CH_3), 4.04 (s, 3H, NCH_3), 4.55 (s, 2H, OCH_2), 5.03 (s, 1H, =CH), 5.18 (s, 1H, =CH), 7.02 (d, 1H, *J* = 3.6 Hz, Pyr-H), 7.19 (d, 1H, *J* = 10.4 Hz, Ph-3-H), 7.55 (d, 1H, *J* = 6.4 Hz, Ph-6-H); MS (APCI, positive) *m/z*: 349 ($[M+H]^+$).

Anal

Calcd. for $C_{15}H_{13}ClF_4N_2O$: C, 51.66; H, 3.76; N, 8.03. Found: C, 51.43; H, 3.85; N, 7.95.

3-[4-chloro-2-fluoro-5-(ethoxyl)methoxyphenyl]-1-methyl-5-trifluoromethyl-1H-pyrazole (7f)

This compound was obtained as white solid (alcohol), yield 83%; mp 71.0~72.5 °C; ^{1}H NMR (400 MHz, $CDCl_3$), δ: 1.26 (t, 3H, J = 6.8 Hz, OCH_2CH_3), 3.82 (q, 2H, J = 6.8 Hz, OCH_2CH_3), 4.04 (s, 3H, NCH_3), 5.33 (s, 2H, OCH_2O), 7.01 (d, 1H, J = 3.2 Hz, Pyr-H), 7.19 (d, 1H, J = 10.4 Hz, Ph-3-H), 7.79 (d, 1H, J = 7.2 Hz, Ph-6-H); MS (APCI, positive) *m/z*: 353 ($[M+H]^+$).

Anal

Calcd. for $C_{14}H_{13}ClF_4N_2O_2$: C, 47.67; H, 3.71; N, 7.94. Found: C, 47.95; H, 3.82; N, 7.73.

N,N-diethyl 2'-[2-chloro-5-(4-chloro-1-methyl-5-trifluoromethyl-1H-pyrazole-3-yl)-4-fluorophenoxy]acetamide (7g)

This compound was obtained as white solid (alcohol), yield 89%; mp 117.5~118.5 °C; IR (KBr) ν (cm^{-1}): 1667 (C=O); ^{1}H NMR (400 MHz, $CDCl_3$), δ: 1.13 (t, 3H, J = 6.8 Hz, NCH_2CH_3), 1.22 (t, 3H, J = 6.8 Hz, NCH_2CH_3), 3.4 (m, 4H, $CONCH_2$), 4.05 (s, 3H, NCH_3), 4.77 (s, 2H, OCH_2CO), 7.16 (d, 1H, J = 4.4 Hz, Ph-6-H), 7.26 (d, 1H, J = 7.6 Hz, Ph-3-H); MS (API-ES, positive) *m/z*: 442 ($[M+H]^+$), 464 ($[M+Na]^+$).

Anal

Calcd. for $C_{17}H_{17}Cl_2F_4N_3O_2$: C, 46.17; H, 3.87; N, 9.50. Found: C, 45.98; H, 3.94; N, 9.61.

4-chloro-3-[4-chloro-2-fluoro-5-(2'-methyl)allyloxyphenyl]-1-methyl-5-trifluoromethyl-1H-pyrazole (7h)

This compound was obtained as white solid (alcohol), yield 95%; mp 58.0~58.5 °C; IR (KBr) ν (cm^{-1}): 1663 (C=C); ^{1}H NMR (400 MHz, $CDCl_3$), δ: 1.85 (s, 3H, CH_3), 4.06 (s, 3H, NCH_3), 4.50 (s, 2H, OCH_2), 5.02 (s, 1H, =CH), 5.15 (s, 1H, =CH), 7.04 (d, 1H, J = 6.0 Hz, Ph-6-H), 7.25 (d, 1H, J = 9.2 Hz, Ph-3-H); MS (APCI, positive) *m/z*: 383 ($[M+H]^+$).

Anal

Calcd. for $C_{15}H_{12}Cl_2F_4N_2O$: C, 47.02; H, 3.16; N, 7.31. Found: C, 47.21; H, 3.23; N, 7.25.

Methyl 2'-[2-chloro-5-(4-chloro-1-ethyl-5-trifluoromethyl-1H-pyrazole-3-yl)-4-fluoro phenoxy]acetate (7i)

This compound was obtained as white solid (alcohol), yield 83%; mp 92.0~93.0 °C; IR (KBr) ν (cm^{-1}): 1749 (C=O); ^{1}H NMR (400 MHz, $CDCl_3$), δ: 1.52 (t, 3H, J = 7.2 Hz, NCH_2CH_3), 3.80 (s, 3H, OCH_3), 4.33 (q, 2H, J = 7.2 Hz, NCH_2CH_3), 4.72 (s, 2H, OCH_2COO), 7.04 (d, 1H, J = 4.4 Hz, Ph-6-H), 7.27 (d, 1H, J = 8.8 Hz, Ph-3-H); MS (API-ES, positive) *m/z*: 415 ($[M+H]^+$), 437 ($[M+Na]^+$).

Anal

Calcd. for $C_{15}H_{12}Cl_2F_4N_2O_3$: C, 43.40; H, 2.91; N, 6.75. Found: C, 43.69; H, 3.02; N, 6.86.

Ethyl 2'-[2-chloro-5-(4-chloro-1-ethyl-5-trifluoromethyl-1H-pyrazole-3-yl)-4-fluoro phenoxy]acetate (7j)

This compound was obtained as white solid (alcohol), yield 91%; mp 75.0~77.0 °C; IR (KBr) ν (cm^{-1}): 1746 (C=O); ^{1}H NMR (400 MHz, $CDCl_3$), δ: 1.29 (t, 3H, J = 7.2 Hz, CH_2CH_3), 1.52 (t, 3H, J = 7.2 Hz, CH_2CH_3), 4.27 (q, 2H, J = 7.2 Hz, CH_2CH_3), 4.35 (q, 2H, J = 7.2 Hz, CH_2CH_3), 4.71 (s, 2H, OCH_2COO), 7.04 (d, 1H, J = 6.4 Hz, Ph-6-H), 7.28 (d, 1H, J = 9.2 Hz, Ph-3-H); MS (API-ES, positive) m/z: 429 ($[M+H]^+$), 451 ($[M+Na]^+$).

Anal

Calcd. for $C_{16}H_{14}Cl_2F_4N_2O_3$: C, 44.78; H, 3.29; N, 6.53. Found: C, 44.49; H, 3.43; N, 6.44.

Isopropyl 2'-[2-chloro-5-(4-chloro-1-ethyl-5-trifluoromethyl-1H-pyrazole-3-yl)-4-fluoro phenoxy]acetate (7k)

This compound was obtained as white solid (alcohol), yield 96%; mp 99.5~101.0 °C; IR (KBr) ν (cm^{-1}): 1744 (C=O); ^{1}H NMR (400 MHz, $CDCl_3$), δ: 1.27 (d, 6H, J = 6.4 Hz, $CH(CH_3)_2$), 1.52 (t, 3H, J = 6.8 Hz, NCH_2CH_3), 4.34 (q, 2H, J = 6.8 Hz, NCH_2CH_3), 4.68 (s, 2H, OCH_2COO), 5.14 (sept, 1H, J = 6.4 Hz, OCH), 7.03 (d, 1H, J = 6.0 Hz, Ph-6-H), 7.28 (d, 1H, J = 8.8 Hz, Ph-3-H); MS (APCI, positive) m/z: 443 ($[M+H]^+$), 465 ($[M+Na]^+$).

Anal

Calcd. for $C_{17}H_{16}Cl_2F_4N_3O_2$: C, 46.07; H, 3.64; N, 6.32. Found: C, 46.41; H, 3.46; N, 6.25.

N,N-diethyl 2'-[2-chloro-5-(4-chloro-1-ethyl-5-trifluoromethyl-1H-pyrazole-3-yl)-4-fluorophenoxy]acetamide (7l)

This compound was obtained as white solid (alcohol), yield 81%; mp 73.0~75.0 °C; IR (KBr) ν (cm^{-1}): 1664 (C=O); ^{1}H NMR (400 MHz, $CDCl_3$), δ: 1.14 (t, 3H, J = 7.2 Hz, $CONCH_2CH_3$), 1.23 (t, 3H, J = 7.2 Hz, $CONCH_2CH_3$), 1.51 (t, 3H, J = 7.2 Hz, Pyr-NCH_2CH_3), 3.3~3.6 (m, 4H, $CONCH_2$), 4.35 (q, 2H, J = 7.2 Hz, Pyr-NCH_2), 4.77 (s, 2H, OCH_2CON), 7.16 (d, 1H, J = 6.0 Hz, Ph-6-H), 7.26 (d, 1H, J = 8.8 Hz, Ph-3-H); MS (APCI, positive) m/z: 456 ($[M+H]^+$), 478 ($[M+Na]^+$).

Anal

Calcd. for $C_{18}H_{19}Cl_2F_4N_2O_3$: C, 47.38; H, 4.20; N, 9.21. Found: C, 47.26; H, 4.32; N, 9.34.

Crystal Structure Determination

Saturated solution of **7h** in EtOAc was covered with *n*-hexane, and stand in air at room temperature to give single crystals. The data were obtained on a Bruker SMART APEX CCD diffractometer with graphite monochromated Mo Kα radiation (λ = 0.71073 Å). Empirical absorption corrections were performed using the SADABS program [27]. Structures were solved by direct methods and refined by full-matrix least-squares based on all data using F^2 using shelx97 [28]. All of the non-hydrogen atoms were refined anisotropic ally. All of the hydrogen atoms were generated and refined in ideal positions. Crystallographic parameters of **7h**: empirical formula: $C_{15}H_{12}Cl_2F_4N_2O$; formula weight: 382.03; crystal system: Monoclinic; space group: *P*2(1)/*c*; a = 13.9177 (7) Å, b = 14.0499 (7) Å, c = 8.4389 (5) Å; β = 93.326 (3) deg; V = 1647.38 (17) $Å^3$; Z = 4; Dcalcd. = 1.549 g/cm^3; T = 273(2) K; μ = 0.71073 mm^{-1}; $F(000)$ = 780; θ: 2.06 to 25.32 deg; limiting indices: $-16 \leq h \leq 16$, $-16 \leq k \leq 16$, $-10 \leq l \leq 9$; Reflns collected/unique: 11231/2984 [*R(int)* = 0.0407]; $GOF(F^2)$: 1.000; R_1/wR_2 [I>2sigma(I)] = 0.0450/0.1183, R_1/wR_2 (all data) = 0.0776/0.1448; Largest diff. peak and hole: 0.291 and -0.461 e.A^{-3}.

Post-Emergence Herbicidal Activity Test

Compounds were formulated as 22 g/L emulsible concentrates, which were diluted with water to the required concentration and applied to pot-grown plants in a greenhouse.

Seeds of assayed weeds (*Echinochloa crusgalli*, *Setaria viridis*, *Abutilon theophrasti* and *Acalypha australis*) were germinated in water at 30 °C under dark conditions for 48 h. The germinated seeds were placed in a pot (0.1 m^2) as 10 seeds per-pot. While *Echinochloa crusgalli* was in the third-leaf stage, *Setaria viridis* was in the second-leaf stage, *Abutilon theophrasti* and *Acalypha australis* had two or three leaves, the diluted formulation was applied. Fifteen days after treatment, the upper-soil parts of the plants were cut off, and their weights were measured freshly. The degree of weeds control by the test compounds were scored as 0 - 10, 0 means no activity while 10 means complete death of weeds.

Each test was repeated three times.

Pre-Emergence Herbicidal Activity Test

The test solutions of the compounds and the seeds of assayed weeds were prepared as above. The diluted formulation test solutions were applied for pre-emergence treatment 24 h after weeds were sown. Twenty days after treatment, the degree of weeds control by the test compounds were scored as above.

ACKNOWLEDGMENTS

This work was supported financially by the National Natural Science Foundation of China (No. 20606005), the Doctor Foundation of Liaoning Province, China (No. 20031068).

REFERENCES

[1] Matringe, M; Camadro, JM; Labbe, P; Scalla, R. Protoporphyrinogen oxidase as a molecular target for diphenyl ether herbicides. *Biochem J* 1989, *260*, 231–235.

[2] Camadro, JM; Matringe, M; Thome, F; Brouillet, N; Mornet, R; Labbe, P. Photoaffinity labeling of protoporphyrinogen oxidase, themolecular target of diphenylether-type herbicides. *Eur J Biochem* 1995, *229*, 669–674.

[3] Smith, AG; Marsh, O; Elder, GH. Investigation of the subcellular location of the tetrapyrrole-biosynthesis enzyme coproporphyrinogen oxidase in higher plants. *Biochem J* 1993, *292*, 503–508.

[4] Hess, FD. Light-dependent herbicides: an overview. *Weed Sci* 2000, *48*, 160-170.

[5] Hao, GF; Zuo, Y; Yang, SG; Yang, GF. Protoporphyrinogen oxidase inhibitor: an ideal target for herbicide discovery. *Chimia* 2011, *65*, 961-969.

[6] Gressel, J; Levy, AA. Agriculture: the selector of improbable mutations. *P Natl Acad of Sci USA* 2006, *103*, 12215-12216.

[7] Shaner, DL. Herbicide safety relative to common targets in plants and mammals. *Pest Manag Sci* 2004, *60*, 17-24.

[8] Jacobs, JM; Jacobs, NJ; Sherman, TD; Duke, SO. Effect of diphenyl ether herbicide on oxidation of protoporphyrinogen to protoporphyrin in organellar and plasma membrane enriched fractions of barley. *Plant Physiol* 1991, *97*, 197–203.

[9] Zhou, YH; Miao, WR. Cheng, LB. Wang, DX; Bai, ZS. Progress on protoporphyrinogen oxidase-inhibiting herbicides. *Nongyaoxue xuebao* 2002, *4*, 1-8 (in Chinese).

[10] Li, X; Nicholl, D. Development of PPO inhibitor-resistant cultures and crops. *Pest Manag Sci* 2005, *61*, 277-285.

[11] Theodoridis, G. Preparation of herbicidal (sulfonylaminophenyl)triazolinones. WO 8703782 (1987).

[12] Poss, KM. Preparation of 1-[(carboxyalkyl)phenyl]triazolin-5-ones and analogs as herbicides. WO 9002120 (1990).

[13] Menges, M; Hamprecht, G; Menke, O; Reinhard, R; Schafer, P; Zagar, C; Westphalen, KO; Otten, M; Walter, H. Preparation of benzothiazolylpyridines and related compounds as herbicides, plant desiccants, and defoliants. WO 9906394 (1999).

[14] Hwang, IT; Hong, KS; Choi, JS; Kim, HR; Jeon DJ; Cho KY. Protoporphyrinogen IX-oxidizing activities involved in the mode of action of a new compound N-[4-chloro-2-fluoro-5-{3-(2-fluorophenyl)-5-methyl-4,5-dihydroisoxazol-5-yl-methoxy}-phenyl]-3,4,5,6-tetrahydrophthalimide. *Pest Biochem and Phy* 2004, *80*, 123-130.

[15] Hamper, BC; Mao, MK; Phillips, WG; Preparation of substituted 3-Aryl-5-haloalkyl-pyrazoles having herbicidalactivity. US 6121458 (2000).

[16] Huang, MZ; Huang, KL; Ren, YG; Lei, MX; Huang, L; Hou, ZK; Liu, AP; Ou, XM. Synthesis and herbicidal activity of 2-(7-fluoro-3-oxo-3,4-dihydro-2H-benzo[b][1,4]oxazin-6-yl)isoindoline-1,3-diones. *J Agric Food Chem* 2005, *53*, 7908-7914.

[17] Macias, FA; Marin, D; Oliveros-Bastidas, A; Castellano, D; Simonet, AM; Molinillo, AMG. Structure-activity relationship (SAR) studies of benzoxazinones, their

degradation products, and analogues. Phytotoxicity on problematic weeds Avena fatua L. and Lolium rigidum Gaud. *J Agric Food Chem* 2006, *54*, 1040-1048.

[18] Huang, MZ; Luo, FX; Mo, HB; Ren, YG; Wang, XG; Ou, XM; Lei, MX; Liu, AP; Huang, L; Xu, MC. Synthesis and herbicidal activity of isoindoline-1,3-dione substituted benzoxazinone derivatives containing a carboxylic ester group. *J Agric Food Chem* 2009, *57*, 9585-9592.

[19] Zhang, L; Tan, Y; Wang, NX; Wu, QY; Xi, Z; Yang, GF. Design, syntheses and 3D-QSAR studies of novel N-phenyl pyrrolidin-2-ones and N-phenyl-1H-pyrrol-2-ones as protoporphyrinogen oxidase inhibitors. *Bioorg Med Chem* 2010, *18*, 7948-7956.

[20] Jiang, LL; Tan, Y; Zhu, XL; Wang, ZF; Zuo, Y; Chen, Q; Xi, Z; Yang, GF. Design, Synthesis, and 3D-QSAR Analysis of Novel 1,3,4-Oxadiazol-2(3H)-ones as Protoporphyrinogen Oxidase Inhibitors. *J Agric Food Chem* 2010, *58*, 2643-2651.

[21] Zhou, Y; Xue, N; Wang, G; Qu, J. Synthesis, Structure and Herbicidal Activity of Substituted Phenyl Pyrazole Derivatives. *J Chem Res* 2010, *34*, 684-688.

[22] Zhou, Y; Chen, Y; Miao, W; Qu, J. Synthesis, Structures and Herbicidal Activity of Isoxazole Derivatives. *J Heterocyclic Chem* 2010, *47*, 1310-1316.

[23] Xue, N; Zhou, Y; Wang, G; Miao, W; Qu, J. Syntheses and herbicidal activity of pyrazolyl benzoxazole derivatives. *J Heterocyclic Chem* 2010, *47*, 15-21.

[24] Dong, X; Zhou, Y; Xu, S; Qu, J. Synthesis and Herbicidal Activity of 5-Arylmethoxy Phenylpyrazole Derivatives. *Chin J Org Chem* 2011, *31*, 1020-1026. (in Chinese)

[25] Meazza, G; Bettarini, F; Porta, PL; Piccarfi P; Signorini E; Portoso D; Fornara L. Synthesis and herbicidal activity of novel heterocyclic protoporphyrinogen oxidase inhibitors. *Pest Manag Sci* 2004, *60*, 1178-1188.

[26] Nandihalli, UB; Duke, SO. Structure–Activity relationships of protoporphyrinogen oxidase inhibiting herbicides, in Duke, SO; Rebeiz, CA, editor. Porphyric pesticides, ACS Symposium Series No 559, American Chemical Society, Washington, DC, 1994, 133–146.

[27] Sheldrich, GM. SADABS, Program for Empirical Absorption Correction; University of Göttingen: Germany, 1996.

[28] Sheldrich, GM. SHELX97, Program for Crystal Structure Deter-mination; University of Göttingen: Germany, 1997.

In: Handbook on Herbicides
Editors: Daiki Kobayashi and Eito Watanabe

ISBN: 978-1-62948-053-4

Chapter 14

BEHAVIOR OF HERBICIDES IN PADDY WATER AND SOIL AFTER APPLICATION

Kuniaki Kawata** and Tomohiro Kose
Faculty of Applied Life Sciences, Niigata University of Pharmacy and Applied Life Sciences, Akiha-ku, Niigata, Niigata, Japan

ABSTRACT

Degradation and distribution of herbicides in paddy water and paddy soil after application to paddy fields including experimental paddy plots are reviewed to examine the half-life and the organic carbon normalized soil sorption coefficient (K_{OC}). The target compounds are 10 herbicides: bromobutide, pyrazolynate, mefenacet, simetryn and thiobencarb, and two degradation products, bromobutide-debromo and destosyl pyrazolynate. Their half-lives in paddy water and soil after application are assessed. Their respective K_{OC} in paddy fields (K_{OCP}) are calculated based on their reported concentrations in paddy water and soil. The variations of K_{OCP} values are evaluated and discussed through comparison with the reported K_{OC} using laboratory experiments and calculations.

Keywords: Herbicide, Degradation product, Paddy field, Water, Soil, Half-life, Soil sorption coefficient

INTRODUCTION

Paddy rice farming has played an important role in crop production throughout the world. The FY2012 production of rice, the chief grain of Japan, was 8,519 million metric tons. That year, paddy fields covered a total area of 2.47 million ha, accounting for 54.3% of all cultivated area (4.55 million ha) in Japan: 1.58 million ha. More than five hundred compounds are registered as pesticides for agricultural use in Japan. In FY2011, 230 metric

* Correspondence to: Kuniaki Kawata E-mail address, kawata@nupals.ac.jp.

kilotons of pesticides of more than 200 varieties were manufactured. Pesticides are the environmental contaminants causing the greatest concern (Kawata 2009). Numerous studies have assessed their distributions and behaviors in rivers (Iwafune et al. 2010, Mitobe et al. 1999, Tanabe et al. 2001, Tanabe and Kawata 2004, Tanabe and Kawata 2009) and paddy fields (Kawata et al. 2005, Shiota et al. 2006). Approximately 70 herbicides are applied directly to the surface water of paddy fields in Japan, most of which are applied by ground application (Kubo et al. 2012, Morohashi et al. 2012b). Some herbicides applied to paddy fields flow into rivers via drainage channels. The runoff events of the applied herbicides occur by the drainage of paddy water and by rainfall (Morohashi et al. 2012b). The runoff ratios of pesticides including herbicides depend on the paddy fields themselves and water control as well as application methods and weather conditions (Kawata and Kose 2012).

In this chapter, we reviewed behaviors of some common herbicides after application to paddy fields, including experimental paddy plots. First, their half-lives in paddy water and soil are reviewed. Secondly, we evaluate their partitions between the paddy water and soil. For that purpose, their values of organic carbon normalized soil sorption coefficient (K_{OC}) of applied herbicides in paddy fields (K_{OCP}) are calculated using their reported concentrations in paddy waters and soils. They are evaluated and discussed through comparison with their reported K_{OC} values.

Properties of Target Compounds

The 10 target herbicides are common in Japan and detected from river water as well as paddy waters and soils. The herbicides are presented in Table 1. Their chemical structures are presented in Figure 1. Pyriminobac-methyl and simetryn are applied mainly to paddy fields during 7–18 days after planting rice seedlings in paddy fields. However, the other eight herbicides are applied mainly to paddy fields within seven days after planting.

The two targeted degradation products, bromobutide-debromo and destosyl pyrazolynate (DTP), are presented in Table 1. Their chemical structures are portrayed in Figure 1. Bromobutide is known to degrade to bromobutide-debromo via photochemical reaction (Takahashi et al. 1985) and metabolism (Isobe et al. 1984). Bromobutide-debromo, or desbromo-bromobutide, was detected in river water at 0.073–0.108 μg/L at maximum (Iwafune et al. 2010) and 0.05–0.19 μg/L at maximum (Mitobe et al. 1999). DTP is the herbicidal entity of pyrazolynate, or pyrazolate (Yamaoka et al. 1987). Pyrazolynate is hydrolyzed rapidly in water to DTP (Kubo et al. 2012), which has been detected in river water at maximum concentrations of 0.073–0.108 μg/L (Iwafune et al. 2010).

The water solubility and logarithm of the octanol/water partition coefficient (K_{OW}) of each target compound are presented in Table 2. Simetryn is the most water soluble of the 10 herbicides; pyrazolynate is the least water soluble one, although DTP is more than 7,400 times as soluble as pyrazolynate. The log K_{OW} values are from 0.61–0.71 for bensulfuron-methyl to 2.58 (25°C) – 5.06 for pyrazolynate.

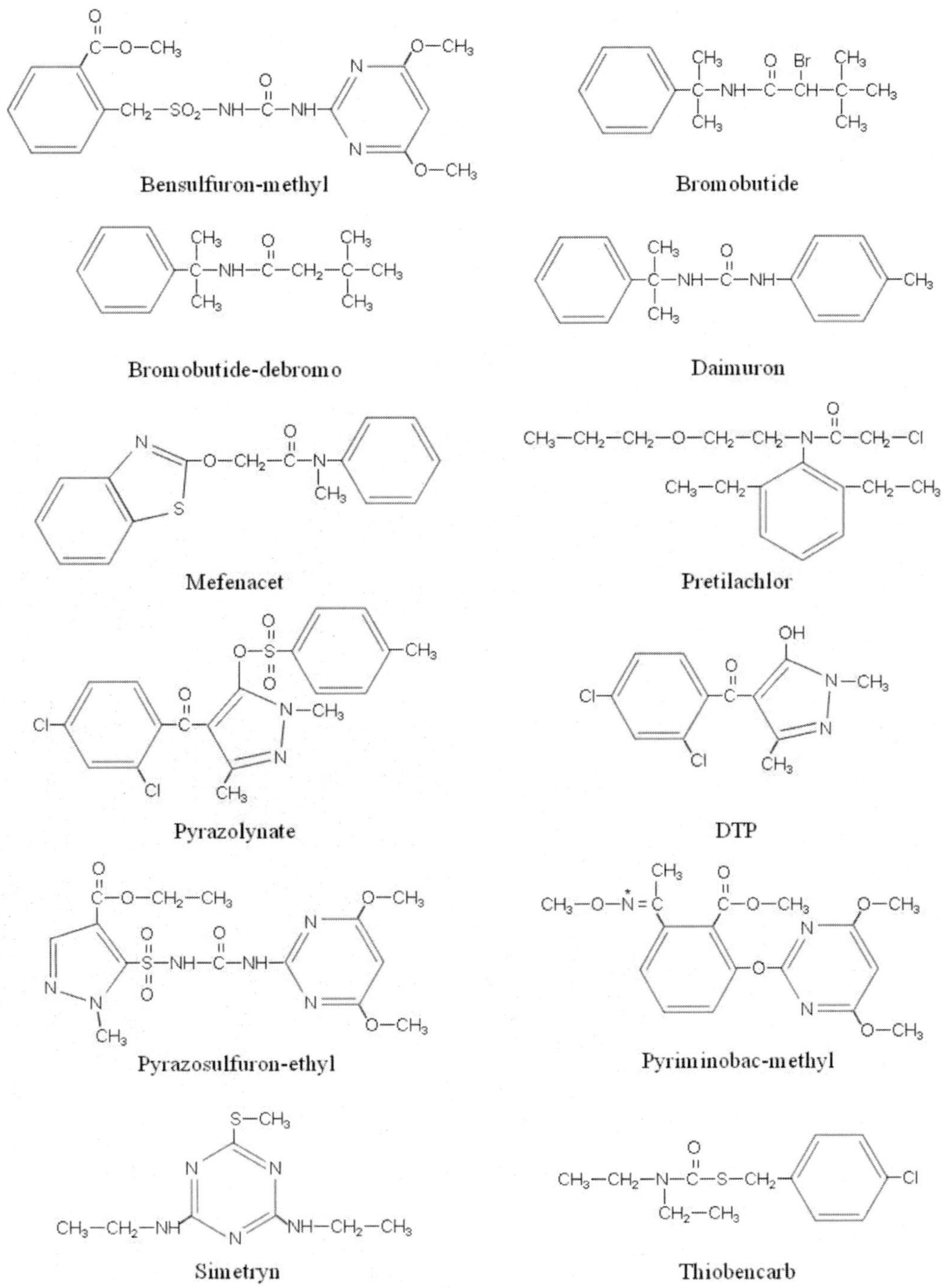

Figure 1. Chemical structures of target herbicides and their degradation compounds.

Table 1. Target herbicides and their degradation compounds

Common name	CAS No.	IUPAC Name CAS Name	Substance group
Bensulfuron-methyl	83055-99-6	Methyl α-[(4,6-dimethoxypyrimidin-2-ylcarbamoyl)sulfamoyl]-o-toluate Methyl 2-[[[[[(4,6-dimethoxy-2-pyrimidinyl)amino]carbonyl]amino]sulfonyl]methyl] benzoate	Sulfonyl-urea
Bromobutide	74712-19-9	(*RS*)-2-Bromo-3,3-dimethyl-*N*-(1-methyl-1-phenylethyl)butyramide 2-Bromo-3,3-dimethyl-*N*-(1-methyl-1-phenylethyl)butanamide	Amide
Bromobutide-debromo	75463-73-9	*N*-(α,α-dimethylbenzyl)-3,3-dimethylbutyramide 3,3-Dimethyl-*N*-(1-methyl-1-phenylethyl)butanamide	Amide
Daimuron	42609-52-9	1-(α,α-Dimethylbenzyl)-3-p-tolylurea 1-(4-methylphenyl)-3-(2-phenylpropan-2-yl)urea	Urea
Mefenacet	73250-68-7	2-(1,3-Benzothiazol-2-yloxy)-*N*-methylacetanilide 2-(2-Benzothiazolyloxy)-*N*-methyl-*N*-phenylacetamide	Oxy-acetamide
Pretilachlor	51218-49-6	2-Chloro-2',6'-diethyl-*N*-(2-propoxyethyl)acetanilide 2-Chloro-*N*-(2,6-diethylphenyl)-*N*-(2-propoxyethyl)acetamide	Chloro-acetamide
Pyrazolynate	58011-68-0	4-(2,4-Dichlorobenzoyl)-1,3-dimethylpyrazol-5-yl toluene-4-sulfonate (2,4-Dichlorophenyl)[1,3-dimethyl-5-[[(4-methylphenyl)sulfonyl]oxy]-1H-pyrazol-4-yl]methanone	Benzoyl-pyrazole
Destosyl pyrazolynate (DTP)	58010-98-3	4-(2,4-Dichlorobenzoyl)-1,3-dimethyl-5-hydroxypyrazole (2,4-Dichlorophenyl)[1,3-dimethyl-5hydroxy-1H-pyrazol-4-yl]methanone	Pyrazole
Pyrazosulfuron-ethyl	93697-74-6	Ethyl 5-[(4,6-dimethoxypyrimidin-2-ylcarbamoyl)sulfamoyl]-1-methylpyrazole-4-carboxylate Ethyl 5-[[[[(4,6-dimethoxy-2-pyrimidinyl)amino]carbonyl]amino]sulfonyl]-1-methyl-1H-pyrazole-4-carboxylate	Pyrazole
Pyriminobac-methyl	136191-64-5	Methyl (*EZ*)-2-(4,6-dimethoxypyrimidin-2-yloxy)-6-(1-methoxyiminoethyl)benzoate Methyl 2-[(4,6-dimethoxy-2-pyrimidinyl)oxy]-6-[1-(methoxyimino)ethyl]benzoate	Pyrimidinyl-oxybenzoic acid
(*E*)-Pyriminobac-methyl	147411-69-6	Methyl 2-(4,6-dimethoxy-2-pyrimidinyloxy)-6-(*E*)-(1-methoxyiminoethyl)benzoate	
(*Z*)-Pyriminobac-methyl	147411-70-9	Methyl 2-(4,6-dimethoxy-2-pyrimidinyloxy)-6-(*Z*)-(1-methoxyiminoethyl)benzoate	
Simetryn	1014-70-6	N^2,N^4-Diethyl-6-methylthio-1,3,5-triazine-2,4-diamine *N*,*N*'-Diethyl-6-(methylthio)-1,3,5-triazine-2,4-diamine	Triazine
Thiobencarb	28249-77-6	*S*-4-Chlorobenzyl diethyl(thiocarbamate) *S*-[(4-Chlorophenyl)methyl] diethylcarbamothioate	Thio-carbamate

Table 2. Properties of target herbicides and their degradation compounds

Common name	Molecular Weight	Molecular Formula	Water solubility [a] (mg/L)	log K_{OW}
Bensulfuron-methyl	410.4	$C_{16}H_{18}N_4O_7S$	12 (25°C, pH 6), 120 (25°C, pH 7)	0.61, 0.62, 0.78 (pH 7)
Bromobutide	312.3	$C_{15}H_{22}BrNO$	3.54 (25°C)	3.62
Bromobutide-debromo	233.4	$C_{15}H_{23}NO$		
Daimuron	268.4	$C_{17}H_{20}N_2O$	0.79 (20°C), 1.2 (20°C), 120 (20°C)	2.7, 3.01
Mefenacet	298.4	$C_{16}H_{14}N_2O_2S$	4 (20°C)	3.23
Pretilachlor	311.9	$C_{17}H_{26}ClNO_2$	50 (20°C), 74 (25°C)	4.08
Pyrazolynate	439.3	$C_{19}H_{16}Cl_2N_2O_4S$	0.056 (25°C)	2.58 (25°C), 5.06
DTP	285.1	$C_{12}H_{10}\ Cl_2\ N_2\ O_2$	415 (25°C)	
Pyrazosulfuron-ethyl	414.3	$C_{14}H_{18}N_6O_7S$	9.76- 9.96 (20°C), 14.5 (20°C)	1.3, 3.13
(*E*)-Pyriminobac-methyl	361.4	$C_{17}H_{19}N_3O_6$	9.25 (20°C)	2.98 (21.5°C)
(*Z*)-Pyriminobac-methyl	361.4	$C_{17}H_{19}N_3O_6$	175 (20°C)	2.70 (20.6°C)
Simetryn	213.3	$C_8H_{15}N_5S$	400 (20°C), 482 (20°C)	2.6
Thiobencarb	257.8	$C_{12}H_{16}ClNOS$	30 (20°C), 167 (20°C)	3.42

[a] Water solubilities and log K_{OW} values were referred from the British Crop Protection Council 2000, Inao et al. 2011, Kanazawa 1996 and Yamaoka et al. 1988.

HALF-LIVES IN PADDY FIELDS

The concentrations of applied herbicides in paddy water increased to maximum within 24 h after application in cases of pyrazolynate (Kubo et al. 2012) and thiobencarb (Phong et al. 2008) to three days after applications in cases of bromobutide (Morohashi et al. 2012b), mefenacet (Inao et al. 2003), pretilachlor and pyrazosulfuron-ethyl (Ishii et al. 2004). The maximum concentrations were reported as observed within 24 h after application in cases of bromobutide (Morohashi et al. 2012b) and pyrazolynate (Kubo et al. 2012) and pyriminobac-methyl (Inao et al. 2009) to four days after application in cases of bensulfuron-methyl (Ishii et al. 2004). The ratios of the herbicide amounts in the paddy soils increase according to the decrease of paddy water volume (Kawata and Kose 2012).

The decrease of an herbicide in the paddy water and soil can often be interpreted using first-order reaction kinetics (Iwashita et al. 2008, Morohashi et al. 2012b, Qin et al. 2004) in the following equation [1].

$$\ln A = k\,t + C \tag{1}$$

Therein, A stands for the concentration or the amount of the herbicide in the paddy soil, k signifies the decreasing rate (/day) or the rate constant, t denotes the elapsed time after the application (day), and C is a constant. Figure 2 shows the natural logarithm of mean bromobutide concentrations in paddy water (μg/L) and soil (μg/kg dry) versus elapsed time after application with regression lines using the least-squares method. The decreasing plots of paddy waters and soils were well fitted ($p<0.01$) to the first-order reaction, given as equation [1]. The respective half-lives of bromobutide in the water and soil were estimated using equation [2] with the obtained k values.

$$t_{1/2} = -\ln 2 / k \quad (2)$$

Therein, $t_{1/2}$ is the half-life in paddy water or soil.

Morohashi et al. (2012a and 2012b) reported the behavior of bromobutide in paddy water and soil after application to three paddy fields (paddy fields 1, 2 and 3) in Niigata, Japan. Niigata produced 0.656 million metric tons of rice, accounting for 7.7% of total rice production in FY2012 in Japan.

The mean of the obtained half-life value in the paddy water was 2.7 ± 0.34 days during 1–18 days after application (Table 3), which was only 0.035 of that in paddy water without paddy soil under solar irradiation, 11 weeks (Takahashi et al. 1985). The obtained mean half-life values in the paddy soil were 6.9 ± 2.6 days (1–18 days) and 16 ± 4.5 days (18–104 days), which were 0.11–0.22 and 0.25–0.52, respectively, of that in soil, 31–64 days (Kanazawa 1996). These marked decreases of bromobutide both in water and in soil during 1–18 days after application in the study were attributable mainly to the distribution of bromobutide between water and soil as well as the runoff from the paddy field (Morohashi et al. 2012).

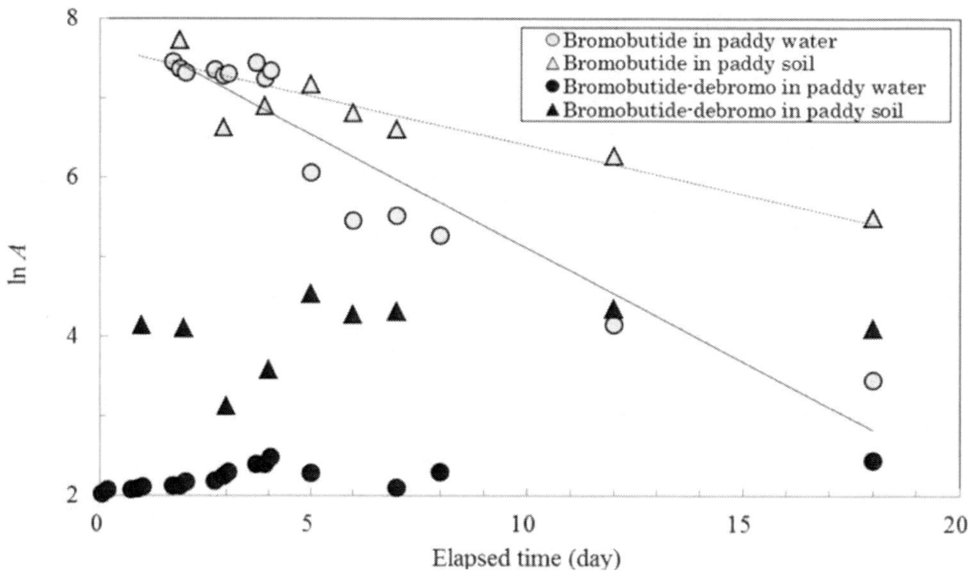

Figure 2. Natural logarithms of mean bromobutide and bromobutide-debromo concentrations in paddy water and soil vs. elapsed time with regression lines obtained using the least-squares method (plotted based on Morohashi et al. 2012b).

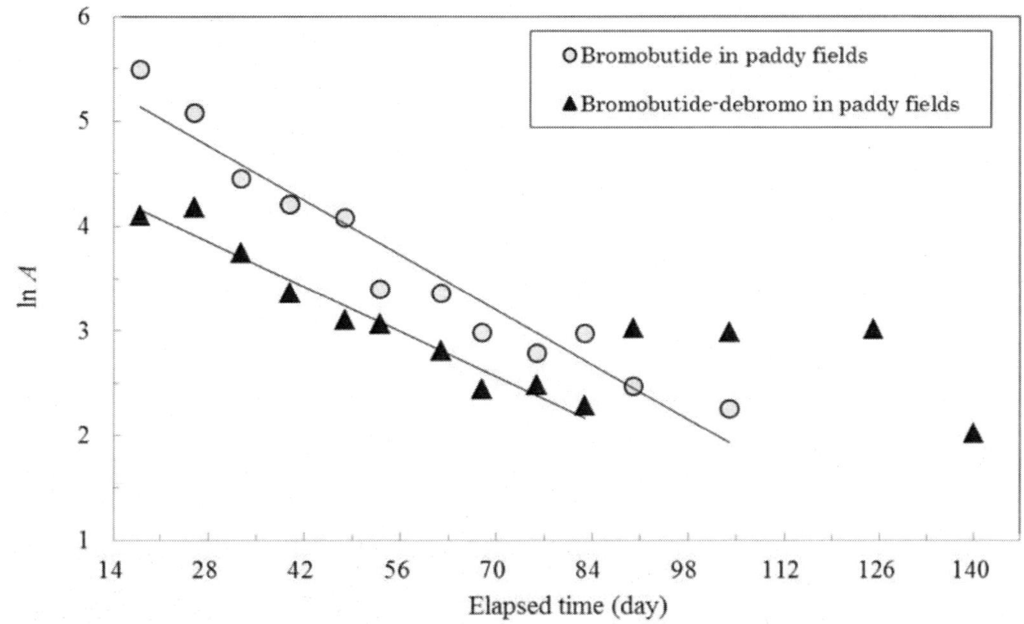

Figure 3. Natural logarithms of mean bromobutide and bromobutide-debromo concentrations in paddy soil during periods without paddy water vs. elapsed time with regression lines obtained using the least-squares method (plotted based on Morohashi et al. 2012a).

Regarding bromobutide-debromo, its concentration in the paddy water or soil represents the remainder of the formation amount from bromobutide and the decreased amount to other degradation compounds. The natural logarithms of mean bromobutide-debromo concentrations in paddy water and soil versus elapsed time after the application are shown in Figures 2 and 3. No significant correlation was found between the natural logarithm of the bromobutide-debromo amount and the elapsed time during 18 days after application (Morohashi et al. 2012b) and after 83 days. The decreasing plot of paddy soils during 18–83 days (Fig. 3) was well fitted (p<0.01) to the first-order reaction given as equation [1] with k = –0.0307 and C = 4.71. The respective half-lives of bromobutide-debromo in the paddy water and soil were estimated using equation [2]. The obtained half-life was 23 ± 5.2 days.

The reported half-life values of target compounds in paddy fields are presented in Table 3 with the reported half-lives in water and soil. Among the target compounds, pyrazolynate has the shortest half-life (0.87 ± 0.091 day) in paddy water. DTP (17 ± 1.4 day) and simetryn (17.7 day at maximum) have the longest half-lives. Moreover, half-life values differed from the reported values, indicating that they were affected by the conditions of paddy fields, farming practices, and weather. In most cases, these half-life values obtained experimentally in paddy fields differed greatly from those found in the laboratory. This is attributed to the runoff from the paddy field, migration to the ground, and accelerated decomposition under weather conditions. Therefore, experiments must be undertaken to evaluate decreases of herbicides in paddy fields after application.

Table 3. Half-lives of target herbicides and their degradation compounds

Common name	Half-life [a] (day)		Half-life in paddy field (day)		Reference [b]
	Water	Soil	Paddy water	Paddy soil	
Bensulfuron-methyl	41–58	11–143	2.8 [3-14 days]; 7.4 [14-51 days]	4.6 [1-21 days]	Ishii et al. 2004
Bromobutide	143 (pH 7, 25°C)	31–64	2.7 ± 0.34 [1-18 days]	6.9 ± 2.6 [1-18 days]	Morohashi et al. 2012b
			–	16 ± 4.5 [18–104 days]	Morohashi et al. 2012a
Bromobutide-debromo	–	–	–	23 ±5.2 [18–83 days] [d]	
Daimuron	2.68	9–77	2.1 [3-14 days]; 9.6 [14-51 days]	30 [3-86 days]	Ishii et al. 2004
Mefenacet	–	23–223	1.8 [3-14 days]; 7.7 [14-51 days], 3.4 [2-14 days]; 5.7 [14-43 days]	21 [1-86 days], 12 [1-43 days]	Ishii et al. 2004
			4.3	8.2	Watanabe et al. 2006
Pretilachlor	14, >200 (pH1-9)	8–50	–	257[e]	
Pyrazolynate	–	10–30	0.87 ± 0.091 [0-5 days]	2.2 ± 0.70 [0-6 days]	Kubo et al. 2012
DTP			17 ± 1.4 [6-76 days]	26 ± 2.1 [6-112 days]	Kubo et al. 2012
Pyrazosulfuron-ethyl	5.6	3–28	1.9 [2-7 day], 7.6 [7-27 days]	11 [1-27 days]	Ishii et al. 2004
(*E*)-Pyriminobac-methyl	224 [c]	2.0-133 [c]	–	–	
(*Z*)-Pyriminobac-methyl	187 [c]	3.5-76 [c]	–	–	
Simetryn	Stable	52–179	1.3-1.4 [within 7 days], 16.2-17.7 [after 7 days]	11.5	Phong et al. 2008
Thiobencarb	30 (pH5-9, 21°C)	10–240	2.1-2.2 [within 14 days], 9.2- 9.7 [after 14 days]	7.2, 8.1	Phong et al. 2008

[a] Half-life values were referred from the British Crop Protection Council 2000, Inao et al. 2011, Kanazawa 1996 and Matsui et al. 2006.

[b] References for half-life values in paddy field.

[c] Calculated based on Inao et al. 2009.

[d] Calculated based on Morohashi et al. 2012a.

[e] Calculated based on Watanabe and Takagi 2000.

K_{OC} VALUES IN PADDY FIELD

Sorption processes play a major role in determining the behavior of organic compounds in the water/soil environment. The most frequently used parameter to indicate the soil mobility of an organic compound is the organic carbon normalized soil sorption coefficient (K_{OC}), which is probably the key descriptor of the distribution of organic compounds between the soil and aqueous phases of the environment. K_{OC} is defined according to the following equation [3].

$$K_{OC} = (C_S / C_W) \times (100 / C_{OC}) \quad (3)$$

In that equation, C_S and C_W respectively represent the concentrations of the herbicide in the paddy soil (μg/kg) and the paddy water (μg/L), and C_{OC} is the concentration of the organic carbon (%). The C_{OC} values are obtained experimentally in the laboratory using several soils containing various organic carbons (Mamy and Barriuso 2005, Piwowarczyk and Holden 2013). C_{OC} are also predicted using regression equations, e.g. log K_{OC} vs. log K_{OW} (Wen et al. 2012), and models (Huuskonen 2003, Schüürmann et al. 2006).

Herbicides applied to paddy water diffuse in paddy water and adsorb to paddy soil. Through diffusion and adsorption processes, the herbicide concentration in water phase and that in soil phase can be in equilibrium. The K_{OC} values of applied herbicides in paddy fields (K_{OCP}) were calculated using equation [3] with pesticide concentrations in paddy water and soil (Iwashita et al. 2008, Morohashi et al. 2012b). The K_{OCP} values for bromobutide and bromobutide-debromo are reported based on their concentrations in three paddy fields (Morohashi et al. 2012b). Variations in the calculated K_{OCP} values in paddy fields after application are presented in Figure 3. Calculated K_{OCP} values of bromobutide increased to the maximum values (100, 160 and 140 mL/g in paddy fields 1, 2 and 3, respectively, at 18, 12 and 18 days after application). The calculated K_{OCP} values in the paddy fields during days 12–18 after application were 93–160 mL/g and 120 ± 25 mL/g in mean ± standard deviation (Morohashi et al. 2012b). Although the reported K_{OC} values were 163–10430 mL/g (Table 4), which include both the experimentally obtained ones and the calculated ones, the obtained values, 93–160 mL/g, were less than the reported values. Therefore, the applied bromobutide could be not distributed between the soil and water in the ratio at the equilibrium inferred from the K_{OC} values during the investigated period, existing more in the paddy water than under the equilibrium condition (Morohashi et al. 2012b).

Kubo et al. (2012) reported the behavior of pyrazolynate in paddy water and soil after application to three paddy fields (paddy fields 1, 2 and 3) in Niigata, Japan. The K_{OCP} values for pyrazolynate and DTP are calculated based on their concentrations in three previously reported paddy fields (Kubo et al. 2012). Variations in the calculated K_{OCP} values in paddy fields after application are given in Figure 4. Calculated K_{OCP} values of pyrazolynate were at maximum at 2 days (79 mL/g in paddy field 2 and 126 mL/g in paddy field 3) and 3 days after application (96 mg/L in paddy field 1). However, those of DTP at maximum were 283 mL/g in paddy field 1 at 14 days after application, 39 mg/L in paddy field 2 at 6 days, and 157 mL/g in paddy field 3 at 1 day. The mean calculated K_{OCP} values of pyrazolynate during 2–5 days after application were 43–98 mL/g (63 ± 39 mL/g in mean ± standard deviation). Those of DTP during 1–14 days after application were 23–80 mL/g (56 ± 67 mL/g in mean ±

standard deviation), as shown in Table 4. Yamaoka et al. (1988) reported the soil sorption coefficient (K_d) values of DTP as 4.1–95.7 mL/g based on an adsorption experiment using 10 varieties of soils. The K_{OC} values of DTP were calculated based on K_d values using the organic matter contents as C_{OC}. The calculated value, 1080 ± 1580 mL/g (Table 4), was 19 times K_{OCP} (63 ± 39 mL/g) in the paddy fields. This tendency is the same as that of bromobutide described above.

Table 4. Calculated K_{OCP} in paddy fields and reported values

Common name	Reported K_{OC} [a]	Calculated K_{OCP} [b]	Reference [b]
Bensulfuron-methyl	257, 499, 5400	285 ± 56	Ishii et al. 2004
Bromobutide	163, 306, 652, 653, 10430	120 ± 25 [c]	
Bromobutide-debromo		147 ± 55	Morohashi et al. 2012b
Daimuron	186, 940, 959, 6855	19100 ± 5730	Ishii et al. 2004
Mefenacet	3036	216 ± 169	Inao et al. 2003
		356 ± 152	Ishii et al. 2004
		406 ± 57	Watanabe et al. 2006
Pretilachlor	254, 628, 1613	318 ± 124	Watanabe and Takagi 2000
Pyrazolynate	2855, 7855, 29830	63 ± 39	Kubo et al. 2012
DTP	1080 ± 1580 [d]	56 ± 67	Kubo et al. 2012
Pyrazosulfuron-ethyl	10, 209, 371, 455	212 ± 163	Ishii et al. 2004
(*E*)-Pyriminobac-methyl	425, 741, 1270	346 ± 118	Inao et al. 2009
(*Z*)-Pyriminobac-methyl	215, 372, 636	176 ± 52	Inao et al. 2009
Simetryn	333, 6915	119 ± 29	Phong et al. 2008
Thiobencarb	676, 1602, 3170	1250 ± 489	Phong et al. 2008

[a] K_{OC} values were referred from the Food Safety Commission 2008, Inao et al. 2011, Kanazawa 1996 and Matsui et al. 2006.

[b] K_{OCP} values were calculated based on references except for bromobutide.

[c] Values reported by Morohashi et al. 2012b.

[d] Calculated based on Yamaoka et al. 1988.

K_{OCP} values for the other target compounds in paddy fields were also calculated based on the reported concentrations of target compounds in paddy water and soil. The calculated K_{OCP} values in paddy fields are shown in Table 4 together with the reported K_{OC} values. The mean K_{OCP} values showed no statistically significant correlation with those of K_{OC}. In addition, neither of the means of K_{OCP} and K_{OC} values showed a significant correlation with those of K_{OW}. The reported K_{OC} values of an herbicide differ greatly among reports, e.g. 257–5400 for bensulfuron-methyl and 163–10430 for bromobutide (Table 4). These differences might result from differences in experimental conditions such as the soils used (Yamaoka et a. 1988) or because they include values calculated from models for K_{OC} prediction. It is noteworthy that the applied herbicide was distributed between the paddy water and soil in the ratio at equilibrium assumed by the K_{OCP} values during the investigated period. However, K_{OCP}

signifies the distribution in an actual paddy field including the runoff event from the paddy field, migration from the surface to the deeper ground, and decomposition under weather conditions. Therefore, K_{OCP} is used effectively for evaluation of herbicide behavior in paddy fields instead of K_{OC}. In fact, K_{OCP} is a useful indicator for evaluation of the distribution of an herbicide between paddy water and soil.

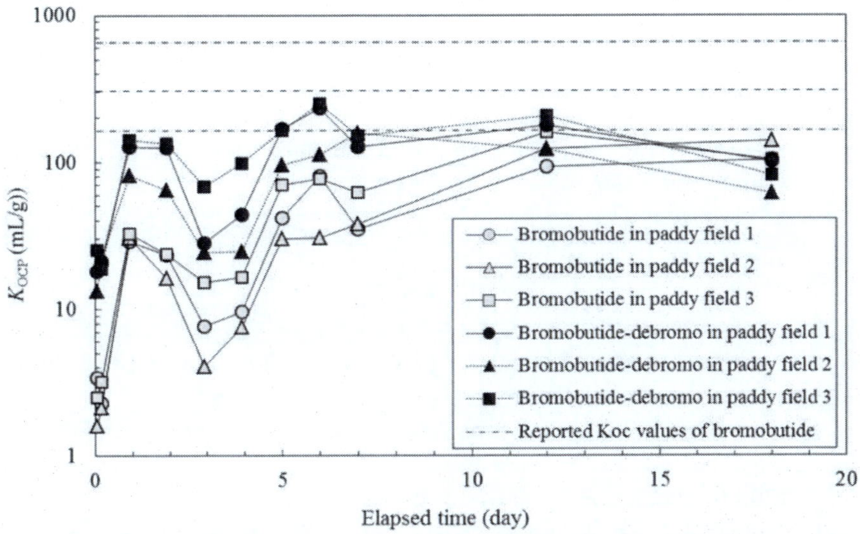

Figure 4. Variations in calculated K_{OCP} values of bromobutide and bromobutide-debromo in paddy fields.

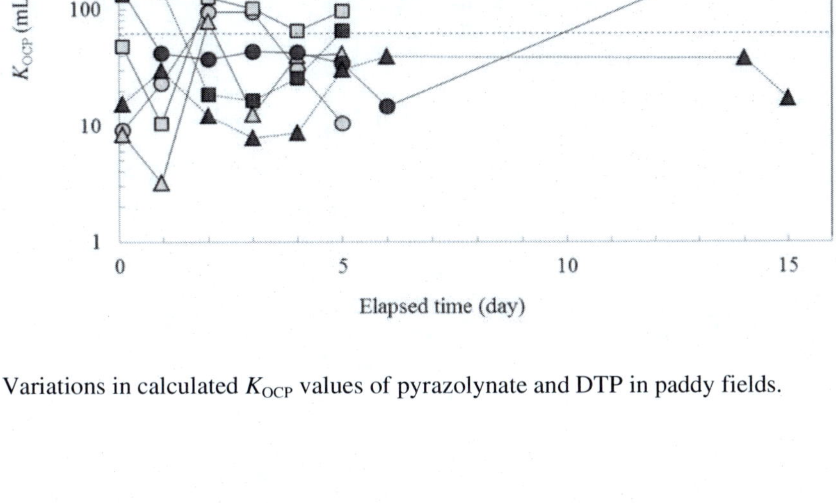

Figure 5. Variations in calculated K_{OCP} values of pyrazolynate and DTP in paddy fields.

CONCLUSION

In this chapter, we reviewed the half-lives and K_{OCP} of 12 compounds, 10 common herbicides in Japan, and 2 degradation products after application in paddy fields including experimental paddy plots. First, their half-lives in paddy water and soil are reviewed. In most cases, the half-life values of herbicides obtained experimentally in paddy fields differ greatly from those obtained in laboratory experiments. Therefore, experiments must be conducted to evaluate decreases of herbicides in paddy fields after application.

Secondly, K_{OCP} values are calculated for the 12 target compounds using their reported concentrations in paddy waters and soils. No significant correlation was found between mean K_{OCP} values and those of K_{OC}. K_{OCP} can be a useful indicator for evaluation of the herbicide distribution between paddy water and soil.

REFERENCES

British Crop Protection Council (2000). *The pesticide manual,* 12/e, ed. by Tomlin CDS, British Crop Protection Council, Surrey, UK.

Food Safety Commission (2008*). Evaluation Report,* bromobutide. Food Safety Commission, Tokyo, Japan.

Huuskonen, J. (2003). Prediction of soil sorption coefficient of a diverse set of organic chemicals from molecular structure. *Journal of Chemical Information and Computer Sciences,* 2003, 43, 1457–1462.

Inao, K., Ishii, Y., Kobara, Y. & Kitamura, Y. (2003). Landscape-scale simulation of pesticide behavior in river basin due to runoff from paddy fields using pesticide paddy field model (PADDY). *Journal of Pesticide Science,* 28, 24–32.

Inao, K., Hojyo, T., Annoh, H., Miyazaki, S., Saito, T. & Park, H.-D. (2011). Predicting the behavior of paddy pesticides in a river basin using a simulation model (PADDY-Large): Application to a tributary of the Chikuma River under rice cultivation. *Journal of Pesticide Science,* 36, 413–427.

Inao, K., Mizutani, H., Yogo, Y. & Ikeda, M. (2009). Improved PADDY model including photoisomerization and metabolic pathways for predicting pesticide behavior in paddy fields: Application to the herbicide pyriminobac-methyl. *Journal of Pesticide Science,* 34, 273–282.

Ishii, Y., Inao, K. & Kobara, Y. (2004). Dissipation of some herbicides in a flooded rice field and increase of water-holding times after application of herbicides. *Bulletin of National Institute for Agro-Environmental Sciences, 23*, 15–25.

Isobe, N., Matsuo, M. & Miyamoto, J. (1984). Comparative metabolism studies on N-(1-methyl-1-phenylethyl)-2-bromo-3,3-dimethylbutanamide (bromobutide) and its debromonated derivative in rats and mice. *Journal of Pesticide Science,* 9, 105–115.

Iwafune, T., Inao, K., Horio, T., Iwasaki, N., Yokoyama, A. & Nagai, T. (2010). Behavior of paddy pesticides and major metabolites in the Sakura River, Ibaraki, Japan. *Journal of Pesticide Science,* 35, 114–123.

Iwashita, M, Maeda, T, Hori, T, Asada, T, Oikawa, K, & Kawata, K (2008). Decrease of 4,5,6,7-tetrachlorophthalide in paddy field soil after aerial application. *Bulletin of Environmental Contamination and Toxicology,* 81, 383–386.

Kanazawa, J. (1996). *Environmental character and toxicity data of pesticides.* Godo-shuppan, Tokyo, Japan.

Kawata, K. (2009). Behavior of aerial applied 4,5,6,7-tetrachlorophthalide in air, water and soil, in *"Fungicides: Chemistry, Environmental Impact and Health Effects",* De Costa, P. and Bezerra, P. (ed.), pp.1–12, Nova Science Publishers, Hauppauge, NY, USA.

Kawata, K., Asada, T., Tanabe, A. & Oikawa, K. (2005). Runoff of clomeprop and oxaziclomefone from a paddy field. *Bulletin of Environmental Contamination and Toxicology,* 75, 922–928.

Kawata, K. & Kose, T. (2012). Behavior of pesticides and their transformation products in river water in Japan, in *"The Impact of Pesticides",* Jokanović, M. (ed.), pp. 203–224, Publishing Services LLC, Cheyenne, WY, USA.

Kubo, T., Ohno, M., Nagasawa, S., Kose, T. & Kawata, K. (2012). Behavior of herbicide pyrazolynate and its hydrolysate in paddy fields after application. *Bulletin of Environmental Contamination and Toxicology, 89,* 985–989.

Mamy, L. & Barriuso, E. (2005). Glyphosate adsorption in soils compared to herbicides replaced with the introduction of glyphosate resistant crops. *Chemosphere,* 61, 844–855.

Matsui, Y., Narita, K., Inoue, T. & Matsushita, T. (2006). Screening level analysis for monitoring pesticide in river water using a hydrological diffuse pollution model with limited input data. *Water Science and Technology,* 53, 173–181.

Mitobe, H., Ibaraki, T., Tanabe, A., Kawata, K., Sakai, M. & Kifune, I. (1999). Variation of pesticides in a river flowing through paddy fields area. *Journal of Environmental Chemistry,* 9, 311–320.

Morohashi, M., Nagasawa, S., Enya, N., Ohno, M., Suzuki, K., Kose, T. & Kawata, K. (2012a). Decrease of herbicide bromobutide and its debromo metabolite in paddy field soil during 24 weeks after application. *Bulletin of Environmental Contamination and Toxicology,* 89, 176–180.

Morohashi, M., Nagasawa, S., Enya, N., Suzuki, K., Kose, T. & Kawata, K. (2012b). Behavior of bromobutide in paddy water and soil after application. *Bulletin of Environmental Contamination and Toxicology,* 88, 521–525.

Piwowarczyk, A. A. & Holden, N.M. (2013). Phenoxyalkanoic acid herbicide sorption and the effect of co-application in a Haplic Cambisol with contrasting management. *Chemosphere,* 90, 535–541.

Phong, T.K., Watanabe, H., Hien, T.Q., Vu, S.H., Tanaka, T., Nhung D.T.T. & Motobayashi, T. (2008). Excess water storage depth—a water management practice to control simetryn and thiobencarb runoff from paddy fields. *Journal of Pesticide Science,* 33, 159–165.

Qin, S., Gan, J., Liu, W. & Becker, J.O. (2004). Degradation and adsorption of fosthiazate in soil. *Journal of Agricultural and Food Chemistry,* 52, 6239–6242.

Schüürmann, G., Bert, R.-U. & Kühne, R. (2006). Prediction of the sorption of organic compounds into soil organic matter from molecular structure. *Environmental Science & Technology,* 40, 7005–7011.

Shiota, N., Hori, T., Nissato, K., Asada, T., Oikawa, K. & Kawata, K. (2006). Runoff of aerially applied phthalide from paddy fields. *Bulletin of Environmental Contamination and Toxicology,* 77, 508–515.

Takahashi, N., Mikami, N., Yamada, H. & Miyamoto, J. (1985). Photodegradation of the herbicide bromobutide in water. *Journal of Pesticide Science,* 10, 247–256.

Tanabe, A. & Kawata, K. (2004). Determination of triazine pesticides and related compounds in environmental water by liquid chromatography – mass spectrometry. Analytical *Sciences,* 20, 227–230.

Tanabe, A. & Kawata, K. (2009). Daily variation of pesticides in surface water of a small river flowing through paddy field area. *Bulletin of Environmental Contamination and Toxicology*, 82, 705–710.

Tanabe, A. Mitobe, H., Kawata, K., Yasuhara, A. & Shibamoto, T. (2001). Seasonal and spatial studies on pesticide residues in surface waters of the Shinano River in Japan. *Journal of Agricultural and Food Chemistry,* 49, 3847–3852.

Watanabe, H. & Takagi, K. (2000). A simulation model for predicting pesticide concentrations in paddy water and surface soil II. Model Validation and Application. *Environmental Technology,* 21, 1393–1404.

Watanabe, H., Takagi, K. & Vu, S.H. (2006). Simulation of mefenacet concentrations in paddy fields by an improved PCPF-1 model. *Pest Management Science, 62*, 20–29.

Wen, Y., Su, L.M., Qin, W.C., Fu, L., He, J. & Zhao, Y.H. (2012). Linear and non-linear relationships between soil sorption and hydrophobicity: model, validation and influencing factors. *Chemosphere,* 86, 634–640.

Yamaoka, K., Nakagawa, M. & Ishida, M. (1987). Hydrolysis of the rice herbicide pyrazolate in aqueous solutions. *Journal of Pesticide Science,* 12, 209–212.

Yamaoka, K., Tohjigamori, M., Tsujino, Y., Nakagawa, M. & Ishida, M. (1988). Adsorption and desorption of DTP, the herbicidal entity of pyrazolate, by soils and vertical mobility of pyrazolate and DTP in paddy fields. *Journal of Pesticide Science,* 13, 261–268.

In: Handbook on Herbicides
Editors: Daiki Kobayashi and Eito Watanabe

ISBN: 978-1-62948-053-4

Chapter 15

A LABORATORY LYSIMETER FOR PESTICIDE TRANSPORT STUDIES

A. A. Piwowarczyk* and N. M. Holden
UCD Bioresources Research Centre / Biosystems Engineering,
UCD School of Biosystems Engineering , University College Dublin,
Belfield, Dublin, Ireland

ABSTRACT

The transport of pesticides through the soil is one of many processes controlling their final distribution in the environment. The presence of pesticides in soil, groundwater, surface water and air may cause considerable adverse effects on ecosystems and human health. Human health may be affected by pesticide residues in food and drinking water, while ecosystems may be affected by loss of biodiversity and decreases in populations of sensitive organisms. A preliminary investigation into the use of a laboratory lysimeter apparatus to study transport of pesticides through undisturbed soil columns was undertaken. Two phenoxyalkanoic acid herbicides (MCPA and mecoprop-p) were applied simultaneously to a soil subject to long-term tillage and grassland management. The amounts leached from the tillage soil accounted for 3.5 mg of MCPA and mecoprop-p, which was approximately 66% and 68% of the initially applied amount, respectively. The amounts leached from the grassland soil accounted for 3.6 mg of MCPA (approximately 69% of initial application) and 3.0 mg of mecoprop-p (approximately 58% of initial application). Even though the quantities leached were comparable, the peak concentration appeared much sooner in the soil with the greatest saturated hydraulic conductivity. This observation may be important for pesticides with a long degradation half-life, as the time they need to degrade to a less harmful form (i.e. CO_2) may not be sufficient before reaching groundwater. The study showed the usefulness of the lysimeter apparatus for studying pesticide transport through undisturbed soil columns under controlled laboratory conditions.

Keywords: Pesticide transport, undisturbed soil column, saturated hydraulic conductivity, soil management

* Corresponding authors: agnieszka.piwowarczyk@ucd.ie, nick.holden@ucd.ie.

INTRODUCTION

Water is a solvent capable of carrying significant quantities of dissolved pesticides and other agro-chemicals. Pesticides can be found in ground- and surface waters due to runoff and/or leaching [1,2]. The transport of pesticides through the soil is a function of water movement mechanisms, the chemical properties of the soil and the properties of the pesticide itself (degradation rate and adsorption-desorption characteristics). The physics of soil water and solute movement can be used to determine the fate of pesticide compounds, although the physico-chemical interactions of the solute with soil particles will vary depending on the nature of the solute as well as the solid [3]. In many pesticide studies these interactions are characterized by pesticide specific sorption isotherms that are related to soil properties such as organic carbon content, soil texture, pH and cation exchange capacity [4-6]. Both disturbed and undisturbed soil columns have been used to study the transport of pesticides in the soil system, for example Köhne et al. [7] and Dousset et al. [8]. When a disturbed column is used, the focus of the study is on the role of soil constituents and pesticide interactions, while undisturbed soil columns permit consideration of the role of pore structures on transport kinetics. Studying undisturbed soils also permits evaluation of the role of soil water status on transport, but systematically controlling soil water status can be quite difficult.

Pesticide transport studies can be conducted in the field [9], in field lysimeters [10] or in laboratory lysimeters [7]. Field scale studies have the advantage of revealing the consequences of interactions with climate and environment, but are complicated by the fact that it becomes difficult to control the consequences of variable temperature effects on degradation rates.

Studies to understand pesticide kinetics in soils require laboratory experiments in order to have a controlled environment as one of the variables / treatments. This paper describes a laboratory lysimeter facility that was designed to permit: (i) undisturbed soil columns to be evaluated for pesticide transport experiments with controlled boundary conditions, and (ii) a test case of saturated transport of two phenoxyalkanoic acid herbicides in one soil under contrasting long-term management.

MATERIALS AND METHODS

Description of Soils and Sampling

One typical agricultural soil from Ireland (Haplic Cambisol, Oakpark series) was used for this study. Full details of the site location, management history and sampling strategy can be found in the recently published work of Piwowarczyk and Holden [11]. Undisturbed soil columns were sampled by driving stainless steel cylindrical columns (10 cm i.d. and 20 cm height) into the soil with three replicates. Once filled with soil, the columns were excavated and sealed with foil and placed into air- and water-tight plastic bags. The ends of the columns were supported to ensure as little transport disturbance as possible. The columns were stored at 4 °C until used.

The Laboratory Lysimeter Unit

The lysimeter apparatus (Figure 1) was designed to enable examination of both saturated and unsaturated transport of agrochemicals through undisturbed soils, although only saturated transport is presented here. The apparatus was operated in an air-conditioned laboratory (ambient temperature of 20 °C ± 2) at the Department of Agriculture, Food and the Marine Pesticide Control Laboratory (Celbridge, Co. Kildare, Ireland; ISO 17025). The undisturbed soil columns (10 cm i.d. and 20 cm height) were held between thin layers of sand, kept in place by a nylon mesh, a glass microfiber filter (Ø 90 mm, Whatman) and glass wool (to minimize the dead-end volume and for better distribution of the pesticide solution), and clamped by end plates and steel rods. The columns were prepared with a clean surface on the top and the bottom tightly held against the sand to ensure hydraulic continuity throughout the system. The end caps were sealed in place using molten paraffin wax. Above the soil columns there were two reservoirs, one contained 0.01 M $CaCl_2$ solution and one contained herbicides dissolved in 0.01M $CaCl_2$. The pressure head to drive the transport was by gravity for this experiment. The apparatus could be set up using either a peristaltic pump or vacuum pump to provide the pressure head, although these are not necessary for some studies.

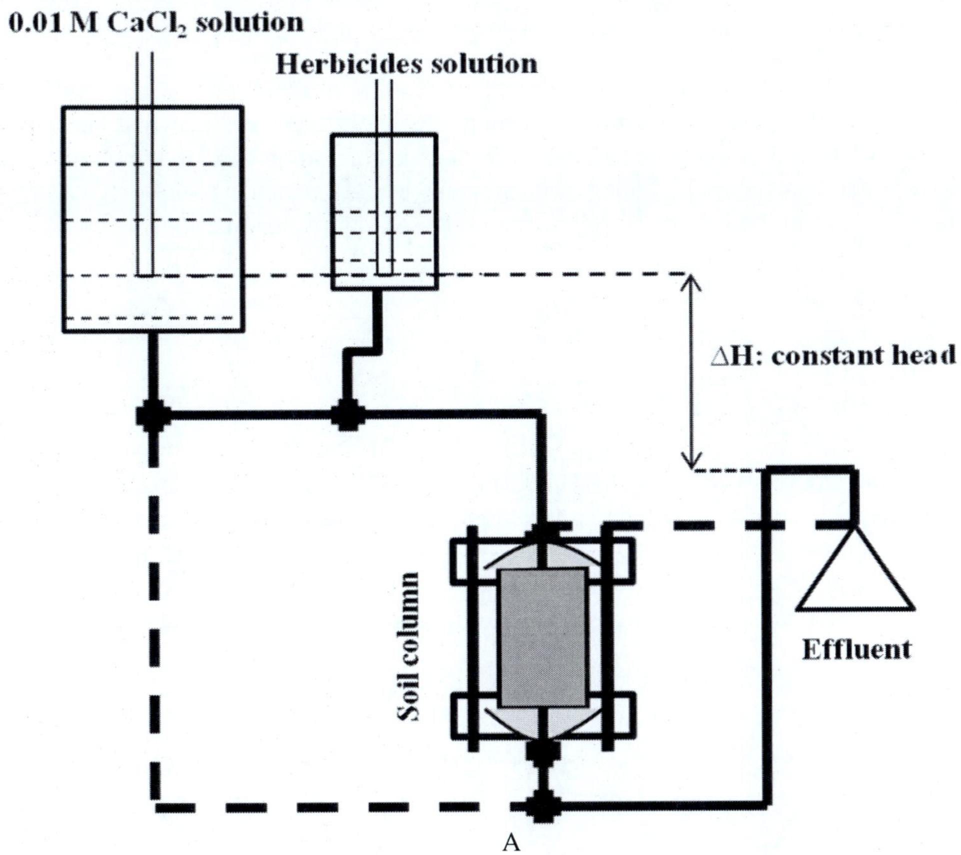

Figure 1. (Continued).

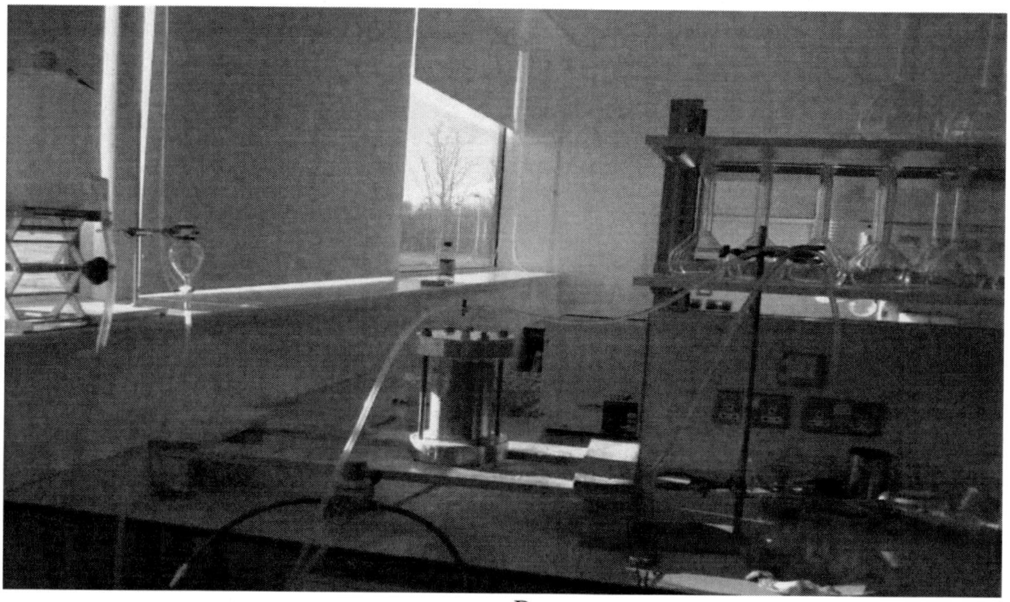

B

Figure 1. The lysimeter apparatus used in the experiment. A: Schematic; B: Photograph.

Before the experiment commenced, the soil columns were saturated in a water tank filled with 0.01 M $CaCl_2$ solution for a period of about 48 hours, and later clamped between the endplates. 0.01 M $CaCl_2$ solution was passed through each column from the bottom to allow air to escape for an additional 18 hours. The system was then switched to allow 0.01M $CaCl_2$ solution to pass from the top of the column and this was run for another 48 hours in order to achieve equilibrium and a steady flow (Figure 1b). The saturated hydraulic conductivity (K_{sat}) was calculated by Darcy's law:

$$K_{sat} = \frac{\Delta VL}{A\Delta t\Delta H} \quad (1)$$

where A is the cross area of the soil column (m^2); L is the length of soil column (m); ΔH is the pressure head (cm water) in the soil column. The herbicides were injected as a pulse. Three soil columns were employed with one being used as a control treatment in order to check whether there were any pesticide residues in the soils collected. Additional samples were collected from all columns 4 hours prior herbicides application to check for any residues in the leachate.

Preparation of Agrochemicals Applied to the Soil Columns and Sample Collection

Once a steady flow was achieved, a pesticide solution of 519.7 mg L^{-1} MCPA and 506.6 mg L^{-1} MCPP-p was injected into the system it was run thereafter for about 28 hours. The mass applied to each column was approximately 5.2 mg of MCPA and 5.1 mg of mecoprop-

p. These amounts were greater than those calculated from the recommended field rates for grassland (personnal communication with DAFM) and given soil column dimensions (1.4 mg of MCPA and 2.4 mg of MCPP-p). The use of higher amount was justified to ensure samples were above the limit of detection for the analytical methods. The same amount of each herbicide was applied to the tillage soil for comparison. The samples were collected in 50 mL tubes for determination of residues in the leachate.

Determination of Herbicide Residues in Leachate

A full desciption of the extraction and analytical methods used to determine MCPA and mecoprop-p concentrations in the collected leachets can be found in Piwowarczyk and Holden [11].

RESULTS AND DISCUSSION

Soils

Selected physical and chemical properties of the grassland and tillage soils used in the study are presented in Table 1. Soil properties varied by management and depth. The organic carbon and the cation exchange capacity (CEC) were higher in the samples taken from the upper layer (0-10 cm) of both profiles, although greater values were found in the grassland soil. The soil pH of the tillage soil was greater at 0-10 cm depth compared to the grassland soil, probably due to liming [12]. The sand content was greater in the upper layer in both cases with some increase in clay content with depth. The soil bulk density of the tillage samples did not differ by depth while a small increase was observed at 10-20 cm depth in the grassland soil, probably as a result of compaction caused by grazing [13]. The saturated hydraulic conductivity (K_{sat}) was greater in the grassland soil, probably due to a better pore distribution and structure [14], also reflected in lower bulk density values obtained from the grassland soil taken from 0-10 cm depth.

Table 1. Characteristics of the representative soil used in the study sampled under tillage (OT) and grassland (OG) management

Soil	Depth (cm)	OC (%)	pH (0.01 M $CaCl_2$)	CEC (cmol kg^{-1})	Sand (%)	Silt (%)	Clay (%)	Bulk density (g cm^{-3})	K_{sat} (cm day^{-1})
OT	0-10	3.8	5.8	16.0	66	28	6	1.34	69.0
	10-20	3.0	5.1	14.9	55	30	12	1.33	62.7
OG	0-10	4.4	5.0	18.3	65	26	9	1.18	98.3
	10-20	3.7	5.3	14.2	46	37	19	1.35	109.5

Transport Study

There was no MCPA or mecoprop-p observed in the leachate from the control columns or in the control samples collected prior to herbicide application from any of the soil columns.

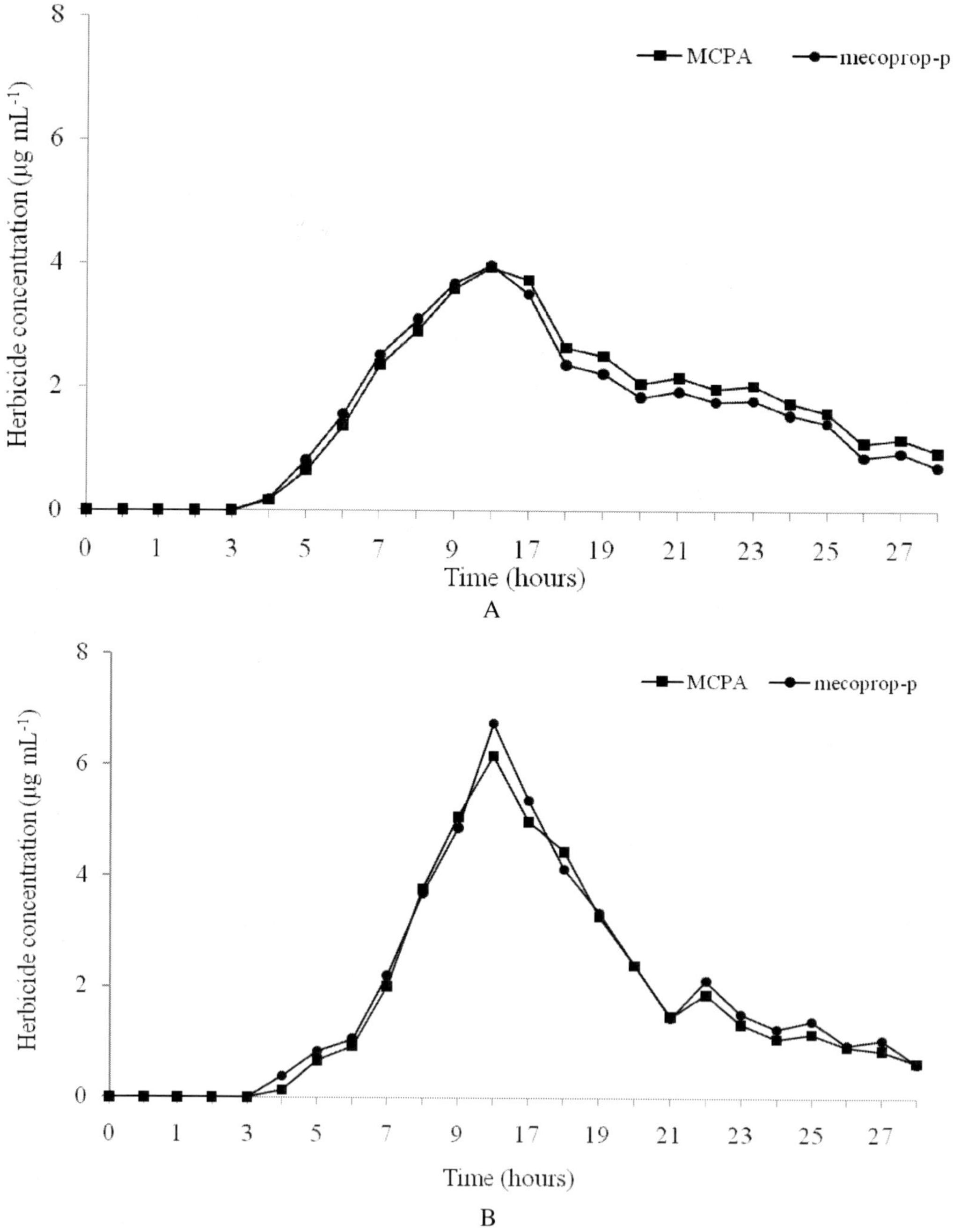

Figure 2. Breakthrough curves of MCPA and mecoprop-p. A: column 1 and B: column 2 (tillage soil).

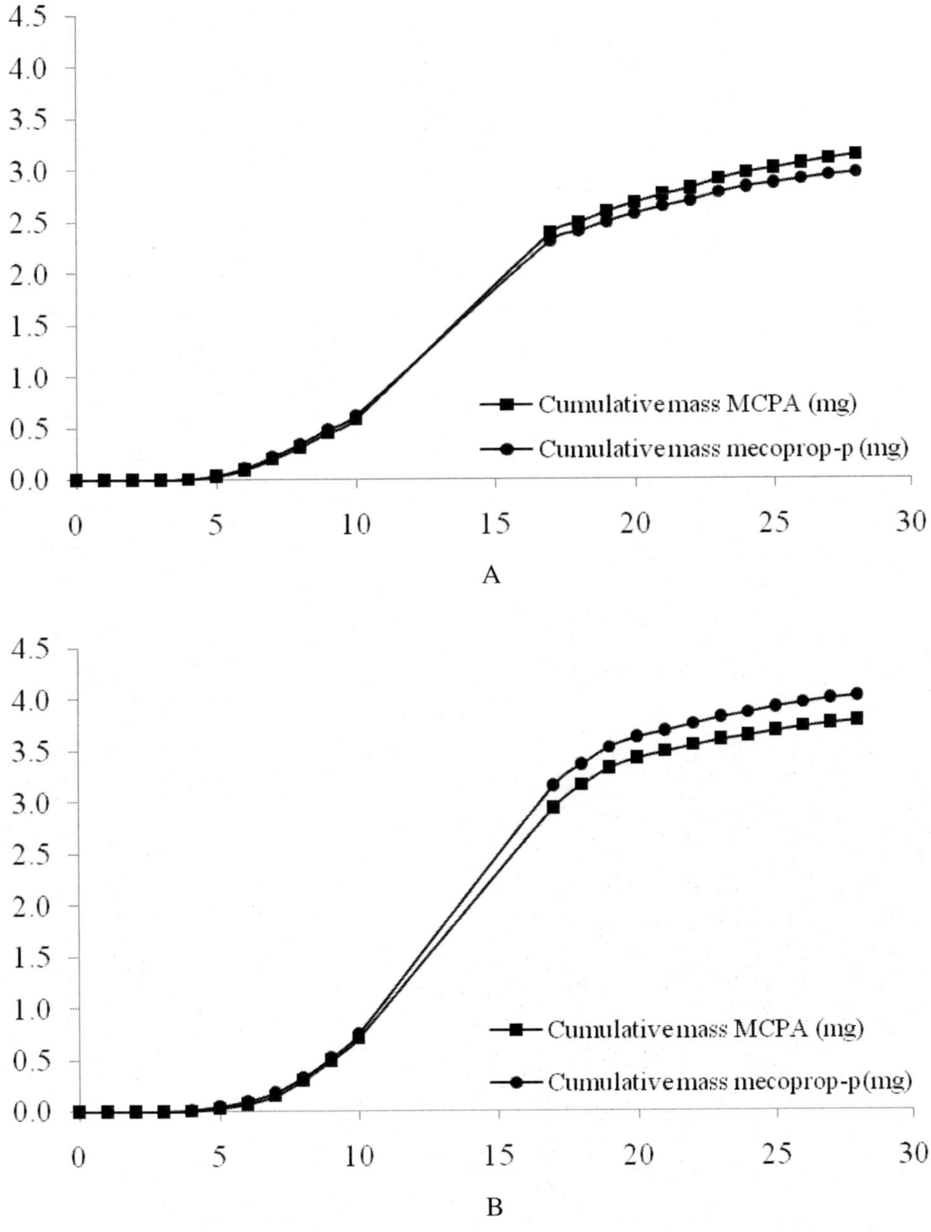

Figure 3. The cumulative mass (Y axis, mg) of MCPA and mecoprop-p leached from column 1 (A) and from column 2 (B) over 28 hour experimental period (X axis, hours) in the tillage soil experiment.

The average volume collected over the experimental period from the tillage soil columns was 39.5 mL hour^{-1} (SD = 5.0) and 41.0 mL hour^{-1} (SD = 4.2). Pressure heads (ΔH) in the tillage column experiment were 3.5 cm and 4.0 cm of water. The average volume collected over the experimental period in the grassland soil columns was 41.8 mL hour^{-1} (SD = 4.60) and 39.4 mL hour^{-1} (SD = 4.5). Pressure heads (ΔH) were 2.6 cm and 2.2 cm of water. The breakthrough curves of MCPA and mecoprop-p observed in the tillage soil are presented in Figure 2 (a, b). The cumulative mass of both herbicides leached from column 1 is shown in

Figure 3 (a) and from column 2 in Figure 3 (b). Approximately 60% (3.14 mg) of MCPA and 58% (2.97 mg) of mecoprop-p applied to the tillage soil was leached from colum 1, while approximately 73% (3.79 mg) of MCPA and 79% (4.03 mg) of mecoprop-p was leached from column 2. The maximum concentration of both herbicides was observed at about 10 hours post application while the first herbicide appearance in leachate was around 4 hours.

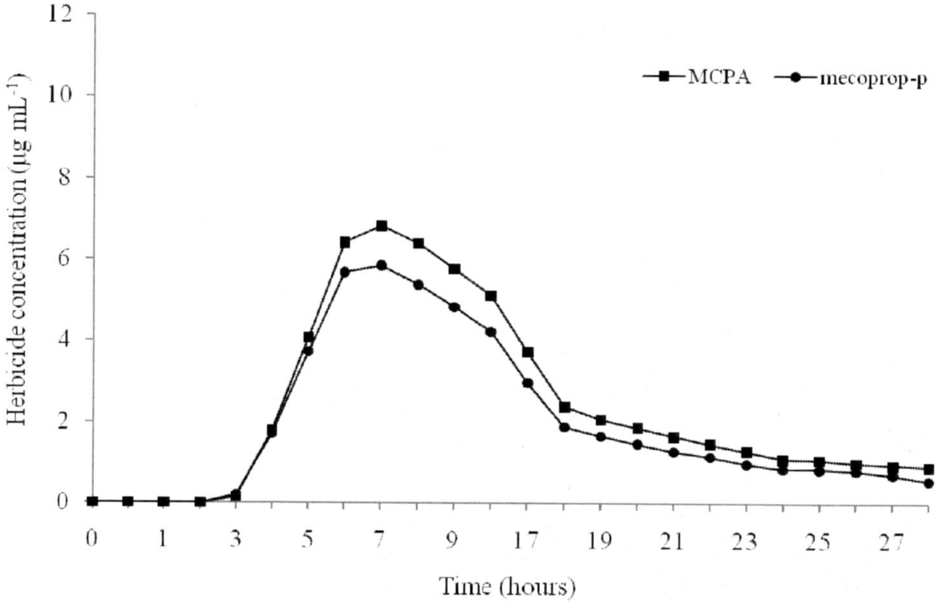

A

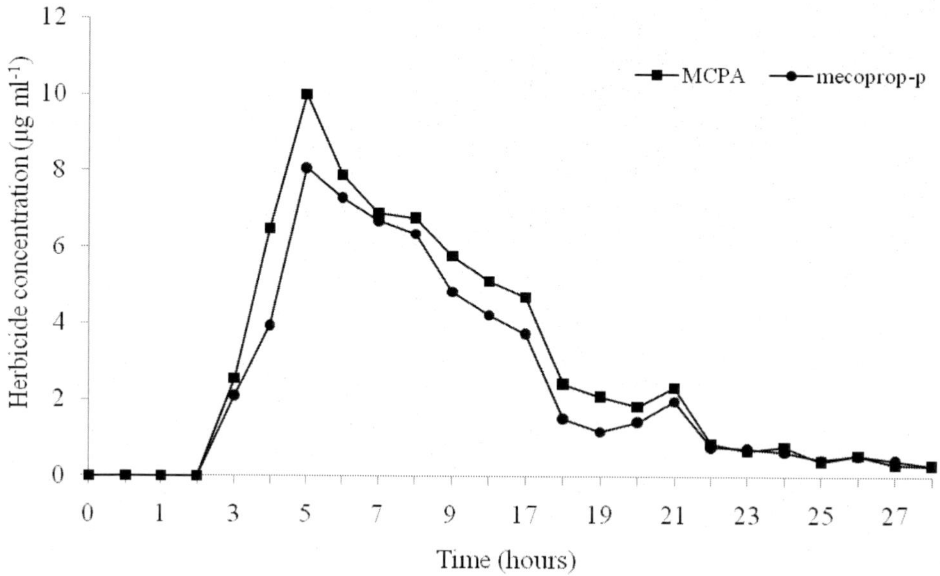

B

Figure 4. Breakthrough curves of MCPA and mecoprop-p. A: column 1 and B: column 2 (grassland soil).

Figure 4 presents the breakthrough curves of the two herbicicides studied in the grassland soil. Unlike in the tillage soil, the peak concentration occurred sooner at between 4 and 7 hours after application. This earlier maximum occurrence in the grassland soil was probably a result of better soil structure and porosity, as the hydraulic conductivity in the grassland soil was found to be greater than in the tillage soil (Table 1).

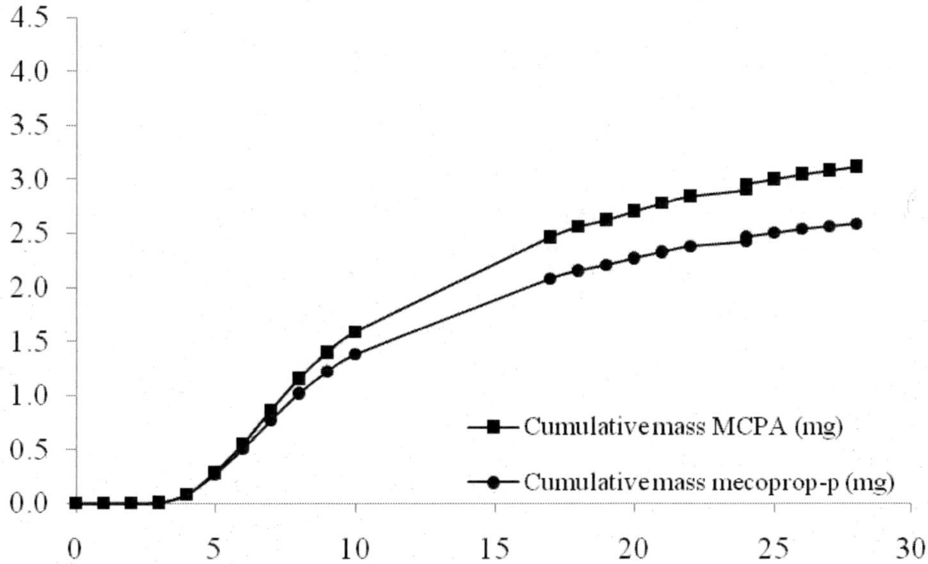

A

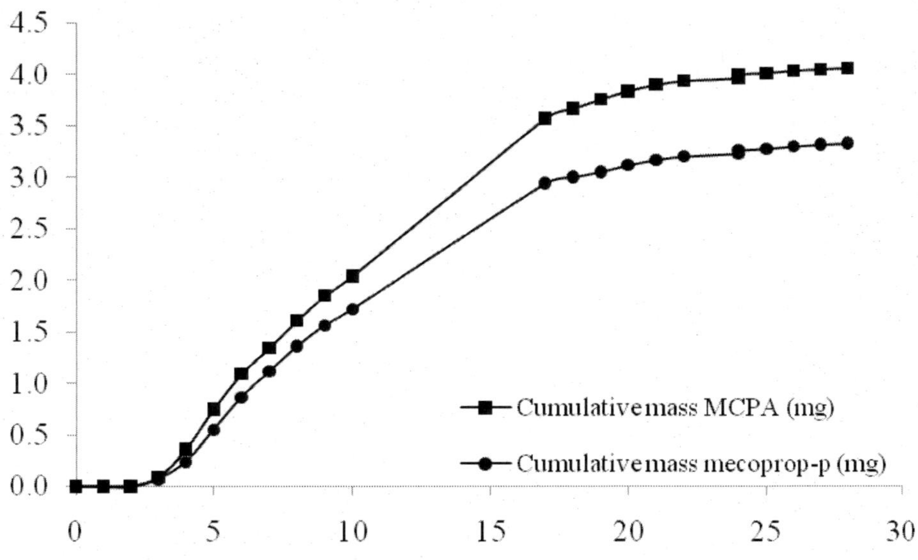

B

Figure 5. The cumulative mass (Y axis, mg) of MCPA and mecoprop-p leached from column 1 (A) and from column 2 (B) over 28 hour experimental period (X axis, hours) in the grassland soil experiment.

Approximately 60% (3.12 mg) of MCPA and 51% (2.59 mg) of mecoprop-p was leached from column 1, and approximately 78% (4.06 mg) of MCPA and 65% (3.33 mg) of mecoprop-p was leached from column 2. The cumulative mass of MCPA and mecoprop-p leached from column 1 is shown in Figure 5 (a) and from column 2 in Figure 5 (b). The first herbicide appearance was around 3 hours, which was one hour earlier than from the tillage soil columns.

In Piwowarczyk and Holden [11] was reported that the organic carbon normalised coefficients (K_{oc}) of the two herbicides ranged from 30.45 to 49.94 L kg^{-1} (single application), and from 25.42 to 52.74 L kg^{-1} (mix application), indicating high leaching potential. Such behaviour was supported by observations from the intact soil column experiment, in which both herbicides were leached in relatively large quantities (between 51 and 71% of initially applied herbicide was leached within 28 hours).

To fully describe the transport characteristics the CDE (convection-dispersion equation) would be used in conjunction with bromide or chloride tracer data. The CDE is commonly used to describe solute transport through the soil matrix assuming: (i) the flow regime is steady state, (ii) the soil water content is uniform, (iii) the soil is homogenous and isotropic, (iv) Darcy's law for water flow applies and (v) the solute flux law applies. The CDE has been used by many researchers for pesticide transport studies with soil columns, i.e. Beigel and Pietro [15], Meyer-Windel et al. [16], Abu-Zreig and Abu-Ashour [17], Magga et al. [18], and could be used with this experimental apparatus. In addition, loss of pesticide due to degradation could be determined in a full mass balance, although for MCPA and mecoprop-p it can be assumed that no degradation took place or it was negligible [19,20].

Conclusion

It was possible to examine the herbicides transport through undisturbed soil columns with the lysimeter apparatus. It has the potential to allow a full mass balance, co-application of conservative tracer and control of boundary conditions (so that other than saturated flow can be studied). The high mobility of MCPA and mecoprop-p estimated in isotherm sorption experiments [11] was supported by the example study where from undisturbed soil columns over 50% of both herbicides was leached within 28 hours. The laboratory column apparatus has potential to allow controlled environment studies of pesticide transport in a way not possible with conventional columns or field lysimeters.

Acknowledgments

Authors wish to acknowledge the Department of Agriculture, Food and the Marine Pesticide Control Laboratory (Celbridge, Co. Kildare, Ireland) to facilitate carrying out the chemical analysis, and in particular Dennis Carr and Noel Cosgrove for their technical support as well as other PCL technical and office staff. This work was funded by the Irish Government through Research Stimulus Fund contract RSF 07 544.

REFERENCES

[1] Goss, DW. (1992). Screening procedure for soils and pesticides for potential water quality impact. *Weed Technology*, *6*, 701-708.

[2] Holvoet, KMA; Seuntjens, P; Vanrolleghem, PA. Monitoring and modelling pesticide fate in surface waters at the catchment scale. *Ecological Modelling*, 2007, 209, 53-64.

[3] Leeds – Harrison, PB. 1995. The movement of water and solutes to surface and groundwaters. In: Walker A. et al. editors. *Pesticide Movement to Water*, (Monograph No 62, pp. 3-12). Surrey, UK: British Crop Protection Council.

[4] Gao, JP; Maguhn, J; Spitzauer, P; Kettrup, A. Sorption of pesticides in the sediment of the Teufelsweiher Pond (Southern Germany). In: Equilibrium assessments, effect of organic carbon content and pH. *Water Research*, 1998, 32, 1662-1672.

[5] Wauchope, RD; Yeh, S.; Linders, JBHJ; Kloskowski, R; Tanaka, K; Rubin, B; Katayama, A; Kördel, W; Gerstl, Z; Lane, M; Unsworth, B. Review Pesticide soil sorption parameters: theory, measurement, uses, limitations and reliability. *Pest Management Science*, 2002, 58, 419-445.

[6] Picton, P. and Farenhorst, A. (2004). Factors influencing 2,4 – D sorption and mineralization in soil. Journal of Environmental Science and Health Part B- Pesticides, *Food Contaminants, and Agricultural Wastes*, *B39*, 367-379.

[7] Köhne, JM; Köhne, S; Šimunek, J. Multi-process herbicide transport in structured soil columns: Experiments and model analysis. *Journal of Contaminant Hydrology*, 2006, 85, 1-32.

[8] Dousset, S; Thevenot, M; Pot, V; Šimunek, J; Andreux F. Evaluating equilibrium and non-equilibrium transport of bromide and isoproturon in disturbed and undisturbed soil columns. *Journal of Contaminant Hydrology*, 2007, 94, 261-276.

[9] Højberg, AL; Engesgaard, P; Bjerg, PL. Pesticide transport in an aerobic aquifer with variable pH- Modelling of a field scale injection experiment. *Journal of Contaminant Hydrology*, 2005, 78, 231-255.

[10] Winton, K. and Weber, J.B. (1996). A review of field lysimeter studies to describe the environmental fate of pesticides. *Weed Technology, 10*, 202-209.

[11] Piwowarczyk, A.A and Holden, N.M. (2013). Phenoxyalkanoic acid herbicide sorption and the effect of co-application in a Haplic Cambisol with contrasting management. *Chemosphere,* 90, 535-541.

[12] Webb, J; Loveland, PJ; Chambers, BJ; Mitchell, R; and Garwood T. The impact of modern farming practices on soil fertility and quality in England and Wales. *Journal of Agricultural Science*, 2001, 137, 127-138.

[13] Drewry, JJ; Cameron, KC; Buchan, GD. Pasture yields and soil physical property responses to soil compaction from treading and grazing: a review. *Australian Journal of Soil Research*, 2008, 46, 237–256.

[14] Hiller, D. Flow of water in saturated soil. In: *Environmental Soil Physics*. San Diego, CA: Academic Press, 173-202; 1998.

[15] Beigel, C. and Pietro, L.D. (1999). Transport of triticonazole in homogenous soil columns: influence of nonequilibrium sorption. *Soil Science Society of America Journal, 63*, 1077-1086.

[16] Meyer-Windel, S; Lennartz, B; Widmoser, P. Bromide and herbicide transport under steady-state and transient flow conditions. *European Journal of Soil Science*, 1999, 50, 23-33.

[17] Abu-Zreig, A. and Abu-Ashour, J. (2004). Chloride and atrazine transport through saturated soil column. *Toxicological and Environmental Chemistry, 86*, 181-190.

[18] Magga, Z; Tzovolou, DN; Theodoropoulou, M.A; Dalkarani, T; Pikios, K; Tsakiroglou, CD. Soil column experiment used as a means to assess transport, sorption, and biodegradation of pesticides in groundwater. *Journal of Environmental Science and Health Part B*, 2008, 43, 732-741.

[19] EEC. (2003). Review report for the active substance: Mecoprop-p. SANCO/3065/99. European Commission.

[20] EEC. (2004). Review report for the active substance: MCPA. SANCO/4062/2001. European Commission.

INDEX

A

absorption spectra, 248, 250, 252, 253
absorption spectroscopy, xii, 245, 246, 247, 250
access, 38, 134, 241
accessions, 43
accounting, 277, 282
acetic acid, vii, xi, 125, 194, 231, 232, 244, 268
acetone, 153, 270
acetonitrile, 156, 234
acetylcholine, 98, 99, 145
acetylcholinesterase, 137
acetylcholinesterase activity, ix, 86, 103, 104, 106, 108
acidic, 13, 15, 19, 57, 63, 69, 71, 76, 116, 214, 232, 268
acidity, 226
active oxygen, 210
active site, 23, 33, 38, 99, 223
active transport, 75, 76
adaptation, vii, 1, 19, 22, 25, 27, 117
adenine, 192
adenocarcinoma, 140
adenosine, 63
adenosine triphosphate, 63
adipose tissue, 142
adsorption, vii, xi, 89, 90, 102, 103, 105, 114, 116, 126, 127, 213, 215, 217, 220, 221, 222, 223, 224, 225, 226, 227, 228, 231, 232, 233, 234, 235, 236, 237, 238, 239, 240, 241, 242, 243, 244, 285, 286, 289, 292
adsorption coefficients, xi, 231, 237
adsorption isotherms, 116, 220, 221, 222, 223, 227, 243
adults, 136
advancement, 86
adverse effects, xii, 87, 92, 192, 193, 216, 232, 291
agar, 3, 4, 8
age, 49, 61, 64, 187
agricultural market, 246
agriculture, ix, xi, 30, 44, 56, 88, 105, 111, 180, 186, 213, 215, 219, 232, 233
agrobacterium, 22
airborne particles, 131
alanine, 38
alcohols, 152, 158
aldehydes, 152
alfalfa, 42, 178, 179
algae, 91, 95, 128
algorithm, 7
alkylation, xii, 35, 263, 265
ALS, 32, 33, 36, 37, 38, 40, 43, 45, 46, 49, 50, 54
alters, 99, 107, 108
aluminium, 227
ambient air, 130
amine(s), 91, 120, 215, 225, 227
amine group, 120, 225
amino, 3, 33, 38, 41, 50, 56, 57, 89, 135, 194, 224, 280
amino acid(s), 3, 33, 38, 41, 50, 56, 89, 194
ammonium, 5
amphibia, 136, 209
amphibians, 91, 95, 103, 146
amplitude, 60, 62, 64, 65, 66, 70, 71, 72
androgen, 138
aneuploidy, 123, 132
aniline, 117, 132, 229
anoxia, 82
anthocyanin, 50
antibiotic, 2, 45
antibiotic resistance, 2, 45
antioxidant, ix, 35, 86, 96, 97, 98, 101, 103, 106, 109, 135, 136, 137, 138, 145, 193, 194, 200, 204, 207, 208
antisocial behavior, 127
apoptosis, 137, 142, 143, 149, 196
apoptotic mechanisms, 143

aquatic systems, 40
aqueous solutions, 155, 156, 232, 290
Arabidopsis thaliana, 35, 37, 45
Argentina, viii, xi, 85, 87, 88, 91, 93, 101, 102, 103, 105, 106, 107, 108, 213, 214, 216, 219, 226, 228, 229
arginine, 40
aromatic compounds, 251
aromatic rings, 34, 251
aromatics, 131
ARS, 244
ascorbic acid, 96
asparagus, 134
aspartate, 49
aspartic acid, 38
assessment, 87, 102, 108, 126
assimilation, 34, 57
atherosclerosis, 136
atmosphere, 87, 125, 218
atmospheric pressure, 119
atomic force, 227
atoms, 214, 274
ATP, 38, 57, 58, 59, 63, 74, 75, 143
autoimmunity, 83
automate, 118
avoidance, 108, 183
awareness, x, 175, 184, 186

B

background noise, 120
bacteria, 23, 24, 26, 57, 58, 77, 81, 91, 117, 121, 129, 162, 171
bacterial strains, 27, 116
bacterium, 117, 153
banks, 187
barriers, 33, 64
base, 41, 138, 176, 224, 225, 248, 251, 255
base pair, 248, 251, 255
behaviors, 35, 278
benefits, 116
benign, 91
bicarbonate, 206, 228
binding energies, 225
bioaccumulation, 88, 92, 95, 109, 116, 117, 192
bioassay, x, xii, 151, 153, 163, 164, 165, 169, 170, 263, 268
bioavailability, 128, 153, 154, 164, 171
biochemistry, 36, 242
biodegradation, 112, 113, 115, 116, 302
biodiversity, xii, 87, 92, 107, 183, 216, 291
bioindicators, ix, 86
biological activity, vii, 171, 267
biological fluids, 135
biological responses, 139
biological systems, 81, 92, 113
biomarkers, ix, 86, 95, 96, 101, 103, 106, 109, 121, 122, 123, 137, 210
biomass, 5, 7, 8, 30, 187
biomolecules, ix, 133, 136
biomonitoring, 93, 131
biosynthesis, 33, 34, 36, 38, 50, 51, 81, 89, 104, 275
biosynthetic pathways, 49
biotic, 114, 115
bleaching, 40
blood, 104, 136, 137, 139, 141, 142, 148, 194
body weight, 134, 138, 140
bonding, 255
bonds, 81
bone marrow, 123, 139, 140, 141, 147, 148
brain, 99, 103, 106, 136, 145, 246
Brazil, ix, 131, 133
breakdown, 92, 214, 243
breast cancer, 137
breeding, 93
bursa, 37
by-products, 152, 153, 156, 160, 162, 164, 171

C

Ca^{2+}, 143
cadmium, 104, 228
calcium, 206
calcium carbonate, 217
calibration, 120, 156
cancer, vii, 77, 83, 121, 130, 131, 132, 136, 140, 141, 148, 194, 209, 214, 246, 259
cancer cells, 77, 194, 209
cancer therapy, 77
CAP, 178
capillary, 59, 118, 119, 120, 129, 130
carbon, xi, xii, 3, 17, 19, 30, 75, 81, 115, 117, 118, 152, 231, 233, 238, 240, 241, 243, 277, 278, 285, 292, 295, 300, 301
carbon dioxide, 30
carboxyl, 38, 89
carcinogen, 140, 192
carcinogenesis, 146
carcinoma, 194
carotene, 96
carotenoids, 50, 135
cartoon, 76
cascades, 143
case study, 54, 128
caspases, 143
catabolism, 26, 118

catalysis, 24, 27, 106, 214
catalytic activity, 20, 48
catalytic effect, 117
catecholamines, 96
category a, 77
catfish, 104, 105
cation, xi, 191, 192, 224, 233, 241, 292, 295
cattle, ix, 133
causal relationship, 141
CD95, 143
cDNA, 256
CEC, 235, 295
cell culture, 42, 53, 112, 121, 125, 126
cell cycle, 126
cell death, x, 133, 138, 142, 143, 149, 249
cell line(s), ix, 42, 53, 111, 112, 123, 124, 125, 126, 127, 132
cell metabolism, 57, 80
cellular environment, vii, 1
cellular signaling pathway, 142
central nervous system, 192
challenges, 188
charge density, 229
chemical degradation, 114, 128, 214
chemical interaction, 104, 292
chemical properties, vii, ix, 112, 235, 292, 295
chemical structures, 96, 134, 278
chemiluminescence, 120
chemisorption, 215
children, 192
Chile, 83, 180
China, 263, 274
chlorination, xii, 263, 270
chlorine, xii, 154, 263, 265, 267, 268
chlorophyll, xii, 40, 42, 50, 51, 52, 263, 264
chloroplast, viii, 34, 38, 40, 41, 42, 55, 56, 58, 59, 63, 64, 73, 75, 76, 77, 134
CHO cells, 112, 123
cholesterol, 137
choline, 65
cholinesterase, 109
chromatid, 122, 126, 132, 148, 195, 202, 209, 210
chromatographic technique, 120
chromatography, 5, 119, 129, 130, 220, 229, 243
chromosomal alterations, 258
chromosomal damage-enhancement-mechanism, xi, 191
chromosome, 2, 26, 122, 123, 125, 126, 194, 195, 202
circulation, 61
city(ies), 121, 130, 131, 216
classes, xii, 32, 33, 50, 76, 136, 263, 264
classification, vii, ix, 31, 36, 111, 131, 141, 142, 147, 233, 235, 240, 243
clay minerals, 223, 225, 227
cleanup, 119, 120
cleavage, viii, 2, 5, 10, 12, 17, 18, 20, 22, 27, 117, 171
climate, 56, 215, 292
climate change, 56
closure, xii, 177, 263
clustering, 41
C-N, x, 151
CNS, 99
CO_2, xiii, 125, 291
coding, 2, 6, 15, 16
codon, 40, 41
coenzyme, 29, 47, 50
collaboration, 186
color, 217
combined effect, 240
combustion, 233
commercial, 26, 59, 86, 89, 91, 93, 102, 107, 123, 219, 264
common groundsel, viii, 29, 30
communication, 295
community(ies), 23, 27, 87, 92, 107, 109, 188
compaction, 295, 301
companion cell, 34
competition, 45, 51, 56, 223
competitiveness, 188
complement, 154
complexity, 120, 126, 184
compliance, 234
composition, 41, 42, 53, 95, 113, 114, 115, 116, 117, 137, 153, 159, 160, 161, 217, 229, 242
compost, 116, 128
condensation, xii, 263
conditioning, 234
conductivity, xiii, 220, 291, 294, 295, 299
conflict, 127
conjugation, 4, 22, 34, 35, 36, 136, 140
consent, 112
conservation, 178, 179
constituents, x, 87, 91, 104, 135, 151, 158, 292
consumption, ix, 3, 5, 9, 10, 17, 18, 20, 21, 22, 118, 133, 208
consumption patterns, 9
consumption rates, 3, 18
containers, 157, 214
contaminant, 130
contamination, 116, 127, 136, 214, 243
control group, 93, 94, 98, 99, 100
cooling, 7
coordination, 24, 89, 215, 225

copolymer(s), 193
copper, 89, 109, 215, 228, 229
correlation(s), 13, 158, 221, 238, 283, 286, 288
correlation coefficient, 158
cost, viii, 30, 45, 55, 56, 118, 184, 187
cost effectiveness, 184
Costa Rica, 210
cotton, 47, 87
cotyledon, 52
covalent bonding, 99
covering, 264
creep, 182
criticism, 154
Croatia, 146, 260
crop production, xii, 78, 116, 179, 189, 263, 277
crop residue, 116
cross-fertilization, 180
cultivation, 3, 4, 8, 9, 10, 11, 15, 16, 17, 18, 20, 25, 86, 91, 107, 288
cultivation conditions, 18
cultural practices, 184
culture, vii, 1, 4, 7, 8, 10, 12, 42, 46, 126, 138, 154, 194, 210
culture medium, 126, 194
cuticle, 33, 91, 154, 155, 158, 159, 163
cutin, 158
cycles, 6
cycling, viii, 6, 29
Cyprus, 132
cysteine, 38
cytochrome(s), 29, 45, 106, 135, 145, 192, 264
cytokinesis, 126, 132
cytoplasm, 33, 76
cytoskeleton, 137
cytotoxicity, 147, 192
Czech Republic, 176

D

damages, 123, 126, 136, 139
danger, x, 120, 162, 171, 175
data collection, 61, 63
database, 244
DCA, 125
decay, 5, 8, 156
decomposition, 117, 193, 283, 287
decoupling, 268
defects, 214, 246
defense cellular mechanisms, x, 133
deficiencies, 64
deforestation, 87
deformation, 137, 217
degradation process, 115, 117, 159
degradation rate, 116, 152, 292
Delta, 6, 175
denaturation, xi, 6, 245, 247, 248, 249, 251, 255, 256, 260, 261
density values, 295
deoxyribonucleic acid, 260
Department of Agriculture, 233, 234, 241, 293, 300
depression, 215
depth, vii, xi, 213, 217, 289, 295
derivatives, vii, xii, 3, 25, 47, 112, 113, 117, 119, 128, 136, 146, 263, 264, 265, 267, 268, 276
desiccation, 33
desorption, vii, xi, 107, 127, 231, 232, 233, 234, 236, 237, 239, 241, 242, 243, 244, 290, 292
destosyl pyrazolynate, xii, 277, 278
detachment, 125
detectable, 10, 31, 57
detection, viii, 8, 10, 55, 58, 64, 67, 77, 82, 109, 112, 119, 120, 129, 130, 131, 142, 153, 171, 235, 295
detoxification, viii, 29, 30, 35, 36, 44, 48, 49, 77, 186
developing countries, 45, 56, 211
deviation, 70, 286
DFT, 23
diacylglycerol, 42
diamonds, 12
diarrhea, 214
diffraction, 218
diffusion, 34, 61, 66, 75, 118, 148, 285
digestion, 6
dimethylformamide, 269
diploid, 132
discrimination, 60
discs, 82, 157
diseases, 136, 213, 246
dispersion, 118, 119, 129, 217, 232, 300
displacement, 61, 87
disposition, 16, 18, 20, 23
dissociation, xi, 231, 237, 241
dissolved oxygen, 4
distilled water, 206
distribution, viii, xii, 1, 13, 22, 35, 76, 79, 93, 95, 96, 116, 158, 162, 180, 185, 189, 220, 226, 232, 233, 237, 240, 277, 282, 285, 287, 288, 291, 293, 295
diversity, x, 23, 72, 77, 175, 188
DNA, vi, vii, ix, xi, 6, 7, 41, 96, 103, 104, 106, 122, 131, 133, 138, 139, 140, 141, 142, 143, 145, 146, 147, 148, 245, 246, 247, 248, 249, 250, 251, 252, 253, 254, 255, 256, 257, 258, 259, 260, 261
DNA breakage, 122, 138
DNA damage, 96, 103, 131, 138, 139, 140, 143, 145, 146, 147, 148, 246, 249, 258
DNA sequencing, 41
DNA strand breaks, 106

DOI, 78
donors, 26
dosage, 74, 214
double helix, 251, 255, 258
drainage, 217, 278
dream, 187
drinking water, xii, 148, 153, 232, 291
drought, 183
drug interaction, 251
drug resistance, 36, 77
drugs, 143, 255
drying, 7, 194, 233
DSC, 128
dumping, 214
dyes, 260

E

ecology, 45, 187, 242
economic performance, 187
economic problem, 184
ecosystem, 92, 102, 112, 113, 116, 129, 139
ecotoxicological, 92, 99, 101, 142
editors, 26, 301
EEA, 242
effluent(s), 5, 6, 115
effluents, 115
electrochemistry, 246
electrolyte, 250, 252, 253, 254, 257
electron(s), 30, 35, 36, 42, 53, 134, 135, 192, 198, 200, 202, 204, 205, 207
electron paramagnetic resonance, 53
electrophoresis, vii, 1, 6, 13, 120, 129, 130, 131, 139, 148, 246, 260
electrophoretic separation, 3
electrospray ionization, x, 151
elongation, 168
elucidation, 21, 58, 139
emission, 153, 158, 186
employment, 87
enantiomers, vii, 1, 7, 8, 10, 17, 18, 20, 25, 28
encoding, 2, 23, 27
endocrine, 92, 95, 115, 137, 142, 246
endocrine system, 138
energy, 3, 17, 19, 57, 59, 74, 116, 118, 142, 156, 209, 224
England, 240, 301
entropy, 248, 249, 255
environment(s), vii, viii, ix, xii, 1, 2, 23, 24, 42, 57, 65, 85, 86, 87, 88, 89, 91, 92, 93, 95, 105, 109, 112, 113, 114, 115, 116, 120, 126, 127, 153, 154, 162, 171, 192, 214, 219, 232, 246, 285, 291, 292, 300
environmental aspects, 127
environmental change, 92
environmental conditions, xi, 2, 87, 231
environmental factors, 63, 109, 240, 241
environmental impact, 86, 87, 115
Environmental Protection Agency (EPA), 91, 109, 134,141, 144, 147, 259
environmental quality, 116
environmental stress, ix, 86, 92, 95, 136
epidemiologic studies, 148, 246
epidermis, 33
epigenetics, 142
epithelial cells, 131, 136
epithelium, 136
equilibrium, xi, 112, 116, 220, 222, 231, 232, 234, 235, 237, 248, 260, 285, 286, 294, 301
equipment, 118, 176, 183
erythrocytes, 103, 136, 137, 139, 140, 145
ESI, x, 120, 151, 160
ester, 160, 229, 233, 276
ethers, 40
ethyl acetate, 155, 269, 270
eukaryotic, 50, 121
Europe, vi, vii, x, 45, 80, 175, 176, 178, 182, 184, 188, 242
European Commission, 302
European Community, 258
European Union EU, 232, 233, 246
evacuation, 66
evidence, 2, 20, 21, 36, 59, 64, 68, 69, 71, 75, 77, 79, 91, 115, 128, 142, 143, 180, 205
evolution, x, 2, 24, 27, 30, 32, 34, 36, 43, 44, 45, 51, 54, 88, 162, 175, 180, 182, 183, 186, 187
EXAFS, 228, 229
excitotoxicity, 143
exclusion, viii, 33, 55, 57, 59, 64, 73, 76, 77, 180
excretion, 10, 135, 136
execution, 143
experimental condition, 95, 223, 286
experimental design, 94, 157
exporters, 77, 86
exposure, ix, 23, 30, 33, 58, 75, 86, 94, 96, 97, 98, 99, 101, 104, 107, 108, 112, 117, 120, 121, 130, 131, 136, 137, 138, 139, 142, 143, 145, 149, 176, 208, 214, 219, 246, 258, 259
extinction, 19
extraction, 4, 6, 74, 112, 118, 119, 120, 126, 129, 130, 153, 234, 237, 295
extracts, viii, 5, 6, 10, 11, 13, 17, 36, 55, 130, 140, 269, 270

F

factories, 127
FAD, 40
families, 40, 214
farmers, x, 44, 87, 121, 131, 175, 177, 184, 210, 233, 246, 259
farmland, 183
farms, 87, 127
FAS, 143
fatty acids, 34, 38, 193, 196, 197
fertility, viii, 29, 116, 301
fertilization, 93, 210
fertilizers, 101, 136, 146
fiber(s), 56, 138
fibroblasts, 209
fidelity, 74
field crops, 47
financial, 44, 226, 233, 258
financial support, 44, 226, 233, 258
fish, vii, 86, 91, 92, 93, 95, 96, 97, 98, 99, 101, 102, 103, 104, 105, 106, 108, 109, 137, 139, 143, 146, 147
fitness, 30, 182, 188
fixation, 30, 194
flame, 129
flavonoids, 36, 38, 135
flight, x, 151
flora, 117, 214
flora and fauna, 214
fluctuations, 22
fluid, 42, 118, 119, 137
fluid extract, 118
fluorescence, 6, 22, 31, 42, 119, 120, 128, 139, 246, 258, 259
FMC, 264
food, viii, xii, 29, 56, 77, 88, 90, 92, 115, 129, 139, 209, 210, 214, 261, 291
food additive(s), 261
food chain, 88, 90, 92, 115, 139, 214
food production, 56
food products, 210
food security, 77
force, 267
formation, x, xi, 4, 5, 9, 12, 19, 20, 40, 47, 78, 89, 90, 96, 114, 137, 138, 139, 140, 151, 152, 156, 161, 162, 171, 191, 193, 204, 205, 207, 210, 215, 224, 225, 283
formula, 161, 274
fragile site, 202
fragments, 123, 139, 160
France, x, 175, 176, 177, 178, 179
free energy, 222
free radicals, 96, 193, 200
freshwater, ix, 86, 93, 101, 103, 106, 107, 108, 109, 116
fructose, 4
fruits, 87, 177, 178
FTIR, 102, 225, 226, 228

G

GC-FID, 129
gel, vii, 1, 3, 6, 13, 131, 139, 148, 246
gene amplification, 30, 42, 43, 45, 79
gene expression, 3, 7, 15, 16, 18, 21, 23, 36, 123, 138, 142
gene pool, 182
gene transfer, 2, 23
genes, vii, 1, 2, 3, 4, 6, 15, 16, 18, 20, 21, 22, 26, 27, 40, 41, 42, 45, 57, 59, 138, 176, 182
genetic background, viii, 2
genetic diversity, 77
genetic engineering, viii, 55, 89
genetic information, 2, 17, 22
genetics, 180
genome, 42
genotyping, 50
genus, 27, 117
geography, 41
Germany, 1, 4, 6, 7, 27, 155, 233, 242, 276, 301
germination, x, 151, 153, 154, 157, 163, 164, 165, 166, 171, 178
Gibbs energy, 249
gill, 98, 137, 139, 140, 146
gland, 106
glomerulus, 137
glucose, 34
glutamate, 80, 143
glutathione, 29, 34, 35, 48, 49, 59, 96, 135, 136, 140, 145
glycerol, 4
glycine, xi, 38, 56, 89, 213, 214, 228
glyphosate resistance, x, 41, 42, 44, 45, 48, 52, 53, 54, 56, 57, 58, 59, 68, 71, 74, 77, 78, 79, 80, 82, 83, 175, 176, 177, 178, 180, 181, 182, 184, 185, 186, 187, 188, 189
graphite, 274
grass(es), x, xi, 33, 35, 48, 50, 65, 72, 89, 134, 151, 152, 154, 156, 163, 164, 167, 170, 171, 179, 219, 233, 245
grasslands, 87, 215, 233
gravity, 156, 217, 293
grazing, 243, 295, 301
Greece, x, 175, 176, 177, 178, 179, 185, 188
greenhouse, 31, 58, 60, 64, 71, 73, 154, 274

groundwater, ix, xi, xii, 87, 90, 127, 133, 153, 214, 231, 232, 240, 241, 291, 302
growth, vii, xi, 1, 3, 4, 5, 7, 8, 10, 12, 13, 15, 16, 17, 18, 20, 22, 24, 25, 26, 30, 31, 34, 36, 42, 45, 58, 60, 65, 66, 74, 82, 89, 92, 93, 108, 112, 113, 116, 117, 125, 146, 149, 153, 154, 157, 158, 167, 168, 177, 178, 182, 183, 245
growth rate, vii, 1, 3, 5, 7, 8, 15, 17, 18, 19, 20, 22, 24, 25, 30
guidelines, 43, 178, 242

H

habitats, 87, 185
half-life, x, xii, xiii, 151, 157, 158, 214, 240, 241, 277, 282, 283, 284, 288, 291
halogen, 117, 267
harvesting, 125, 264
hazards, xi, 147, 245
health, vii, xii, 57, 58, 127, 130, 131, 142, 148, 149, 192, 193, 246, 259, 291
health effects, 142, 246
heat shock protein, 137
heating rate, 248
height, 60, 65, 182, 249, 292, 293
heme, 24, 25, 27, 40, 51, 137, 264
hemoglobin, 264
hepatocytes, 108
herbicide uptake, viii, 33, 55, 74
heterogeneity, 214, 221
hexane, x, 151, 155, 158, 159, 162, 274
hippocampus, 136
history, 41, 176, 180, 292
homeostasis, 137
homogeneity, 57
hormone(s), 108, 142
horses, 243
host, vii, 1, 2, 3, 12, 17
household sector, 246
HPLC-UV, 120
hub, 66
human, viii, ix, xii, 29, 36, 56, 91, 111, 112, 120, 121, 127, 133, 137, 140, 146, 148, 192, 193, 208, 209, 213, 214, 216, 246, 258, 259, 291
human body, 214
human exposure, 127
human health, ix, xii, 91, 111, 112, 121, 127, 148, 192, 213, 216, 246, 291
humidity, 233
hybrid, 23, 156
hydrazine, 265, 268
hydrocarbons, 130, 152
hydrogen, xii, 96, 192, 193, 195, 200, 202, 208, 210, 251, 255, 263, 267, 268, 274
hydrogen atoms, 274
hydrogen bonds, 251
hydrogen peroxide, 96, 192, 193, 195, 200, 202, 208, 210
hydrolysis, 34, 95, 98, 116, 118, 152, 159, 214, 227, 229
hydroperoxides, 193
hydrophobicity, 290
hydroxide, 59
hydroxyl, 96, 192, 193, 210, 214, 215, 224
hypertension, 136
hypothesis, 20, 21, 41, 59, 61, 62, 73, 204
hysteresis, 239, 244

I

ibuprofen, 132
ID, 186
ideal, 63, 72, 152, 274, 275
identification, ix, x, 52, 81, 86, 111, 115, 119, 121, 126, 151, 156
illumination, 157
imbalances, viii, 2, 10
imidazolinone, 32, 38
immersion, 125, 126
immune response, 92, 104, 123
immunity, 83
in vitro, ix, xi, 47, 78, 92, 111, 112, 121, 123, 124, 125, 136, 139, 140, 141, 143, 147, 148, 153, 191, 194, 196, 260
in vivo, viii, 47, 55, 57, 58, 60, 71, 72, 73, 74, 76, 77, 78, 80, 81, 82, 121, 123, 132, 136, 138, 139, 140, 147, 148, 192
incidence, 148, 246
incomplete combustion, 130
incubation period, 71
incubation time, 70
incubator, 154
individual character, 21
individual characteristics, 21
individual differences, 5
individuals, 30, 32, 41, 43, 44, 176, 180, 182, 183, 188, 206
inducer, 207
induction, 123, 137, 138, 195, 197, 205, 207
industrial chemicals, 120
industrialization, 86
industry, 259
inequality, 86
infection, 105
ingestion, ix, 111, 121, 133

ingredients, 91, 246
inheritance, 41
inhibition, ix, 25, 35, 36, 40, 47, 49, 53, 58, 63, 64, 65, 67, 69, 74, 78, 81, 86, 89, 92, 99, 108, 109, 111, 134, 138, 143, 154, 168, 196, 200, 249
inhibitor, 41, 43, 49, 51, 52, 54, 56, 91, 263, 275
initiation, 7, 157, 177
injury, 163, 164, 168, 192, 208
inoculation, 24
inoculum, 4
insecticide, 214, 228
insertion, 138
interface, 228, 258
interference, 47, 51
intermediaries, 96
international division of labor, 86
interphase, 121, 123
intervention, 36
intestinal tract, 134
intestine, 135
invertebrates, 99, 104
ion uptake, 137
ion-exchange, 209
ionization, x, 119, 151
ions, 89, 152, 161, 163, 215
Iowa, 44
IR spectra, 268
IRC, 7
Ireland, 45, 155, 231, 233, 234, 240, 241, 243, 291, 292, 293, 300
iron, 23, 24, 27, 89, 137, 215, 217, 220, 224, 225, 226, 227
irradiation, 152, 153, 155, 156, 157, 158, 159, 160, 162, 193, 194, 282
irrigation, 91, 177, 183, 214
islands, 26
isolation, 20, 116, 129, 260
isoleucine, 36, 38, 50
isomerization, x, 151, 160, 171
isomers, 160
isotherms, xi, 213, 231, 232, 233, 234, 235, 237, 239, 240, 241
isozymes, 49, 89
issues, 60, 86, 147, 183
Italy, x, 46, 111, 131, 175, 176, 177, 179, 185

J

Japan, 191, 204, 205, 277, 278, 282, 285, 288, 289, 290
Jordan, 79, 186

K

karyotype, 194, 202
KBr, 270, 271, 272, 273
ketones, 152
kidney, 96, 135, 136, 142, 145, 148, 192
kinetic model, 116
kinetic parameters, 13, 19
kinetics, 12, 13, 19, 25, 128, 155, 156, 158, 234, 236, 244, 281, 292
Korea, 46

L

labeling, 62, 80, 275
laboratory studies, 128
laboratory tests, 121
lactose, 194
lakes, 87
land tenure, 87
landscape, 217, 219
larvae, 210
Latin America, viii, 85, 86
leaching, 113, 114, 127, 227, 240, 241, 292, 300
lead, 58, 73, 99, 112, 113, 115, 135, 136, 152, 171, 215
legend, 15, 16, 194, 202
lesions, 138, 210
leucine, 36, 50
leukemia, 246
leukocytes, xi, 141, 191, 195, 196, 197, 198, 199, 200, 201, 202, 203, 205
liberation, 12
ligand, 143, 220, 224, 226
light, xii, 23, 42, 45, 64, 81, 125, 153, 154, 157, 160, 178, 194, 263, 264
limestone, 130
lipid oxidation, 209
lipid peroxidation, xii, 96, 98, 106, 135, 137, 138, 192, 196, 197, 208, 209, 263
lipids, 42, 53, 134, 136, 158, 208
lipoproteins, 210
liquid chromatography, x, 119, 128, 130, 151, 290
liquid phase, 232
liver, 96, 98, 106, 112, 123, 125, 132, 135, 136, 137, 139, 141, 145, 148, 149, 192, 209, 214
liver cells, 139
liver disease, 136
livestock, 87, 216, 235
localization, 24, 52
loci, 30
locus, 180

logging, 216
low-density lipoprotein, 208
LTD, 204, 205
lumen, 34
Luo, 276
lymphocytes, 132, 140, 141, 148
lymphoma, 246, 259
lysis, 6

M

machinery, 96, 214
macromolecules, 96, 224
magnesium, 269, 270
magnet, 72
magnetic field, 57, 66
magnetic resonance, 81, 260
magnetic structure, 228
magnitude, 95, 96, 116, 119, 225, 251
majority, 30, 119, 152, 176, 241
malaria, 83
Malaysia, 41, 52, 79
mammals, 92, 121, 135, 139, 275
man, 113
management, xii, 36, 44, 45, 54, 79, 129, 136, 176, 178, 182, 183, 184, 185, 186, 187, 188, 189, 233, 235, 243, 246, 289, 291, 292, 295, 301
manipulation, 74
manufacturing, 148
marketing, 86
marrow, 140
Maryland, 36, 185
mass, x, 5, 7, 21, 34, 71, 119, 120, 128, 130, 151, 156, 160, 161, 290, 294, 297, 299, 300
mass spectrometry, x, 119, 120, 128, 130, 151, 156, 290
materials, 154
matrix(s), 42, 112, 118, 120, 121, 126, 129, 274, 300
matter, 44, 89, 115, 130, 131, 149, 215, 218
measurement(s), 13, 70, 72, 160, 161, 226, 228, 243, 247, 258, 261, 301
meat, ix, 133
media, 4, 36, 46, 59, 61, 74, 75, 76, 89, 95, 101, 152, 153, 158, 160, 162, 180, 214, 215, 224
medical, 26
medicine, 258
Mediterranean, 177, 183, 184, 186, 188
Mediterranean countries, 183, 184
meiosis, 139
melatonin, 145
melt, 6
melting, xii, 245, 249, 254, 255, 260, 261
melting temperature, 248, 251
membrane permeability, 209
membranes, 40, 42, 53, 61, 83, 134, 135, 137, 143
mercury, 22, 128
meristem, 56, 60, 64, 66, 67, 68, 82
MES, 59
mesophyll, 34
meta-analysis, 146
metabolic, 148, 243
metabolic changes, 92
metabolic disorder, 116
metabolic pathways, 67, 89, 288
metabolism, 10, 21, 22, 25, 30, 33, 35, 45, 47, 48, 49, 50, 57, 58, 77, 80, 81, 117, 135, 136, 145, 148, 188, 278, 288
metabolites, ix, 35, 58, 74, 85, 115, 119, 129, 130, 133, 134, 135, 136, 137, 140, 145, 288
metabolized, 33, 34, 35, 43, 71, 77, 116, 134, 136, 140
metabolizing, 96, 145
metal ion(s), 67, 89, 193, 219, 225, 228
metals, 57, 89, 215
metaphase, 121, 122, 125, 194
metaphase plate, 126
meter, 57
methanol, x, 125, 151, 155, 156, 158, 159, 160, 161, 162, 194, 233, 234, 265, 268
methyl group, 117, 267, 268
methylation, xii, 142, 263
methylene blue, 260
Mexico, 98, 131
mice, 92, 139, 141, 142, 147, 149, 288
microbial communities, 107, 115
micromycetes, 127
micronucleus, 103, 123, 131, 132, 139, 147, 148
microorganisms, 23, 25, 89, 90, 115, 116, 127
microscope, 139
microscopy, 194, 227
microwaves, 118
migration, 87, 139, 232, 283, 287
mineral water, 155, 157, 159
mineralization, 116, 117, 128, 129, 130, 219, 227, 301
miniaturization, 118
mitochondria, 41, 50, 96, 209, 210
mitosis, 139
mitotic index, 126
modelling, 242, 301
models, 23, 38, 105, 128, 136, 138, 146, 152, 157, 158, 220, 224, 285, 286
modifications, viii, 2, 19, 246
modules, 26
moisture, 157, 240, 241
molecular biology, 258

molecular oxygen, 96, 192, 208
molecular structure, 116, 140, 228, 243, 255, 288, 289
molecular weight, 57, 60, 96, 214, 260
molecules, xi, xii, 34, 60, 61, 78, 96, 109, 115, 121, 126, 127, 136, 142, 152, 160, 161, 215, 220, 231, 232, 237, 245, 246, 249, 250, 251, 258, 260
monolayer, 53, 222
morphology, 107
mortality, 92, 93, 94, 101, 105, 121, 130, 246, 259
mortality rate, 94
mosaic, 26
mosquitoes, 45
motif, 24
mRNA, 138
mucous membrane(s), 134
multiplication, 18
mussels, 106
mutagen, 123
mutagenesis, 121
mutant, 38, 41, 42, 45, 47, 53
mutation(s), 30, 37, 38, 40, 41, 42, 43, 45, 49, 50, 51, 52, 54, 56, 57, 59, 78, 79, 258, 275

N

Na^+, 137
NaCl, 247
NADH, 207
National Academy of Sciences, 79
natural resource exploitation, 86
natural resources, 86
NCS, 265
necrosis, 89, 136, 142, 143, 149, 214
needy, 56
nervous system, 99, 102, 143
Netherlands, 233
neurotoxicity, ix, 86, 258
neurotransmitter, 98
neutral, 61, 115, 204, 219
New Zealand, 243
NH_2, 90
nicotinamide, 192
Nile, 105
nitric oxide, xi, 191, 205, 208
nitric oxide synthase, xi, 191
nitrite, xi, 191, 204, 205, 206, 207, 208
nitrogen, 4, 80, 81, 204, 205, 210
nitrogen gas, 204, 205
NMR, v, vii, viii, 45, 55, 57, 58, 59, 60, 61, 62, 63, 64, 65, 66, 67, 68, 69, 70, 71, 72, 73, 74, 76, 77, 80, 81, 82, 153, 225, 246, 268, 269, 270, 271, 272, 273
nodules, 80
nonequilibrium, 301
normal development, 210
North America, 58, 105, 179
Northern Ireland, 242
nuclear magnetic resonance, 80, 82, 225
nuclei, 123, 126
nucleic acid, 136, 260, 261
nucleus, 137, 139, 267
nutrient(s), 35, 56, 58, 116, 155, 157, 193
nutrition, 86

O

ODS, 156
OECD, 194, 210, 242
oil, 33, 269
operations, 178
operon, 18
opportunities, 188
optical density, 7
optimization, 222
organ(s), 142, 229
organelle(s), 47, 64, 76
organic chemicals, 116, 288
organic compounds, 113, 240, 243, 285, 289
organic matter, 113, 116, 128, 217, 220, 222, 224, 227, 232, 240, 241, 286, 289
organic solvents, x, 89, 120, 151, 152, 162
organism, ix, 86, 92, 138, 139, 214
osmolality, 137
overlap, 158
overproduction, 58, 68, 208
ovulation, 210
ownership, 87
ox, 59, 61, 62, 63, 65, 67, 68, 69
oxalate, 217, 227
oxidation, 34, 40, 135, 152, 196, 197, 210, 218, 264, 275
oxidative damage, 22, 137, 145
oxidative reaction, 135
oxidative stress, ix, 81, 92, 95, 96, 97, 98, 103, 105, 106, 133, 136, 137, 138, 143, 145, 146, 208, 209, 210
oxygen, 21, 25, 42, 59, 61, 96, 103, 117, 137, 193, 208
oyster(s), 123, 132

P

Pacific, 123, 132
parallel, 98, 182, 248

parameter estimation, 82
parasite(s), 77, 83, 105
partition, 89, 134, 242, 278
pathways, 2, 18, 27, 73, 90, 115, 117, 143, 152, 170, 214
PCR, 6, 7, 18, 23, 26
peer relationship, 127
penicillin, 194
PEP, 56, 57, 67
perfusion, 58, 59, 61, 66, 69, 72, 74, 75, 76
peripheral blood, 132, 139, 140
peripheral neuropathy, 214
permeability, 196, 217
permit, 2, 292
peroxidation, 210
peroxide, 192, 193
personal communication, 233
pest populations, 30
pesticide, vii, xii, 23, 26, 87, 115, 116, 127, 131, 136, 148, 214, 215, 224, 225, 229, 232, 233, 241, 242, 244, 246, 258, 259, 260, 285, 288, 289, 290, 291, 292, 293, 294, 300, 301
pests, 30, 179, 258
petroleum, 130, 131
phage, 26
pharmaceuticals, 132
phenol, 6, 215
phenomenology, 51
phenotype(s), 30, 33, 41
phenoxyacetate herbicides, vii, 1, 26
phenylalanine, 75, 89
phloem, 34, 56, 58, 62, 66, 67, 71, 74, 78
phosphate(s), 33, 34, 52, 56, 57, 62, 63, 64, 69, 72, 74, 78, 81, 89, 99, 106, 108, 135, 192, 194, 215, 223, 224, 227, 229, 247, 250, 252, 253, 254, 256, 257
phosphoenolpyruvate, 56, 57
phosphorous, 58, 69, 74, 228
phosphorus, 57, 58, 65, 217, 226, 229
phosphorylation, 57, 99
photobleaching, 52
photochemical degradation, x, 90, 151, 153, 215
photodegradation, 152, 153, 154, 155, 157, 158, 160, 162, 163, 171, 227
photolysis, vii, x, 151, 152, 153, 154, 157, 158, 162, 170, 214, 228, 229
photosynthesis, ix, 30, 32, 45, 58, 62, 111
photosynthetic performance, 45
phylogenetic tree, 41
physical exercise, 136
physical properties, 72
physics, 292
physiological, 26, 48, 49, 52, 53, 82
physiological mechanisms, 180
physiology, 242
phytoplankton, 92
plant diseases, 258
plant growth, x, 30, 151, 153, 155, 167, 169, 170, 171
plants, vii, ix, 30, 31, 32, 33, 34, 35, 36, 38, 41, 42, 43, 44, 45, 48, 50, 51, 57, 58, 59, 60, 61, 62, 64, 65, 66, 67, 71, 74, 78, 79, 82, 89, 91, 116, 117, 133, 153, 154, 158, 164, 167, 168, 176, 177, 178, 182, 183, 187, 192, 208, 274, 275
plasma membrane, 59, 71, 74, 75, 76, 275
plasmid, vii, 1, 2, 3, 4, 6, 12, 15, 16, 17, 18, 21, 22, 23, 24, 25, 26, 27, 28
plasticity, 180, 217
plastid, 36, 38, 41
platform, 58, 71
point mutation, 36, 37, 43, 58, 68, 121
poison, 20, 125
Poland, 176
polar, 42, 119, 120, 130, 153, 160, 232
polarizability, 116
polarization, 42
pollutants, 27, 95, 112, 116, 118, 121, 126, 132, 142, 143, 192, 210, 214, 244, 246
pollution, ix, 87, 103, 111, 121, 130, 214, 219, 241, 289
polyacrylamide, 26
polyamines, 261
polycyclic aromatic hydrocarbon, 131
polymerase, 6, 15, 53
polymerase chain reaction, 6, 15, 53
polymorphism(s), 50, 52
polyunsaturated fatty acids, 137
ponds, 143, 214
pools, 40
POPs, 142
population, viii, 18, 29, 30, 32, 41, 42, 44, 45, 46, 51, 78, 79, 86, 87, 92, 93, 99, 121, 148, 176, 178, 182, 185, 187, 206, 221, 246
population density, 206
population growth, 86
porosity, 217, 299
Portugal, x, 175, 176, 177, 178, 179, 185, 245, 247
posttranslational modification, vii, 1, 22
potassium, 59, 194, 206, 270
potato, 87
precedent, 76
precipitation, 4, 5
preparation, 5, 118, 119, 122, 126, 194
preservation, 57
prevention, 210
principles, 25, 105, 109, 176, 177

probability, 60, 221, 267
probe, 59
producers, 179, 183
profitability, 184
project, 130, 258
prokaryotes, 38
proliferation, 142
proline, 41, 42
propagation, 143
protection, 88, 144, 229
protein synthesis, 89
proteins, ix, 6, 19, 26, 36, 82, 123, 133, 136, 137, 143
prototypes, 187
psoriasis, 136
pumps, 34, 59, 75, 77
pure water, 6, 7
purification, 6, 155, 247, 269
purines, 251
purity, 4, 155, 233
P-value, 195, 196, 197, 198, 199, 201, 203
pyrophosphate, 81

Q

quantification, 26, 58, 115
quartz, 155, 219, 248
quinone, 32

R

race, vii, 1, 3, 10, 18, 26
radiation, 153, 155, 209, 214, 274
radical formation, xi, 191
radicals, 96, 193, 207, 208, 210
radio, 57, 59
radiography, 64, 74
rainfall, 215, 278
raw materials, 86
reaction center, 32, 41
reaction mechanism, 21, 25
reaction rate, 116
reactions, 21, 34, 51, 56, 89, 116, 136, 137, 152, 159, 214, 215, 255
reactive oxygen, vii, xi, 1, 19, 96, 143, 191, 208, 209
reagents, 265
reality, 183
receptors, 143
recognition, 117, 188, 261
recombination, 2, 41
recommendations, 149, 182, 185, 187
recovery, 53, 118, 120, 235, 237
regenerate, 42
regions of the world, 179
regression, 5, 13, 31, 157, 168, 239, 249, 282, 283, 285
regression equation, 31, 168, 285
regression line, 282, 283
regression model, 239
regrowth, 187
regulations, 109, 136
relaxation, 67
relevance, 21, 147
reliability, 186, 221, 301
relief, 215
repair, x, 96, 133
repetitions, 205
replication, 7
repression, x, 133
reproduction, 93, 108, 176
reproductive organs, 33
repulsion, xi, 231, 237
requirements, x, 18, 154, 175
researchers, 42, 154, 264, 300
residues, xii, 38, 102, 105, 112, 119, 132, 153, 162, 171, 209, 234, 243, 290, 291, 294, 295
resolution, 60, 120
resources, 30
respiration, 209
response, vii, 8, 18, 22, 25, 31, 41, 43, 50, 58, 59, 68, 79, 80, 82, 83, 96, 98, 99, 103, 117, 137, 138, 145, 154, 164, 165, 169, 170, 171, 180, 182, 188, 209
restrictions, 58
reticulum, 96, 135, 143
reusability, 120
rice field, 46, 87, 288
rings, 267
risk(s), ix, x, xii, 85, 91, 92, 104, 105, 108, 109, 115, 121, 131, 132, 142, 147, 148, 154, 171, 175, 176, 177, 178, 182, 184, 187, 192, 210, 241, 245, 246, 258, 259
risk assessment, 104, 108, 109, 115, 147, 148
rodents, 121, 139, 140, 147
rods, 293
room temperature, 64, 194, 268, 274
root(s), 33, 35, 49, 54, 56, 60, 66, 67, 75, 80, 89, 140, 148, 153, 154, 157, 158, 163, 164, 165, 167, 168, 169, 170
root growth, 154, 158, 167
root system, 56
routes, 214
routines, 142
runoff, 139, 143, 214, 278, 282, 283, 287, 288, 289, 292

ruptured membranes, 137
rural areas, 43, 112, 121, 246
rural population, 121
Russia, 173
RVS, 64, 65, 68, 69

S

safety, 125, 154, 246, 275
salmon, 109
Salmonella, 123, 132
salts, 83, 89, 215
saturated fatty acids, 42
saturated hydrocarbons, 158
saturation, 13, 220
scavengers, 192, 195, 202
science, 260
SCT, 37, 38
second generation, 31
sediment(s), 91, 107, 116, 117, 128, 129, 130, 215, 234, 243m 244, 301
seed, x, 30, 60, 86, 151, 153, 154, 157, 163, 164, 165, 166, 171, 178, 179, 180, 186, 188
seeding, 86, 101, 157, 177
seedlings, 47, 60, 106, 157, 278
selective agents, ix, 111
selectivity, 35, 40, 119, 120, 153, 176
semen, 138
sensitivity, 26, 38, 41, 42, 43, 50, 64, 65, 67, 71, 73, 74, 75, 76, 77, 82, 93, 101, 118, 119, 120, 154, 163, 189
sequencing, 23, 184
serine, 42, 99
serum, 194
sewage, 128, 130
shear strength, 217
shoot(s), 31, 35, 60, 65, 66, 82, 154, 168, 169, 170
shortage, 56
showing, x, 4, 13, 36, 40, 67, 72, 121, 122, 139, 151, 153, 158, 160, 223, 238
side chain, 117
side effects, xi, 245
SIDS, 194, 210
sieve element, 34
signaling pathway, 138, 149
signalling, 24
signals, ix, 72, 86, 95, 143
silver, 104, 105, 206
simulation, 288, 290
single crystals, 274
SiO_2, 219
sister chromatid exchange, ix, 111, 122, 125, 140, 148
skin, ix, 111, 121, 134, 135, 192, 214, 246
skin cancer, 208, 209
skin diseases, 210
Slovakia, 243, 244
sludge, 115, 120, 128, 130
sodium, xi, 155, 156, 191, 194, 204, 206, 228, 247, 265, 268, 269, 270
sodium hydroxide, 269, 270
software, 5, 6, 13
soil erosion, 178
soil particles, xi, 90, 226, 231, 237, 241, 292
soil pollution, 216
soil seed bank, 30
soil type, 90
solid phase, 118, 119, 129, 153, 232, 237
solubility, 89, 120, 134, 136, 153, 155, 219, 233, 278, 281
solvents, x, 118, 151, 158, 160, 162, 171
sorption, xi, xii, 113, 116, 117, 130, 227, 231, 232, 233, 234, 237, 240, 241, 242, 243, 277, 278, 285, 286, 288, 289, 290, 292, 300, 301, 302
sorption experiments, 300
sorption isotherms, xi, 231, 292
sorption process, 240
South Africa, 79, 180
South America, 45, 68, 80, 93, 105
sowing, 91, 107, 216
soybeans, 42, 87, 91, 105, 185
Spain, x, 47, 48, 49, 130, 151, 175, 176, 177, 178, 179, 185, 186, 188, 189, 229
specifications, 104
spectroscopy, vii, 57, 80, 81, 246, 248, 251, 256, 258, 260, 261
sperm, 138, 259, 260, 261
spermatogenesis, 138
spindle, 123, 138
spleen, 148
Sprague-Dawley rats, 140
sprouting, 185
stability, xi, 24, 42, 101, 120, 155, 228, 237, 245, 255, 258, 260
stakeholders, x, 175, 184
standard deviation, 66, 70, 72, 235, 236, 237, 239, 240, 285
standard error, 93, 94, 98, 100
state(s), vii, 1, 3, 5, 6, 8, 9, 10, 11, 12, 13, 15, 16, 17, 18, 19, 20, 25, 41, 42, 53, 56, 57, 60, 61, 65, 76, 81, 92, 96, 128, 135, 142, 177, 248, 255, 300, 302
steel, 292, 293
sterile, 6, 7
sterols, 158
stimulation, 99
stimulus, 142

stock, 3, 5, 60, 126, 155, 247
stomata, 183
storage, 3, 5, 289
stress, 42, 81, 95, 96, 136, 137, 138, 142, 143, 146, 183, 186
stress response, 137
stressors, 96
structural changes, 251
structural gene, 2
structure, xi, xii, 21, 23, 35, 38, 40, 48, 51, 52, 64, 66, 88, 92, 99, 113, 117, 134, 136, 139, 153, 161, 217, 225, 245, 247, 251, 261, 263, 267, 295, 299
subacute, 143
substitution(s), 38, 39, 41, 42, 50, 87, 117, 145, 152, 267
substrate(s), vii, 1, 3, 5, 8, 9, 10, 11, 12, 13, 17, 18, 21, 25, 36, 40, 51, 56, 57, 58, 59, 62, 65, 71, 74, 75, 135
sucrose, 56, 59
sugar beet, 152, 153
sugarcane, 116, 128, 134, 228
sulfate, 5, 206, 269, 270
sulfonylurea, 32, 36, 38, 46
sulfuric acid, 206
suppliers, 59
suppression, 77, 80, 142, 177, 192, 209
surface area, xi, 213, 217, 221, 223, 229
surface layer, 233, 235
surface washing, 60
surfactant, 91, 186
surrogates, x, 151, 152
survival, 77, 93, 104, 105, 136, 146, 213
survival rate, 93
susceptibility, 60, 66, 106, 142, 186, 217
suspensions, 4, 194
sustainability, x, 175, 184
swelling, 136
Switzerland, 28
symptoms, 40, 95
synergistic effect, 214
synthesis, 38, 40, 52, 78, 81, 89, 96, 138, 146, 152, 263, 264

T

T cell, 142
tactics, 177
target, vii, viii, ix, xii, 7, 29, 30, 31, 32, 33, 34, 35, 36, 38, 40, 41, 43, 45, 47, 49, 50, 52, 56, 57, 68, 78, 79, 85, 86, 87, 92, 112, 114, 118, 154, 164, 170, 180, 232, 264, 275, 277, 278, 279, 281, 283, 284, 286, 288
technical support, 242, 300
techniques, 5, 60, 74, 112, 118, 119, 126, 140, 153, 232, 243, 261
technology(s), 86, 101
telophase, 139
temperature, 58, 60, 64, 80, 81, 117, 118, 155, 160, 215, 220, 225, 240, 241, 248, 251, 254, 255, 257, 268, 292, 293
terpenes, 34
testing, 31, 95, 147, 242
testis, 246
testosterone, 107, 138
texture, 115, 217, 235, 292
therapeutic targets, 83
thermal stability, 255
thermodynamic parameters, xii, 245, 248, 255
thermodynamics, 255
thermolysis, 159
threonine, 41
thymus, xii, 245, 247, 260
thyroid, 209
time frame, 71
tissue, viii, 33, 42, 55, 56, 57, 58, 59, 60, 61, 62, 63, 64, 65, 66, 67, 68, 69, 70, 71, 72, 74, 75, 76, 82, 98, 99, 101, 136, 138, 142, 192, 208, 246
TNF, 143
tobacco, 42, 51, 80
tocopherols, 135
toluene, 269, 280
torsion, 267
total product, 86
toxic effect, vii, 93, 112, 116, 137
toxic substances, 144, 153
toxicology, 108, 109, 139, 149
training, 136
traits, 2, 22, 27, 77, 180, 187
transcription, 137, 138
transformation, vii, x, 112, 114, 115, 127, 151, 152, 159, 170, 289
transformation processes, 114
transformation product, vii, x, 112, 114, 115, 127, 151, 152, 159, 170, 289
transition metal, 210
transition temperature, xii, 245
translocation, viii, 29, 30, 33, 34, 35, 47, 48, 56, 58, 61, 62, 63, 64, 66, 67, 68, 71, 74, 76, 78, 81, 82, 83, 116, 117, 180, 185, 186, 188
transmission, 155, 177
transport, vii, xii, 30, 35, 53, 61, 62, 75, 76, 83, 112, 128, 210, 291, 292, 293, 300, 301, 302
treatment, 58, 59, 60, 62, 63, 64, 65, 66, 67, 71, 72, 73, 74, 77, 106, 115, 125, 137, 138, 140, 154, 157, 166, 167, 168, 184, 193, 194, 202, 205, 274, 294

treatment methods, 58
triazine herbicide, viii, 29, 30, 35, 47, 48, 134, 136, 138, 139, 140, 141, 144, 147, 148
trifluoroacetate, 265, 268
triggers, 142
trypsin, 6, 24, 125
tryptophan, 38, 89
tumor(s), 36, 140, 143, 192, 246
tumor growth, 192
tumor necrosis factor, 143
Turkey, 188
turnover, 22, 26
two-state model, 255
tyrosine, 89

U

UK, 144, 186, 234, 242, 243, 288, 301
underlying mechanisms, 137
uniform, 60, 220, 300
United Nations, 87
United States (USA), 5, 6, 24, 27, 40, 41, 43, 44, 51 57, 60, 71, 79, 101, 103, 134, 141, 156, 179, 188, 208, 235, 244, 259, 275, 289
urban, 87, 115, 130
urbanization, 86
urea, 32, 53, 112, 113, 115, 117, 119, 129, 132, 280
urine, 136
Uruguay, 102
UV, xii, 119, 120, 126, 155, 158, 160, 245, 246, 247, 248, 249, 250, 251, 252, 253, 256, 258, 259, 261
UV absorption spectra, 158, 248, 251
UV light, 126
UV spectrum, 160, 251

V

vacuole, viii, 34, 55, 57, 58, 59, 63, 64, 66, 67, 68, 69, 71, 72, 73, 74, 75, 76, 77
vacuum, 60, 61, 72, 155, 156, 234, 293
Valencia, 148
validation, 290
valine, 36, 228
valuation, 119
valve, 220
variables, 58, 95, 221, 292
variations, xii, 21, 59, 99, 130, 251, 277
varieties, 278, 286
vascular system, 89
vegetables, 152, 179
vegetation, 170
vertebrates, vii, x, 99, 107, 133, 144
vessels, 220
viscera, 99
viscosity, 57, 258
vision, 115
visualization, 125
vitamin E, x, 40, 145, 191, 194, 209, 210
vitamins, 96
volatility, 56
volatilization, 114, 237
vomiting, 214

W

wastewater, 115, 120
water quality, 109, 301
water supplies, 240, 244
weakness, 74
weather patterns, 56
wells, 214
Western Australia, 186
wild type, 19, 21, 30, 42, 53
wildlife, 192, 193
Wisconsin, 228
wool, 61, 293
workers, 86, 121, 127, 131, 141, 148, 192, 209, 210, 246, 259
World Health Organization (WHO), 90, 91, 95, 104, 109, 131
worldwide, x, 30, 56, 78, 87, 175, 183, 219, 246

X

xenon, 155
X-ray analysis, xii, 263
X-ray photoelectron spectroscopy (XPS), 225
XRD, xi, 213, 217, 218, 225

Y

yeast, 4, 38
yield, 40, 56, 261, 269, 270, 271, 272, 273

Z

zinc, 215
zooplankton, 95